CONSUMERS IN THE COUNTRY

REVISITING RURAL AMERICA

Pete Daniel and Mary C. Neth, *Series Editors*

Mary Neth, *Preserving the Family Farm: Women, Community, and the Foundations of Agribusiness in the Midwest, 1900–1940*

Sally McMurry, *Transforming Rural Life: Dairying Families and Agricultural Change, 1820–1885*

David B. Danbom, *Born in the Country: A History of Rural America*

David Vaught, *Cultivating California: Growers, Specialty Crops, and Labor, 1875–1920*

Ronald R. Kline, *Consumers in the Country: Technology and Social Change in Rural America*

Consumers
in the Country

Technology and Social Change in
Rural America

RONALD R. KLINE

The Johns Hopkins University Press • *Baltimore and London*

© 2000 The Johns Hopkins University Press
All rights reserved. Published 2000
Printed in the United States of America on acid-free paper
9 8 7 6 5 4 3 2 1

The Johns Hopkins University Press
2715 North Charles Street
Baltimore, Maryland 21218-4363
www.press.jhu.edu

Kline, Ronald R.
Consumers in the country : technology and social change in rural America /
Ronald R. Kline
p.   cm. — (Revisiting rural America)
Includes bibliographical references and index.
ISBN 0-8018-6248-5 (alk. paper)
1. Technology—Social aspects—United States.   2. Country life—United States.
I. Title.   II. Series.
T14.5.k578 2000
303.48′3′0973091734—dc21   99-045676

A catalog record for this book is available from the British Library.

For David, Melissa, Michele, Margot, and Maggie

The silence of the Plains, this great unpeopled landscape of earth and sky, is much like the silence one finds in a monastery, an unfathomable silence that has the power to re-form you. And the Plains have changed me. I was a New Yorker for nearly six years and still love to visit my friends in the city. But now I am conscious of carrying a Plains silence within me into cities, and of carrying my city experiences back to the Plains so that they may be absorbed again back into silence, the fruitful silence that produces poems and essays.

—Kathleen Norris, *Dakota*

Contents

Acknowledgments

I dedicate the book to my stepchildren, David, Melissa, Michele, Margot, and Maggie. They, their spouses (Carole, Jon, Mark, Jordan, and Mark), their mother, Marge, my partner for over twenty years, and granddaughter Ruth have become the center of my life. Their support during the preparation of this book has been a blessing.

I want to thank my father, Raymond Orville Kline, for sharing with me his experiences driving, fixing, and modifying automobiles on the farm in the 1930s and selling appliances for rural electrification as a small-town merchant after World War II. My mother, Donna Lee (Dick) Kline, helped me to re-call my childhood experiences with the party-line telephone. Great Uncle Bill Dick and great Aunt Ruth Dick of Newton, Kansas, kindly gave me the eye-opening photograph of a relative washing the clothes with a Model T, which forms the cover for this book. I have especially enjoyed talking with my step-daughter Margot and her partner Jordan about how they have used traditional energy sources, like wood and coal, as well as solar power to live "off the grid" near cloudy Ithaca, New York, for several years.

Trevor Pinch, my colleague in the Science and Technology Studies De-partment at Cornell University, has played a large role in building the theo-retical framework for this book. Trevor and I wrote an article on the rural automobile that forms the basis for chapter 2. In that paper, we incorporated a form of gender power relations into the social construction of technology approach to technology studies, which Trevor co-founded in the mid-1980s, and described a mutual construction of both technology and social groups.

My ideas about how consumers have used resistance to weave new tech-nologies into their lives have benefitted greatly from many conversations with Christina Lindsay, a graduate student in Cornell's S&TS program. Her colleagues Pablo Boczkowski, Julian Kilker, Suzanne Moon, Miranda Paton, Alec Shuldiner, and Jamey Wetmore have also contributed to my thinking about how consumers have created technological and social change. Suzanne conducted more than twenty interviews with farm families in New York State that have added a good deal of social texture to my story. Jonathan Baker, Simon Cole, Sharon Daly, Chris Findlayson, and Miranda Paton helped research the book. Kathleen Babbitt and Scott Crawford provided relevant material from their research on technology and rural life.

Portions of chapters 2, 7, and 9 were presented as conference papers at Cornell University, the University of Trondheim, Norway, and at the annual meeting of the Society for the History of Technology. They later appeared as articles in *Technology and Culture* and *Public Understanding of Science*, and as a chapter in *Rethinking Home Economics* (Ithaca: Cornell University Press, 1997). I thank the publishers of these volumes for permission to reprint, and conference audiences, editors John Staudenmaier, John Durant, Sarah Stage, and Virginia Vincenti, the anonymous referees, and Joan Brumberg, Carolyn Goldstein, Michael Dennis, and Margaret Rossiter, for their helpful comments in the early stages of my research.

This book could not have been written without the generous assistance of several archivists and librarians. I want to thank those at the Henry Ford Museum (Terry Hoover and Bob Casey), Cornell University (Gould Colman), AT&T (Sheldon Hochheiser), the Schenectady Museum (John Anderson), the University of Illinois, the Franklin Delano Roosevelt Library, the Lyndon Baines Johnson Library, and the National Archives in Washington, D.C.

Sheldon Hochheiser, Marge Kline, Nellie Oudshorn, and Trevor Pinch commented on portions of the manuscript. Margaret Rossiter read the entire manuscript, which is much improved as a result. I would also like to thank Mary Neth and an anonymous referee at the Johns Hopkins University Press for their helpful comments. Bob Brugger, Katherine Kimball, and Lee Sioles were exemplary editors.

Much of the research for this book was funded by National Science Foundation Grant Number SBR-9321180 and the Harry and Sue Bovay Program for the Study of the History and Ethics of Professional Engineering at Cornell University.

Introduction

When social analysts ignore the historical dimension, the result is a simplification and schematization of social change that weakens the explanatory power of even the most sophisticated theory. Any understanding of the fate of community in America today or at any time in the past depends upon an expansion of social theory to incorporate the concrete data of historical change into social explanation.
—Thomas Bender, *Community and Social Change in America*

When an editor of the midwestern journal *Wallaces' Farmer* drove his car through the Iowa countryside on a fall evening in 1935, he observed that "too many farm people were living in the dark or the half-dark. . . . The contrast between town and country was the contrast between electricity and the old oil lamp."[1] His view was widely held in the United States. As electricity began to light city streets, stores, and the houses of the wealthy in the 1880s, social commentators and farm people saw it first as an urban luxury and then—when electricity became more common after World War I—as a symbol of the deepening cultural divide separating a "modern" urban life from a "primitive" rural one. An Ohio woman reported in 1913 that "after a visit to her sisters in the city where you push a button and get plenty of good lights instantly . . . the farm woman comes back dissatisfied with her lot."[2] In *Grapes of Wrath*, John Steinbeck's famous novel about Okies escaping the Dust Bowl for the promised land of California, Rose of Sharon tells Ma that she and her husband Connie want to quit farming when they reach the coast. Connie will get a factory job and take a night course in radio. "An' we'll live in town an' go to pitchers whenever an'—well, I'm gonna have a 'lectric iron."[3]

These optimistic views of electricity and other "modern" technologies

1

have pervaded American culture—urban and rural—in the twentieth century. Although critics might point to the dehumanizing effects of the machine age in the 1920s or to widespread technological unemployment in the 1930s, a positive ideology of technological progress seems to have prevailed in the United States for most of the century. In their efforts to reform rural life, promoters viewed four supposedly urbanizing technologies—the telephone, the automobile, radio, and electricity, all of which had been initially marketed as urban luxury goods—as self-evident symbols of modernity and powerful agents of social change. They—and many farmers—inscribed each technology with a largely unquestioned power to transform an old-fashioned rural society into an agrarian version of middle-class, urban consumer culture, just as each technology was thought to have revolutionized the city. During their (largely successful) campaigns to transform rural America in the twentieth century, promoters pushed motorized farming, hybrid seeds, and farm management, along with chemical fertilizers, herbicides, and pesticides, as the means to "industrialize" agriculture. They pushed new communication, transportation, and household technologies as the means to "modernize" rural life.[4] Paradoxically, many reformers wanted to save the family farm by "urbanizing" it, by bringing "city conveniences" to the farmstead and connecting it more closely to town. They hoped the new technologies would make farm life more satisfying, thus halting the worrisome migration to the city.[5]

A wide range of social groups shared these views. Manufacturers, advertisers, rural reformers, government agencies, agricultural professionals, educators, and farm leaders saw the telephone and the automobile, in the early part of the century, then radio and electricity, after World War I, as the latest technology that could solve the long-standing problems of rural life. They predicted that telephones and automobiles would improve the marketing of agricultural products and the purchasing of consumer goods, induce farm families to attend cultural events in the city and town, and turn isolated farm neighborhoods into vibrant rural communities. Automobiles and radios would eliminate regional differences in dress and speech, as well as those between city and country. Radios, electric lights, and electrical appliances would transform dark, drab farmhouses into bright, suburban utopias, where the ancient foes of isolation and drudgery would be banished forever. Above all, promoters claimed that each new technology would keep people, especially talented youth, on the farm. Yet the percentage of the U.S. population living on farms dropped from 42 percent in 1900 to 25 percent in 1930 and 15 percent in 1950 (see table A.1).

An extreme statement of this progress ideology for electricity is reflected in V. I. Lenin's famous dictum, made in 1920, that "communism is socialist power plus the electrification of the whole country." Electricity would not only reform the city and the factory but would also "make it possible to raise the level of culture in the countryside and to overcome, even in the most remote corners of the land, backwardness, ignorance, poverty, diseases, and barbarism."[6] Although few Americans quoted Lenin when pleading for rural electrification, proponents as diverse as utopian novelists, industrialists like Henry Ford, liberal politicians, and conservative farm leaders proclaimed that electricity would revolutionize rural life. John Carmody, head of the New Deal's Rural Electrification Administration, told Congress in 1937, "We believe in the economic wisdom of bringing farm families out of the dark into the light, out of stark drudgery into normal effort, out of a past of unnecessary denial into a present of reasonable convenience."[7] Mrs. Meade Ferguson, chair of the Home Economics Committee of the National Grange, said in a nationwide radio address in 1935, "Adequate electric current for farm homes will mean the emancipation of farm women from the endless household drudgeries which they have borne too long."[8]

The utopian view of rural electrification is dramatically shown in a "scissors picture" created by Country Life reformer Martha Bensley Bruère to illustrate a magazine article published in 1932 (see fig. 1). The top panel depicts the hardships of rural life before electricity. On the left, an anemic-looking cow gives milk to a farm man, who is hunched over from the drudgery of his daily toil. A chicken and rooster listlessly stare at the meager production of one egg. In the house, lit by a kerosene lamp, a tired farm woman scrubs clothes near a hot, menacing stove while her little boy and his dog demand attention. In the second panel, these actors pull back in fear and apprehension as the kilowatts come into their life. The third panel shows life with electricity: The cow, hooked up to a milking machine in a well-lit barn, now dances with joy. The rooster and hen proudly lead a parade of eggs the hen has produced. In the brightly lit house, electricity pumps water, cooks a meal, toasts bread, and makes coffee while a washing machine contentedly burps soap bubbles. Where are all the people and the dog? They are in the bottom panel, relaxing in rocking chairs and cooled by electric fans. Notice some subtle signs of urbanization in the last panel. The boy is now a little scholar with glasses. His mother, fashionably attired, reads the paper, while his father snores in luxurious comfort.[9]

The drawing, of course, exaggerates. It reflects the reformer's ideologies of technological progress and technological determinism (the belief that autonomous technology drives social change) much more accurately than it de-

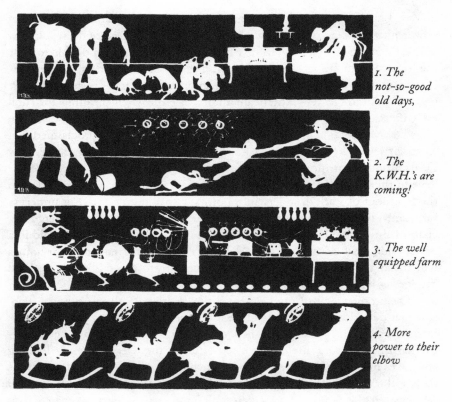

1. The not-so-good old days,

2. The K.W.H.'s are coming!

3. The well equipped farm

4. More power to their elbow

Figure 1. Current Comes to the Farm. A scissors picture by Country Life reformer Martha Bruère, showing the technological optimism of the progressive ideology of electricity in the 1930s.
Source: Morris L. Cooke, "Light and Power: Planning the Electrical Future," *Survey,* 67 (1932): 607–11, on pp. 608, 609.

picts contemporary social conditions. The evidence we have indicates that middle-class farm life was not so technologically deprived before electricity nor so utopian afterward. The so-called golden age of American agriculture— the period from 1909 to 1914, when the purchasing power of farmers was equal to or greater than that of nonfarm workers—had been prosperous enough for large numbers of farm families to buy durable consumer goods. In fact, the Census Bureau reported in 1920 that slightly larger percentages of farm households owned telephones (39 percent) and automobiles (31 percent) than did nonfarm households (34 percent of whom had telephones, and 25 percent

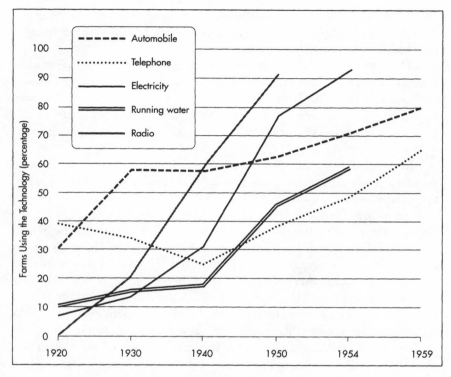

Figure 2. "Urban" Technologies on the American Farm

Source: Data from U.S. Department of Commerce, Bureau of the Census, *U.S. Census of Agriculture, 1959: General Report*, vol. 2, *Statistics by Subjects* (Washington, D.C.: GPO, 1960); *Statistical Abstract* (Washington, D.C.: GPO, 1930); *U.S. Census of Housing: 1950*, vol. 2, *General Characteristics* (Washington, D.C.: GPO, 1953); *U.S. Census of Agriculture, 1954: General Report*, vol. 2, *Statistics by Subjects* (Washington, D.C.: GPO, 1956–57).

automobiles).[10] Although farm income and prices dropped sharply during the depression of the early 1920s and did not fully recover during the decade, the 1930 census reported that nearly 60 percent of farm families owned an automobile, one-third had a telephone, and one-fifth a radio, often battery-operated (see fig. 2). On the other hand, only 13 percent of the country's farms had electricity in 1930, compared with 85 percent of nonfarm households. Yet, as we shall see, many prosperous families used substitutes for electricity—water pumps and washers powered by gasoline engines, gas-mantle lamps, and oil stoves—to ease the hardships depicted in the drawing. Southern farmers, many of whom were poor sharecroppers, lagged behind in all categories (see tables A.2 to A.5). Surveys of middle-class farm women showed that well into the 1960s they worked just as many hours after their houses were electrified as they had before electricity arrived.

The drawing leaves out other important aspects of the "modernization" process. It does not show the vast networks of manufacturers, government agencies, reformers, and farm organizations that tried to use new technologies to transform rural culture. Nor does it indicate the contested nature of that process, except to imply that the fear of electricity was an obstacle to be overcome.

In large measure, this book is a commentary on the scissors picture, a recounting of what is left out of the drawing and a critique of the belief in technological determinism that continues to shape how we think about technology and society. I question whether the telephone, the automobile, radio, and electricity were autonomous social forces that revolutionized rural life in ways predicted by promoters. I explore, instead, the changing relationships among modernizers, farmers, and technological systems—the networks of interactions between groups that promoted these technologies to uplift rural life and farm families who tried to shape them to fit rural culture. Farm life was indeed transformed in the first six decades of the twentieth century, but the transformation was a highly fractured process—both for the promoters and for farm families. Farm men, women, and youth were much more than passive, simple-minded, and grateful recipients of the transfer of technology from the city to the country, as they are depicted in the drawing. They contested efforts to urbanize the farm by resisting each new technology and then weaving it into existing cultural patterns in their own way.

Keywords

The historian Raymond Williams has noted that a word like *urbanization* has diverse meanings that express general social processes, "new kinds of relationships, but also new ways of seeing existing relationships." New terms are invented, and the meanings of old words are adapted, altered, extended, and transferred. "Earlier and later senses coexist, or become actual alternatives in which problems of contemporary belief and affiliation are contested." Several of these protean words, what Williams calls "keywords," are central to my story. Chief among them are *urbanization, modernization, technology, consumer, producer,* and *resistance.*[11]

Throughout the book, I use *urbanization,* as well as *urbanize,* in the sense that many of my historical actors used it, to refer to the cultural activity of investing rural life with an urban character, rather than the demographic process of achieving a population density considered to be urban (defined by the Census Bureau in this period as twenty-five hundred people). In the interwar years, many rural sociologists and historians used *urbanization* in this manner. The rural sociologists Pitirim Sorokin and Carle Zimmerman claimed in 1929 that the "increase of similarities between these [rural and urban] worlds [owing in part to improved transportation and communication systems] has been proceeding more in the way of an 'urbanization' of the rural parts than in the opposite way of 'ruralization' of the cities."[12] But the migration of twenty million people from the country to the city in the 1920s led the rural sociologists Edmund de S. Brunner and John Kolb to conclude in 1935 that "if there has been an urbanization of the country, there has also been a ruralization of the city. . . . Certainly these millions did not shed all of their ideas, attributes, and ways of life at the city gates, as some writers have implied in their great emphasis upon the urbanization of the country."[13]

The historian Paul Johnstone, of the U.S. Department of Agriculture, gave more power to urban influences on rural culture in an influential essay published in 1940. Johnstone claimed that the American "countryside has been undergoing a process of accelerated urbanization for nearly a century." The commercialization of agriculture "was only one phase of the urbanization of farm life and it was in effect merely the means whereby farm people could obtain the products of industry that the absorption of urban culture has taught them increasingly to desire." Good roads, rural free delivery of mail, the telephone, the automobile, and electricity were agents of change. "Each new convenience, every new gadget, has bound the country more closely to the town and made it more like the town."[14] The sustained consumerism that began with the farm prosperity following World War II convinced later historians that Johnstone had been prophetic.[15]

Although more recent historians have also referred to an "urbanization" of rural culture,[16] I take a different tack and study the contested aspects of the attempts to urbanize rural life. In this sense, *urbanization* refers to the twin goals of the reformers: to connect the farm to town and the city through communication and transportation technologies and to bring "city conveniences" to the farmhouse. This was a contested process, briefly described above, in which farm families responded to efforts by promoters to create (what they considered to be) middle-class, urban standards of living in the country through the agencies of new communication, transportation, and household technologies. Farm families adopted many of the material and other characteristics of middle-class, urban consumer culture, but it was an incomplete process in which farm people resisted, modified, and selectively used these technologies to create new rural cultures, new forms of rural modernity—many of which were individual modernities.

I thus use *modernity* to refer to the cultural practices resulting from this contested process of social change. I do not use *modern* or its derivatives in the manner favored by recent social theorists. Many urban and rural people in my story would have agreed with scholars who characterize *modernization* as the development of traits like individualism, rationality, secularism, and tolerance of innovation in capitalist, industrial, urban, mass societies in which technological and social change is the norm. But many of these theories state or imply that *modernization* is an inevitable, monolithic social force rather than a contested, fractured process. Consequently, I use the word in the more restricted sense that most contemporaries used it, as a relative term (relative to time, place, and culture) referring to the process of using new technologies to make life up-to-date, that is, modern.[17] For many contemporaries, urbanization was part of the process of modernization, which included the adoption of all sorts of new practices, including agricultural equipment. Because new communication, transportation, and household technologies were usually seen as urban goods, I tend to use the more specific term *urbanize* (or *suburbanize* for the post–World War II period), rather than *modernize,* to refer to the goals pursued by the promoters of these technologies. But when focusing on attempts to update the farmhouse, I typically use *modernize,* because that term has been employed extensively in this context throughout the twentieth century.

Many of the social groups in my story—such as rural sociologists, agricultural engineers, agricultural economists, and home economists—applied techniques invented to modernize the rural United States before World War II to "undeveloped" countries after the war. What social scientists in the 1960s retrospectively called modernization and viewed as a historical force was actually a contingent, mutual shaping of technology and culture.

In the first part of the book, I use *technology* anachronistically, mainly because it was not a common term in the United States until the 1930s. By that time, its anthropological meaning of the artifacts that characterize a culture had generally replaced its nineteenth-century meaning of knowledge about technical processes. Like many historians and sociologists of technology, I use *technology* in a broad manner, to refer to artifacts, knowledge, skill, and related practices.[18] I use the term *network* to refer to the way these technologies were organized into systems—like the telephone, automotive, radio, and electrical networks—and to the links among mediating agencies and local institutions.

In exploring these networks, *Consumers in the Country* focuses on what the historian Ruth Schwartz Cowan has called the "consumption junction," the mediation—by advertisers, sales people, and others—between groups we call *consumers* of technology and those we call *producers* of technology, such as inventors, engineers, managers, and workers.[19] The line between producers and consumers is, of course, blurred and dependent on one's perspective. One industrial group (such as automobile manufacturers) consumes the products of another industrial group (such as steel companies), workers use technology to make products, and consumers become producers when they engage in paid or unpaid labor. I investigate how people assuming the role of everyday consumers respond to mediators of technology, how consumers help to construct all aspects of a technology by using it, and how the actions of these groups help to create social change. This reciprocal form of social construction examines how farm people reinterpreted four supposedly urbanizing artifacts and systems by using them in alternative ways, often by altering them, and how producers responded to these actions, often by introducing new or modified technologies. The result was the creation of new forms of both technology and rural life.

Resistance was a key element in this process. Rather than viewing resistance to new technology as an irrational behavior exhibited by traditional cultures, I use the term to refer to a common response by a wide variety of urban and rural groups. Forms of resistance have ranged from the Luddite revolts in Britain in the early nineteenth century to organized protests against nuclear power and biotechnology, and resistance by consumers to information technology, in the late twentieth century. Slaves, factory workers, peasants, and other marginal groups have resisted being exploited by technology. In rural America, midwestern farm laborers broke machines during the economic crisis of the 1870s. Farm men and women, both black and white, resisted the imposition of scientific-farming and domestic-science practices before World War II. Midwestern farm men and women tended not to purchase a full complement of modern household appliances. The Old Order Amish in Pennsylvania still refuse to own a telephone or an automobile.[20]

Although recent historians have documented the resistance to the modernization of rural America, most commentators have assumed that farm families welcomed the telephone, the automobile, radio, and electricity with open arms. *Consumers in the Country* shows that the initial resistance to these technologies was widespread and was part of a complex process of contested social change. Farm men and women employed several forms of resistance, from physically attacking a new technology (such as booby-trapping roads to prevent speeding autos from killing farm animals) to using technologies in ways not sanctioned by manufacturers (such as playing music on a telephone party line) to acts of what is usually called "consumer resistance" (such as declining to buy electric ranges).[21] These actions helped to transform both artifacts and social practices, as consumers resisted the imposition of a new technology and its prescribed use on an established way of life.

By analyzing the attempts to urbanize rural life as a contested interaction between producers and consumers, I bridge two historiographic traditions. I join recent scholars in attributing agency to mediators and consumers, in the stories we tell about technology, and to farm women and men, in the stories we tell about rural life. In these accounts, mediators (such as home economists, agricultural implement dealers, utility company salespeople, and chain stores) are no longer simply channels for the direct diffusion of technology and cultural values from producers to consumers; users of technology are more than passive consumers; and farm people resist, to varying degrees, the efforts to modernize agriculture and urbanize rural life. Mediators, urban consumers, and farm people are thus active agents of technological and social change.[22]

Our stories have become more complex in terms of causality because corporations, big government, urban culture, and (for some scholars) technology itself have lost their hegemonic status. These entities are by no means powerless in my account of the transformation of rural life, but neither are farm men and women. My goal is to explore the webs of interactions among promoters, farm people, and new communication and household technologies over the span of sixty years in order to illuminate the contested processes that transformed the family farm in the twentieth century.

Networks of Modernity

Three types of networks map the sociotechnical spaces where promoters and farm people created new forms of rural modernity. Physical systems and related organizations were the means by which farm people wove the telephone, the automobile, radio, and electric light and power into rural life. Government and private agencies formed mediating networks to connect producer and consumer. Finally, farm organizations and communities maintained local networks of family and work cultures.

The attempts to urbanize the countryside varied with each type of technological system and with the time at which it was introduced. Generally, World War I marks the period of rapid growth of corporate and government mediating networks. Before the war, farmers bought telephone equipment and organized local cooperatives to build community telephone systems. They bought automobiles to run on roads they themselves had built and maintained for the horse and buggy. Farm journals, advertisers, and car dealers were the chief mediators for these technologies. The agricultural complex, a government-university-industrial triangle similar to the younger military-university-industrial complex that began in World War I, extensively promoted the industrialization of agriculture at the turn of the century. But its agencies—agribusiness, the U.S. Department of Agriculture (USDA), land-grant colleges, and agricultural experiment stations—did not attempt to urbanize rural life on a large scale until after World War I, when it promoted the radio and rural electrification.[23]

In contrast, farm organizations had mediated between producers and consumers since the beginning of the century in their efforts to improve the quality of rural life. Farmer institutes—a popular form of adult education, usually sponsored by agricultural colleges and state boards of agriculture—addressed these issues from the beginning of the movement in the 1860s. By the 1890s, institutes regularly featured speakers on domestic science and the farm home. One lecturer spoke at fifteen meetings in Missouri on improving and maintaining roads in the 1898–99 season.[24] The National Grange (Order of the Patrons of Husbandry), the main organization of farm people before World War I, was an active agent of modernity. An annual picnic sponsored by the Pennsylvania Grange included displays of sewing machines, washing machines, and then automobiles alongside the more numerous exhibits of agricultural machinery from 1873 to 1916. Electricity began to light the picnic grounds in the 1880s. Women lecturers enjoined Grange members to adopt the methods of domestic science in the nineteenth century and lobbied colleges to teach it. In the early twentieth century, they vigorously promoted the new field of home economics, which had found a home in the agricultural colleges. The National Grange established a Home Economics Committee in 1910, four years before the USDA began to send large numbers of home economists into the field to urbanize rural life.[25]

The interest in the status of farm women drew on a national discourse that had its origins in movements for health reform in the nineteenth century. Although reformers before the Civil War upheld farm women as models of robust health, others worried that their arduous domestic duties and long hours ruined their health. In an article published in the first annual report of the USDA in 1862, Dr. W. W. Hall, a health expert from New York City, called attention to the problem by characterizing farm women as overworked and isolated—and thereby prone to insanity. Hall cited figures from two mental hospitals in Pennsylvania and Ohio to counter the common opinion that rural living was healthier than city life. "In passing through a lunatic asylum the visitor is sometimes surprised to learn that the most numerous class of unfortunates are from the farm. . . . The statistics of the insane in Massachusetts show that the largest numbers of cases were farmers' wives."[26]

The charge that farm women were inclined to insanity—which later analyses showed to be false—lived a long life. A woman Grange lecturer in Wisconsin guessed in 1878 that one-half of the inmates in insane asylums were farm women. Another Wisconsin Granger said in 1880 that her hard work made her understand why farm women went insane. Boards of agriculture in Vermont and Ohio, concerned about abandoned farms and increased migration to the city, wrote in 1877 and 1884 about sick, weak, nervous, and overtaxed women living in isolated farmhouses. Skeptics had only to stop at country graveyards, "part the long grass about the head-stones and read the names of wives who did not live out their days" or to step inside the "insane asylum and see the worse than dead women brought there from country homes." A Pennsylvania woman told her county Grange in 1901 that she had "read one day where there are more farmers' wives who go insane than any other class of people."[27]

The issue of the overworked farm woman infused a rural reform movement that began when President Theodore Roosevelt formed the Country Life Commission in 1908. The main governmental body that promoted the urbanization of rural America before World War I, the all-male commission typified the Progressive Era's faith in scientific experts as instruments of reform. Liberty Hyde Bailey, director of the New York State College of Agriculture at Cornell University, chaired the group, which initially consisted of Kenyon Butterfield, president of the Massachusetts Agricultural College and an early rural sociologist; Henry Wallace, a Presbyterian minister and Iowa pioneer who founded *Wallaces' Farmer*; Gifford Pinchot, head of the Bureau of Forestry at the USDA and the nation's leading conservationist; and Walter Page, the editor of *World's Work* and a proponent of a back-to-the-land movement. W. A. Beard, the editor of the *Great West Magazine*, and C. S. Barrett, a leader in the Farmers' Union in Georgia, were added to give more regional representation. The commission held thirty well-publicized hearings throughout the nation, received almost 100,000 replies from questionnaires sent to farmers and agricultural leaders, and heard from about two hundred meetings held in district schools. Although agriculture seemed to have recovered from the depression of the 1890s—farm prices had risen dramatically, the number of farms had increased from 4.6 million to 5.7 million between 1890 and 1900, and the farm population had grown from 25 million to 30 million in the same period (see table A.1)—the commission found much to improve. It blamed the growing migration from farm to city, a national concern, on the relatively poor economic status of the farmer and the hardships of the farm woman vis-à-vis their urban counterparts. The commission was also concerned that inefficient farming was the cause of high food prices.[28]

To remedy matters, the commission recommended three major programs: regular conferences on rural progress, a national system of farm and home extension work run by state agricultural colleges (which Butterfield and Bailey had long advocated), and a comprehensive survey of country life. The commission praised new communication, transportation, and household technologies as means to improve the quality of farm life. Rural free delivery, good roads, and cooperative telephone systems were "correctives for the social sterility of the open country." The commission urged government and private groups to strengthen efforts in these areas, develop hydropower to electrify farms, improve rural sanitation, establish a federal highway-engineering service, and ease the lot of the overworked farm woman with laborsaving devices.[29] Although Congress refused to allocate money to tabulate the survey and widely distribute the commission's report, the Senate published a limited number of copies in 1909.

The rural response to the report was mixed. Agricultural papers that had championed similar measures praised the commission, but many farm journals resented the moralizing tone of this "rural uplift," especially when slum-infested cities seemed more in need of reform. Several farmers wrote the commission that instead of welfare workers, they needed higher farm prices to enable them to buy more of the telephones, household appliances, and other items recommended by the commission. In this vein, the *Des Moines (Iowa) Homestead* declared that the elevation of country life "will be done by the farmer himself." A Kansas farmer asked Roosevelt to "give us a chance to make money and let the spending to us." The head of the National Grange thought his grassroots organization was a better country life commission than Roosevelt's because it was more attuned to the actual needs and desires of farm people.[30]

These complaints reveal an age-old, urban-rural tension between reformers and farmers that characterized attempts to introduce supposedly urbanizing technologies to the countryside. When faced with resistance to their efforts, promoters—including farm leaders—usually attributed it to the irrational "backwardness" of farmers, who opposed "progress," and forged ahead.

Rarely did they solicit opposing viewpoints or question their reformist beliefs, especially the gospel that science and technology were modernizing forces. Even Liberty Hyde Bailey, who was much more of a traditional agrarian than a technocrat, apparently did not care whether the commission's extensive survey was analyzed. A recent analysis of its replies shows that, contrary to the commission's report, a "*majority* of the total sample and of the farmers were *unqualifiedly satisfied*" with the condition of their homes, rural sanitation, and communications services.[31]

Despite these omissions and rural opposition to the project, the Country Life Commission spawned a vast network of private and public mediating agencies to reform rural life, many of which followed the commission's lead in counting technology as an ally in the movement. Rural reformers—mainly urban-based educators, journalists, social workers, and businesspeople—adopted the report as the blueprint of a Country Life movement and attempted to carry out each of the commission's proposals. As the rural arm of the progressive movement, reformers set up state and national conferences on rural life, which led to the founding of the American Country Life Association in 1919. Numerous Country Lifers, including those from church groups, conducted social surveys on the rural church, school, and standards of living. Charles Galpin, a student of agricultural economist Henry Taylor at the University of Wisconsin, pioneered rural social surveys before World War I and headed the USDA's Division of Farm Life Studies, established in 1919. Following efforts at teaching rural sociology by Butterfield and other educators at the turn of the century, Galpin, Dwight Sanderson at Cornell, John Kolb at Wisconsin, Carle Zimmerman at Minnesota, and their colleagues established rural sociology as an academic discipline in the 1920s.[32]

The major public network that the Country Life movement helped to create was the USDA's Cooperative Extension Service, established by the Smith-Lever Act in 1914. The act institutionalized a cooperative extension system based on the experience of several educational experiments in the field, especially the work of Seaman Knapp to combat the cotton boll weevil at the turn of the century. Instead of relying on farmers to read government pamphlets or attend farmer institutes, Knapp used federal and private money to employ USDA-supervised field agents to establish farmer-operated demonstration farms. These efforts brought readily observable evidence of the practical success of new techniques supplied by agricultural colleges. By 1910, 450 county agents conducted demonstration work in twelve southern states. Several northern and western states had started extension programs before the Smith-Lever Act established a system of cooperation between the USDA and agricultural colleges. The USDA cooperated with the colleges in planning and implementing the program; state or local monies were required to match federal funds; states selected which land-grant colleges would administer the fund; and, generally, state extension directors at the colleges appointed county agents and provided them with program material. The agents were joint federal and state employees. Reflecting the gendered, middle- and upper-class urban ideal of domesticity common at the time, male agents demonstrated agricultural techniques in the field and barn, while female agents demonstrated home economics methods in the house. The Cooperative Extension Service was further divided by region, race, and class. The USDA established separate offices for northern and southern states (which were consolidated in 1921), county agents were segregated in the South, and most agents concentrated on prosperous farmers. Farm and home bureaus—consisting of prominent farm families usually organized by the county agents themselves—reinforced these divisions by providing part of the matching funds for a large number of county agents in the Midwest and Northeast. This arrangement and the marketing of other cooperatives established by farm bureaus further complicated the employment status of county agents, who were supposedly public employees.

The extension system grew rapidly with the federal government's efforts to increase agricultural production and promote the conservation of food in World War I. The number of counties having farm agents increased from about one-half of those in the nation in 1917 to nearly 90 percent in 1918 (2,435 out of about twenty-eight hundred counties). The number of counties having home agents increased in the same period from about one-fourth of all counties to about 60 percent of them. Although four hundred counties dropped their farm agents and nine hundred counties their home agents in 1919, when Congress did not renew their emergency funding, the extension system emerged from the war considerably strengthened. The formation of the American Farm Bureau Federation in 1919 gave the county and state bureaus national leadership and provided much support for a bipartisan "farm bloc" in Congress. Membership in the politically powerful farm bureaus totaled more than 400,000 in 1921. Responding to criticism of the close ties between the private farm bureaus and the public county-agent system, the USDA and the Farm Bureau Federation agreed that county agents would restrict themselves to educational work and not organize or manage co-ops or farm associations—an agreement that proved difficult to enforce.[33]

By 1920, then, the nation's farm families (32 million people living on 6.5 million farms) faced an extensive mediating network of public and private agencies committed to the goal of industrializing agriculture and urbanizing rural life. Farm men, women, and youth encountered these agencies in the country as well as in town, read their exhortations in newspapers and farm journals, listened to their messages on the radio or at meetings of the Grange and local farm bureaus. "Leading" farm families helped run the latter agencies. The process of network building intensified with the agricultural depression of the 1920s and the economic crisis of the 1930s, especially when county agents became the local administrators of New Deal programs like the Agricultural Adjustment Act.[34]

The extent of this urbanization network is clearly shown for the case of rural electrification by a diagram that appeared in an internal "confidential" booklet published by the Rural Electrification Section of the General Electric Company in 1938 (see fig. 3). The other groups depicted in this diagram would probably have drawn different networks from their points of view. General Electric (GE), of course, is at the top of its own diagram; then come layers of mediating agencies between GE and the farm customer. On the second level of the network are communications media, trade associations, professional societies, an industry rural electrification group, government agencies like the REA (Rural Electrification Administration), the USDA, equipment manufacturers, and farm groups. The third level shows those who dealt directly with farm people: land-grant colleges, their extension services, county agents, home economists, leaders of 4-H Clubs for farm youth, private utility companies, cooperatives, and appliance dealers. Finally, we get to the box representing farm men, women, and youth (30 million people in 1938, about one-fourth of the U.S. population), whom a not insignificant number of promoters were trying to convince to buy electricity. A major goal of these networked agencies was to turn farm people into a new group called "farm customer"—that is, consumers of electrical goods—partly by changing their division of labor to the gendered one of the urban domestic ideal. This attempt is symbolized by the label, "Farmer, Wife, and Children," an implied division of labor into male agricultural work and female housework that would have seemed artificial to most farm people in this period.

Resistance was widespread within this network, an indication that the attempts to urbanize rural life were fractured and difficult to achieve, not a monolithic effort by omnipotent government and corporate agencies. Many of the groups depicted in the diagram clashed over how best to reform rural life or transfer technology to the countryside. The early REA held a public-power ideology and criticized private power companies, their trade associations, manufacturers, and the Cooperative Extension Service (before the

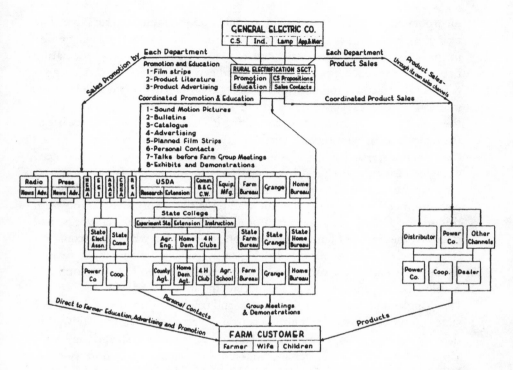

Figure 3. Agencies Influencing the Farm Family. The mediating network of electrical modernization from General Electric's point of view in 1938
Source: General Electric, "Cash In on the Great Farm Market," booklet, 1938, General Electric Collection, Schenectady Museum, Schenectady, New York, p. 9.

REA joined the USDA in 1939) for "skimming the cream" off the market by focusing on wealthy farmers. In turn, the utility industry attacked the REA and its cooperatives as socialistic New Deal organizations. The National Grange accused the younger, business-oriented Farm Bureau of catering to large commercial farmers. Home economists in the USDA criticized REA home economists for doing sales work rather than education. Co-op boards of directors and superintendents resisted implementing the REA's aggressive sales efforts, as did many appliance dealers, who thought most farmers could not afford electrical appliances.

Although there was a large demand for electricity, many farm people resisted joining an REA co-op to obtain electricity and then bought a selected group of electrical appliances, just as farmers had previously resisted and then incorporated the telephone, the automobile, and the radio into rural life. The task of weaving the latter technologies into rural culture was somewhat easier in the early part of the century because the (mostly commercial) mediating networks were much smaller and had not been coupled to the massive governmental and education networks established after World War I. New communication and transportation systems also benefited later agencies such as the Cooperative Extension Service and the REA, which made effective use of the telephone, the automobile, and the radio to reach farm people.

A major reason farm groups in all periods were able to weave these technologies into rural life was that they—like the ethnic workers in interwar Chicago studied by Lizabeth Cohen—encountered new consumer goods in communities having socioeconomic traditions that contrasted sharply with the urban ideal of mass consumption.[35] Farm people, in other words, had their own networks of family and work practices within which they could resist the social implications of technological systems and mediating networks. By the early part of the twentieth century, various regions of the country had developed (agri)cultures based on specializing in different types of farming: New England in dairying, fruit growing, and truck farming; the mid-Atlantic states in truck farming; the South in cotton, tobacco, and rice; the Midwest in dairying, corn, and hogs; the Great Plains in cattle and wheat; and the Pacific Coast in fruits and vegetables, watered by massive irrigation systems. Whereas sharecropping dominated the cotton and tobacco South and huge "bonanza" farms proliferated in the West, smaller family farms, dear to the durable agrarian myth, characterized much of the rest of the country for most of the century. Social networks centered around family, church, and school created structures of mutuality, work sharing, making do, and fellowship, especially in open-country neighborhoods of the Midwest that were a generation or two past the pioneer stage. Roads were poor by urban standards but good enough to support the lively rural institution of "visiting." Farmers may have been physically isolated from the city and other farmers, as urban critics had maintained since the mid-nineteenth century, but they were rarely socially isolated.[36]

Gender structures, identities, and symbolism pervaded farm life. Although the cultural and economic status of farm families varied widely by region and time period, gender relationships remained fairly stable over time. As head of both the farm and the family in the nineteenth century, men were in a position to control the productive and reproductive labor necessary to sustain a large family and, increasingly, to farm on a commercial basis. By the turn of the century, farm women had gained more control over their public and domestic lives as gender relations changed with "modernization," but many traditional sexual divisions of labor remained. On most family farms, men (husband, sons, and hired hands) performed what were regarded as the main income-producing activities in the field, barn, and machine-shop; women (wife, daughters, and hired help) performed "supportive" tasks (from both men's and women's points of view) in the house, garden, and poultry shed. Men and women often shared tasks in the dairy. Although many farm women worked in the field at harvest time and at other periods of labor short-ages, they usually viewed this economic function, as well as their income from selling vegetables, eggs, and dairy products, in terms of "helping out" the man in the field so that the farmstead could stand on its feet economically. For the same reason, women before World War II seem to have accepted the mecha-nization of "men's" jobs in the field before the mechanization of "their" work in the house, but not without some protest.[37]

These and other enduring and flexible aspects of rural culture gave farm people the social and economic spaces within which to resist, modify, and se-lectively use the telephone, the automobile, radio, and electricity. Middle-class farm men and women may have dressed more like city folk when they went to town to buy a car in the 1920s or attend an REA meeting about getting electricity after World War II, but they did so in order to build a better rural life, not the urban or suburban one promoted by modernizers.

Consumers in the Country explores the powerful ideologies that inform the scissors picture and the GE diagram and examines what the drawings leave out: the attempts to urbanize the farm before electricity and the complex in-teractions among technology, promoters, and farmers. Urban and rural folk may have believed that technology drives social change and that moderniza-tion is a monolithic force, but their actions over the course of sixty years chal-lenge these commonly held beliefs. In contesting the social transformation commonly ascribed to up-to-date communication, transportation, and house-hold technologies, farm people created new technologies and new forms of life. They helped to (re)invent rural life in ways not envisioned by Country Life reformers and other promoters. By weaving modified forms of four sup-posedly urbanizing technologies into their lives, they created novel forms of modernity on the American farm.

Part One

URBAN TECHNOLOGY AND
RURAL REFORM

CHAPTER I

(Re)inventing the Telephone

The telephone was the first of our four allegedly urbanizing technologies to be adopted in rural America. As in the case of the automobile, radio, and electrification, inventors developed telephone systems in the third quarter of the nineteenth century, but none of these technologies became widespread until the 1920s. All were initially marketed as expensive business goods or urban luxuries. In the sparsely populated countryside at the turn of the century, where communication and transportation were highly valued, farmers took up the telephone with more relish than the automobile. They found the telephone much easier to control than the dangerous "devil wagon," which roared from the city into the country. But in many other ways, the experience with the telephone established patterns of resistance and assimilation in the use of new consumer technologies to create novel forms of rural modernity before the New Deal.

In the early twentieth century, rural folk often saw the telephone as a sign of urbanization. Telephone instruments, poles, and lines—usually built by farmers or companies independent of the Bell system—connected farm to neighboring farm to town, signifying to many that the countryside had adopted city ways. A newspaper in Galesburg, Michigan, noted in 1904 that "an agent of the Rural Telephone Service has been canvassing our township, and nearly every farmer has subscribed. . . . With rural mail, telephone, etc., the village folks will have to move out into the country to enjoy the city facilities." A newspaper in Garner, Iowa, reported that "something has been doing in the telephone construction business on [Rural] Route 2. Our route is now lined up from Cedar Lake to the Rocky Mountains. With free mail delivery and complete telephone connections we are not slow, even if we don't live in town!" "About sixty new telephones will be put in this division," reported the

Ashland (Ohio) Press in 1904. "Wires are strung from all directions and they make us look kind of citified."[1]

The urban culture portrayed in magazine articles and consumer advertising did much to define which technologies made one "citified." A writer in *New England Magazine* claimed in 1905 that the interurban trolley, rural free delivery of mail (RFD), and the telephone were "urbanizing New England" by connecting the country to the city: The trolley enabled farm women living along its route to go to town more frequently and their children to attend high school. City and town newspapers were brought to the farm by RFD. The telephone allowed farmers to summon the doctor in emergencies and to order supplies; it was "about the greatest urbanizer on record."[2] Stromberg-Carlson, a telephone manufacturing company, ran an advertisement in *Wallaces' Farmer* that year entitled "Modern Country Life" (fig. 4). Drawings of a horse-drawn mail wagon, a wall telephone being used by a well-dressed farm woman, and an interurban trolley herald the new way of life. "The rural mail delivery, the telephone, and the suburban electric railway are working wonderful changes in the life of the farmer's family today," claimed the ad. "The former isolation, which drove many of the young men and women from the farm to the city, has been banished by the many telephone lines now in use all over this country."[3]

Agricultural professionals, imbued with the spirit of the Country Life movement, agreed. The article on rural communication in the 1909 *Cyclopedia of American Agriculture* states that the "striking contrast between city and country life is perhaps today the greatest single cause of that so-called rural depopulation which is attracting the attention of statesmen and reformers the world over. It is reasonable to suppose that it can be done away with by perfecting the agencies of rural communication," especially the telephone. The agricultural economist Thomas Carver praised the telephone in 1914 for improving rural communities. "Probably nothing has done more for country life than the rural telephone. Whenever it has come into general use it has overcome the isolation of farm life as nothing else could have done."[4]

Like most expressions of technological optimism, these statements exaggerate a social problem (farmers were probably no more socially isolated than city dwellers) as well as the ability to solve it. Telephone lines spread throughout the countryside, but the telephone did not keep people on the farm. Neither did the interurban trolley, which the automobile replaced, nor RFD. A federal communications monopoly founded in 1902, RFD became even more commonplace than the telephone, connecting about three-fourths of the nation's 6 million farms to towns and cities in 1910.[5] Despite these improve-

Figure 4. Modern Country Life. Stromberg-Carlson's depiction of the telephone as the latest modernizing agency, joining RFD and the interurban trolley, at the turn of the century
Source: Wallaces' Farmer, Jan. 6, 1905, p. 17.

ments, the percentage of the U.S. population living on farms continued to fall in this period, landing (temporarily) at a level of about 25 percent in 1930 (see table A.1). Some commentators even thought RFD and the telephone, by connecting the farm more closely to the city, increased rural-to-urban migration.

Yet those who stayed on the farm valued the telephone. When the Bell system and the independent telephone companies were slow to bring the new means of communication to the countryside, farmers organized cooperatives to do the job during the agricultural prosperity in the early part of the century. They usually connected their lines to town by hooking up to independent companies or the Bell system. Farm people took matters into their own hands in another way by (re)creating such rural customs as "visiting" on the party line. Bell and the independents fought the practice but finally gave in and designed a telephone system specifically to accommodate it. Using the telephone in these and other ways altered the social geography of rural America, but not always in the manner desired by its promoters or by many farmers.

Cooperating with the Telephone

Farm people played a much larger role in the development of rural telephony than they had in establishing RFD. In fact, rural telephony got off the ground only when farm people drew on the long-standing custom of organizing cooperatives to provide services in the countryside, which telephone companies were reluctant to serve. The reason was economics. The Bell system (which comprised AT&T [American Telephone and Telegraph], Bell operating companies, and manufacturer Western Electric) initially concentrated on the more profitable urban market and sold the telephone as an expensive communications device for businesspeople in large cities. In 1895, nearly all telephones were found in businesses rather than in residences. The largest cities had more than one-half of all telephones, midsize cities nearly one-third, and rural areas a minuscule 3 percent.

This picture changed dramatically after the basic Bell patents expired in 1893 and 1894. Independent companies sprang up across the country to fight the Bell monopoly with populist rhetoric, lower prices, and their own equipment. By starting in the less populous areas neglected by Bell and expanding into large towns, midsize cities, and almost every major city except New York and Chicago, the independents spurred an intense period of competition to hook up the most people to their (usually separate) networks. By 1920, the number of telephones had increased more than fiftyfold since 1895; slightly more telephones were found in residences than in businesses; and—as stated earlier—a larger percentage of farm households had telephones than nonfarm households (39 to 34 percent).[6] The number of telephones on farms varied considerably by region, with the Midwest and Northeast leading the Far West and South (see table A.2). In each of the top four states—Iowa, Kansas, Nebraska, and Illinois, all midwestern states with fairly strong cooperative traditions—more than 70 percent of farms had a telephone. In each of the bottom four states—South Carolina, Louisiana, Florida, and Georgia, all with low farm incomes—fewer than 10 percent of farms had a telephone.

The large increase in rural telephony resulted from the combined efforts of farm men and women (building telephone cooperatives), independent companies, and, eventually, the Bell system. Although the Southern New England (Bell) Telephone Company pioneered in establishing "Club Lines" to connect farmers to its Connecticut exchanges in 1896, independent manufacturers and editors of farm journals in the Midwest promoted rural telephony more extensively than Bell.[7] Firms like Stromberg-Carlson and the Swedish American Telephone Company advertised telephones in farm journals, whose editors praised the telephone's business and social functions. Because telephone companies usually found it too expensive to build lines into sparsely populated areas, promoters advised farm people to buy their own telephone equipment, build their own lines, and create cooperatives to bring phones to the countryside.[8]

Farm men and women turned out to be more innovative than manufacturers had predicted. Ranchers and farm men built many of the early systems as private lines to hook up the neighbors, often using the ubiquitous barbed-wire fences that divided much of the land west of the Mississippi. Although these lines became the butt of urban jokes and yet another symbol of the backwardness of country life, rural (and other) folk correctly saw them as an economical way to bring a modern technology into everyday use. In fact, three of the earliest systems were not farmer lines. In 1900, *Scientific American* celebrated the ingenuity of an Indiana man for using the top wire of a barbed-wire fence to construct an inexpensive fourteen-mile, five-party line between three small towns. Galvanized wire provided continuity at breaks in the fence and at highway and railroad crossings; the earth provided a "ground return," as with other one-wire systems. "The papers often contain stories of western farmers who have used fences of barbed wire for telephone or telegraph lines," observed the *Rural New Yorker* that year. "Many of such stories prove on investigation to be 'fakes,' but J. H. Taylor, of the Rhinehart Cheese Co., Kan., sent us this account of such a telephone." Three and one-half miles of the system used the top wire of a barbed-wire fence; the remaining one and one-half miles consisted of "a single loose wire lying loose in the top of a hedge fence." Two years later, *Telephony*, the Chicago-based magazine of the independent movement, praised a barbed-wire line that connected the ends of an irrigation project through twenty-five miles of Colorado foothills at a cost of only ten dollars.[9]

The practice became so common among midwestern and western farmers at this time that it caught the attention of the *New York Sun,* the urban magazine *Outlook,* and the U.S. Census Bureau. A Bell operating-company president complained to AT&T in 1903 that "The Farmer Lines in the territory of the Pacific States Tel. & Tel. Co. are constructed mainly upon fences." (Even if well made, the grounded circuit did not meet Bell's long-distance standards.) Other telephone men remarked that fence-wire lines were common in Iowa and Nebraska. Edmund Burch, a farm leader in New Mexico, started a rural telephone company in 1904 by installing a barbed-wire system after reading reprints of the *Scientific American* article and similar ones in his local newspaper.[10]

The barbed-wire telephone quickly entered the lexicon of urban humor about farm life—at least in the telephone industry. After praising the irrigation line in Colorado, *Telephony* began reprinting local newspaper stories about barbed-wire systems in its humor section, "In the Rural Line Districts." Farm people in Minnesota, Ohio, Pennsylvania, and South Dakota built lines out of barbed wire.[11] Farmers in Iowa and Indiana used barbed-wire fences as lines and put telephone wires on fence posts as temporary measures while waiting on a regular line to be built.[12] The independent companies called the systems "squirrel lines" (after squirrels running along the top of fences), and one Minnesota group named itself—proudly?— the "Carlisle Barbed Wire Telephone Company."[13] Farm men upgraded two systems by installing a regular telephone line in Indiana and moving the line from fence posts to new poles in Virginia. But instead of praising these actions as examples of modernization, *Telephony* featured the stories as more evidence of the hayseed behavior independents had to contend with in dealing with the necessary evil of "farmer lines."[14]

Farm journals were more understanding of rural conditions. J. W., of Weston, Oregon, asked the *Rural New Yorker* in 1903 if it was practical for twelve farmers to use a barbed-wire fence for part of their ten-mile line. The journal's telephone expert replied, "That's right. Build your own lines and run them, too, as other farmers are doing. Build strictly first-class as far as possible, but build your line by all means in some way. Use barbed wire only when absolutely necessary; 2×4 scantling 12 feet long may be spiked to fence posts every seven or eight rods if posts are well set."[15]

More commonly, telephone cooperatives built lines similar to those used by independent companies for rural areas. In an effort to cut costs, farmers usually built a grounded circuit (with either a fence line or a dedicated telephone line) instead of the more expensive metallic circuit favored by Bell (which used two wires to eliminate interference from interurban trolley and other electric lines). Because of technical difficulties in communicating over rural lines, which were typically longer than urban lines, farm people bought magneto-generator sets powered by local batteries instead of using smaller telephones connected to a large battery at a central office. As companies gradually installed desk-stand and hand-held telephones—made practical by the central-battery system—in towns and cities from the late 1890s to the 1930s, the magneto set, once common in the city and the country, became marked as the old-fashioned "farmer" telephone. Turning a crank extending from the right-hand side of the upright wooden case sent an electrical ringing signal from the magneto generator to the operator or a neighbor on the party line. Coded rings signaled neighbors. Dry-cell batteries provided the current needed to talk on long rural lines with a large number of parties on the line.[16]

The prevalence of party lines in the country can be seen in the Census Bureau's figures for 1907. Independent farmer or rural lines averaged thirty-two telephones per line and one telephone every 1.2 miles, mutual systems had ten phones per line and one every 1.3 miles, and commercial systems hooked up only seven phones on a line, at a rate of one every three-quarters of a mile.[17]

The party line lent itself well to the cooperative form of organization, which had a long history in some regions. In forming telephone cooperatives, farm men and women established mutual associations much like the cooperative grain elevators, creameries, cheese factories, and livestock rings they had set up since the beginning of the Grange movement in the 1870s. At the turn of the century, cooperatives thrived in the Midwest. They operated in Minnesota and Wisconsin, where immigrants had imported the Scandinavian agricultural co-op; in the rich farmlands of Iowa, Illinois, and Indiana; and in the populist strongholds of Kansas and Texas.[18] Like the early cooperatives, which drew on the rhetoric of populism to fight the perceived injustices of middlemen unfairly marketing farm products, the telephone co-ops struck a blow against another monopoly, the Bell "octopus." Independents responded by naming many of their companies the "Home" or "People's" Telephone Company. The Swedish American Telephone Company and Stromberg-Carlson, both founded by Swedish immigrants, drew on that country's cooperative tradition to promote telephone co-ops.[19] The main function of these co-ops, however, was not to engage in populist politics but rather to provide a communications service that Bell and the independents considered unprofitable because of the sparse population in the countryside.

The organization and size of telephone cooperatives varied widely. Later analysts distinguished between "pure mutuals"—which prorated expenses among members, usually did not charge for service, and collectively managed the line—and "stock mutuals"—which issued stock for capital, charged for service, and delegated management to the directors.[20] Although stock mutuals were usually larger, both groups operated under cooperative principles. Stock mutuals were typically composed of groups of farm people, from fifteen to fifty households, each of whom would buy one share per telephone in the cooperative, costing from ten to fifty dollars, and purchase their own telephone sets and wiring at a cost of about fifteen to twenty-five dollars. Annual fees ranged from nothing to eighteen dollars. As part of the gendered division of labor in rural, as well as urban, telephony, men and boys would often cut their own poles, help set them, string the line, and maintain the system for both pure and stock mutuals. The larger cooperatives often paid a woman and her daughter or daughters to operate a switchboard in a farmhouse. Many cooperatives connected their lines to the nearest trade center and other communities through such a switchboard or through one in town, paying a switching fee or negotiating free service because of the large business use of the farm telephone.[21] An Indiana promoter explained the process to AT&T in 1903. "The ordinary method of organization is for a farmer who has a wood-patch to furnish the poles. The other farmers haul and set them. The right of way costs nothing, and, as the work is done in spare time, the pole lines when completed represent no cash outlay; and, from the stand point of the farmer, has cost nothing."[22]

The U.S. Department of Agriculture (USDA) gave organizational advice to some cooperatives in 1915 and published a booklet in 1921 on the status of telephone co-ops and how to organize them. But farm people took the initiative in most cases. Most cooperatives, whether pure or stock mutuals, were neighborhood affairs. A group of twenty-four farms established a twelve-mile line in the early part of the century near Plattsburgh, New York, where a local storekeeper ran the switchboard. In Indiana, the Deer Creek Cooperative Telephone Company incorporated in 1909 and followed standard co-op practices by creating a small board of local directors, limiting the number of shares per member to the number of telephones owned (and limiting that to three), and giving only one vote per share. Three years later, twenty-two California farm men and merchants formed the Mountain Farmers' Telephone Company when a privately owned system doubled its rates. They took nine days to build a metallic line about twenty-five miles long through Ahwahnee Valley, shared the cost equally, provided labor or a substitute, strung most of the wires on trees, cut some poles from a nearby forest, and agreed to maintain the stretch between themselves and their neighbors—all at the low cost of forty-two dollars per mile and twelve dollars per telephone.[23]

In 1913 a farm man wrote *Country Life* that ten new lines had been built in his county during the past year, averaging about fourteen telephones per line. Fourteen farm families had gathered at a schoolhouse to form his co-op, then convinced a telephone company in the nearest town to extend its lines two miles from the central office. The co-op built a metallic circuit to the exchange limits; individuals built spur lines from their farmhouses to the co-op's line. "We built our line during the pleasant days of March before the work of spring crowded us." Two men staked off the poles 160 feet apart, and eight men followed with posthole diggers; three men lifted each pole while one guided it into the hole. Shares cost fifteen dollars, phones eleven dollars; the rate was fifty cents a month. The *Seymour (Ind.) Republican* indicated the local character of most cooperatives in 1912 when it reported that the "directors of the Farmer's Union Telephone Company were out Saturday repairing the telephone lines."[24]

Some cooperatives that began as stock mutuals grew to become large businesses. A mutual telephone company started by a dozen Virginia farmers in 1895 soon merged with other companies in the area. Within ten years, the Rockingham Mutual Telephone Company had sixteen hundred members. An Ohio mutual grew from eight members in the late 1890s to serve three counties in 1901. A Pennsylvania farm man said a co-op established by four families in 1900 "spread worse than measles." By 1908 the grounded-line system served fifteen hundred telephones and paid out six thousand dollars annually to stockholders. Another cooperative started by a few farm families in Indiana in 1903 had six hundred telephones five years later, all on metallic circuits serving two cities, two towns, and seven villages. By 1909, the Farmers' Mutual Telephone Company operated exchanges in forty-one small towns around its headquarters north of Seattle, Washington. Some stock mutuals did not fare nearly as well in this period. Seventy-five Utah ranchers formed the People's Progressive Telephone Company in 1912 but had difficulty maintaining 260 miles of line during severe winters and folded after five years of fitful operations.[25]

More typically, representatives of several healthy mutuals having more than three thousand telephones in Ohio decided in 1906 to connect their lines to an independent company rather than to the regional Bell company. Upon hearing the news, *Telephony* crowed that "90 per cent of all rural lines in the United States are connected with Independent exchanges. The farmer, who is an anti-monopolist, has no use for the Bell system."[26] The attitude was not uncommon in the Midwest—to the consternation of Bell officials. W. O. Pennell, an engineer for the Missouri and Kansas (Bell) Telephone Company, wrote a lengthy internal report in 1906 about how to fight the mutual movement. Pennell observed that mutuals in his territory were often organized into locals (pure or stock mutuals of ten to twenty farms), which combined into districts at a common switching center paid for by the locals. General organizations, consisting of representatives from district officers and each local, standardized equipment and construction methods and built trunk lines between exchanges. In this way, the Cherokee (County) Mutual Telephone Company coordinated a system of twelve hundred phones with free long-distance service in Missouri. The "Mutual Companies seem to be strongest and most rabid in Iowa," noted Pennell. "The starting of many companies in Missouri and Kansas can be directly traced to the initiative of farmers who have come from Iowa, sometimes carrying their telephones with them as part of the household goods."[27]

The mutual movement was strong in southeastern Iowa before World War I. Nearly 90 percent of the 251 telephone companies in this region were mutual companies. These small companies, the vast majority of which served fewer than fifty patrons, accounted for about one-half of the telephones in the region. More than 90 percent of the mutual company presidents were farm men; only two were townspeople. The mutual movement succeeded largely because its rates averaged about $3.50 per year, much lower than the $10.00 to $18.00 per year charged by Bell and the independents.[28]

The nationwide success of the cooperative telephone movement is shown in statistics compiled before World War I. The Census Bureau's special reports on the telephone for 1902 and 1907 divide companies into three categories: "commercial systems" (independent and Bell companies), "mutual systems" (mostly stock mutuals), and "independent farmer or rural lines" (pure and stock mutuals). The reports show that the noncommercial organizations had slightly more rural telephones than commercial systems in 1902 (at about 150,000 each) and held their own against them in 1907 (at about 700,000 phones each). The number of independent farmer or rural lines, most of which did not operate their own switchboards, was impressive, more than 17,000 lines in 1907. (The 1902 figures are not comparable because the bureau changed the way it counted farmer lines in 1907.) The Census Bureau stopped collecting detailed statistics on rural telephony after the 1907 report, and in

1912 it instituted a new category based on revenue. The number of companies with less than five thousand dollars in revenue, which were mostly rural systems, continued to grow from 1912 to 1917. In fact, more than three-fourths of the sixteen hundred telephone cooperatives that survived the Great Depression and responded to a government survey were established between 1900 and 1919, an era of general farm prosperity.[29]

All studies show that the Midwest, the leader in both the independent and the cooperative movements, dominated rural telephony. Iowa, Illinois, Indiana, Missouri, and Ohio led the nation in the number of rural lines in 1902 and, combined, accounted for 70 percent of all rural telephones. The north-central region of the country also dominated the noncommercial category in 1907, with 80 percent of the phones on mutual systems and 70 percent of those on independent farmer or rural lines. Agricultural economists estimated that Minnesota had about six hundred telephone co-ops in 1913, most of which operated on traditional cooperative principles. Another study counted almost five hundred telephone co-ops in Wisconsin during the farm prosperity of World War I.[30]

Telephone cooperatives were not passive recipients or powerless mediators of a new technology that had originated in the city. Local leaders not only built and managed their own lines, they also exercised substantial political power in many areas because of the socioeconomic geography of agriculture in this period. By farming and living in a trading center's hinterland—whose size and shape depended upon forms of transportation, the type of agriculture, the existence of competing trading zones, and so forth—farm men and women made up much of the economy of that center. By all accounts, they were major economic forces, especially in small trading centers.[31]

Bell engineer W. O. Pennell clearly stated this view of farm economics in 1906:

> The power of the mutual company lies chiefly in the fact that the small agricultural town owes its existence and prosperity to the surrounding country. In other words the farmers make the town. Telephone service is not of much value locally in a small village. The people would rather walk to the store or visit their neighbors than telephone. But the telephone is of great value to the farmer as he is separated from his neighbor and the trading point. The merchants, doctors, and lawyers are compelled to take the telephone if they wish the patronage of the farmers.

Some mutuals, in fact, bargained with towns to see who would give them the best connection rates. Merchants in large towns often formed a mutual company, Pennell continued, being "forced into the subscription by the farmers threatening to withdraw their trade to a neighboring town. . . . It is suggestive of the power of the farmers over such towns that through their influence the merchants can be 'held up' to the extent of building a competing exchange," one that competed with the Bell system. In the struggle between Bell and the independents, the "*systems which get the farmers will win out in the town.*"[32]

Farm people exercised this type of power against the independents, as well. An Indiana man wrote the *Rural New Yorker* in 1902 that the demands of a large farmers' mutual for free exchange service with nearby towns at first "seemed unreasonable" to the independent companies. But, "by working through the newspapers and merchants' association, and by the merchants agreeing to discontinue the company's instruments that did not connect with us, we succeeded in getting what we wanted."[33] An independent leader noted in 1903 that farmers often built lines before small towns had established a telephone exchange. In those communities, "managers who have already made every possible concession to these people are now threatened with rival exchanges if they do not give absolutely free service."[34] The independents calculated that whoever connected the most farms to telephone service and served them at affordable rates (often at a loss) would get the more profitable business in town and from tolls. Another independent leader told a West Virginia meeting in 1909, "Any telephone company's strength in the city is largely dependent on its holding in the rural districts. . . . Like the grocer who makes a leader out of some staple article, say sugar or coffee and sells it at about cost, so should we with our rural service make a leader of this class, but never sell it at an actual loss."[35]

The independents and Bell adopted this general strategy of "access competition" for urban and rural areas to increase the value of their networks. The (constructed) phenomenon of access competition helps explain the dramatic increase in telephone ownership outside of large cities in this period and elucidates the factors that drove the development of rural telephony.[36]

Although independents recognized the political and economic importance of farmers, they were extremely ambivalent about serving this "class" of customers. A. B. Fishback told the Michigan Independent Telephone Convention in 1907 that it was "useless to enumerate the difficulties" of dealing with farmer lines because they were so well known to his audience. "Our farmer friends as a class are a peculiar people with which to deal, their independence and remote positions making it all the harder, and our success depends on our ability to get and keep close to them and make them feel that we are their friends indeed and not from purely selfish motives." Independents should not let farmers "be led astray by our enemy," the Bell system, which used "misrepresentations and false statements" to win over the farmer. Farm people should "remember who you owe the credit for making it possible for you to enjoy the privilege of telephone service and cooperate with them and not the people who tried to keep service from you," the Bell system, which ignored farmers during the monopoly years.[37]

Before World War I, *Telephony* published numerous articles like this, articles that drew on the ideologies of populism, the agrarian myth, and rural uplift to portray independents as the true benefactors of farmers. Bowing to the politics of agrarianism, independents acknowledged farming as the basis of the nation's prosperity, then praised the telephone—often in conjunction with RFD and the interurban trolley—for modernizing rural life. They credited the telephone with turning old-fashioned farms into modern business operations, bringing outside help in times of sickness, fire, and other emergencies, and, above all, relieving the "isolation" of rural life, which would help keep women and youth on the farm.[38] Stromberg-Carlson's evocative image of this urbanizing message has already been mentioned. Commenting on the report of the Country Life Commission—which had praised the telephone, RFD, and good roads in much the same manner as the independents praised these technologies—a Peoria man believed the "telephone has already saved many a farmer's wife from going insane on account of isolation from the world." A 1919 article in *Country Gentleman* perpetuated the myth of insanity-prone farm women. "The old-time loneliness on farms, which is said to have often resulted in insanity for the women, has disappeared since party-line telephones were introduced." Despite the growing evidence against the myth (see chap. 3), it lived a remarkably long life in the telephone industry. In 1934, Harry MacMeal, founder and long-standing editor of *Telephony*, praised independents for bringing the telephone to the farm, thus improving all aspects of rural life, especially the incidence of insanity among farm women.[39]

Although the independents portrayed themselves as the farmer's friend, they routinely criticized these very customers in terms much stronger than those Fishback used.[40] As Claude Fischer has shown in his historical study of the telephone, the "independents' internal literature described farmers as ignorant, hard-headed, short-sighted, and tight-fisted—poor customers." *Telephony* complained that farm people "kicked" about rates, avoided bill collectors, "built inferior connecting lines, cared little about maintenance and repair," did not know about finances, demanded free toll service, "listened in on each others' calls . . . and were generally a rude and bumptious lot."[41]

This cultural prejudice pervaded *Telephony*'s regular feature, "In the Rural

Line Districts." Started in the summer of 1904 as farm lines were spreading throughout the country, the full-page section reprinted extracts from local newspapers, often illustrated by "hayseed" cartoons that ridiculed rural life. Entries made fun of the unintended puns of novel telephone metaphors, odd or picturesque names, the parochialism of rural life, barbed-wire lines, strange uses of the telephone, the habit of eavesdropping on party lines, and the popularity of the "hello girls" (operators) in small towns. Replaced in late 1912 by the equally satirical Alfalfa Granger cartoons, the section's deprecating humor served to reinforce urban feelings of superiority about owning modern technology.[42] Telephony's editors probably thought the section would also help independents vent their frustrations in having to deal with unorthodox farmers to win the access competition against Bell.

Many independents operated rural lines at a loss to protect their small exchanges, but others made a profit in the countryside. The Citizens' Company of Grand Rapids, Michigan, constructed lines for 500 farm telephones during the 1901–2 season and had orders on the books for 150 more. The Rural Union Telephone Company of Humboldt, Iowa, got a head start by buying the lines of a mutual and a Bell company. The new firm replaced all equipment with that manufactured by the Swedish American Telephone Company, constructed 92 miles of metallic and 262 miles of grounded rural lines, and installed 500 rural and 475 city phones for new customers—all in the busy year of 1902. Another large system was the Farmers' Telephone and Telegraph Company in Wenatchee Valley, Washington (the presence of "Farmers" in the name of commercial independents was not uncommon). Organized in 1903, the company served about twelve hundred rural subscribers in 1910 in a prosperous valley dotted by apple orchards. The system used a central battery for signaling, but local battery phones were still required in order to talk over the long lines. In 1911, two smaller successful firms in Indiana had about 250 rural subscribers each.[43]

It did not take long for the independents' rural campaign to catch the attention of the Bell system. Frederick Fish, president of AT&T, wrote Bell operating companies in 1903 warning that "this farmers' line question is one that must be solved."[44] The replies to Fish's letter, and to an earlier inquiry in 1902, indicate that the question was how to win the access competition peculiar to rural telephony. "In our small exchanges the business men who are our patrons, insist that farmers' lines are to them a necessity," wrote the president of the Nebraska Telephone Company, "and unless we give them connection with farmers' lines, then they must organize mutual companies in order to protect their trade which might go to some nearby town with which farmers' lines are connected."[45] Angus Hibbard, general manager of the Chicago Telephone Company, said the effect of serving more than two thousand farm customers "has been absolutely to save the local situation for us in a number of exchanges to which they are tributary. In some of these we have driven our competitors entirely from the field; in others we have more than held our own and have held our exchange rates. In addition, our toll business has been greatly developed at these points." Officials of six other Bell companies advocated a similar policy at this time.[46] "Broadly speaking," wrote the president of the Bell Telephone Company of Buffalo, "we have been forced into the building of farmer lines in order to protect our interests where opposition companies maintain exchanges." In 1906, special assistant F. A. Pickernell told Fish "that if we have learned anything in the past three years, we have learned that we must have the farmers connected with our smaller exchanges."[47]

Having more autonomy in these matters under Fish than they later would after Theodore Vail became president of AT&T in 1907, the Bell companies devised numerous schemes to bring telephone service to farm people. Many companies offered three or four types of rural service, involving various combinations of company-built and farmer-built lines. Even when farmers did not build the main line, they usually built spurs to their houses. Rates depended on how many were on the ubiquitous party line, the type of construction, how much of the line was built and maintained by farmers, the rates charged by the competition, and so forth. Made up of rental charges for the telephone set and a switching fee, rates ranged from $12.00 per year on lines having up to twenty parties, a common independent rate, to $24.00 per year for company-owned metallic lines having only eight to ten parties. Toll calls, usually to a town or another farmer line, ran from five to twenty cents a call. Many systems rented Bell instruments for farmer-built lines; the Nebraska Telephone Company rented a complete set at $4.50 per year in 1902 and would switch farmers at $6.00 per year to keep the "opposition exchange out of Neola, Iowa."[48] More innovative were two companies that pioneered in rural telephony. The Southern New England (Bell) Telephone Company provided automatic coin collecting boxes at five cents a call on its Club Line service, established in the mid-1890s. The Cumberland Telephone and Telegraph Company had treated rural areas as part of its residence business since 1890 by creating a rural exchange district in each county.[49]

At least eight Bell companies relegated the generally unprofitable rural business to sublicensees. (Granting sublicenses to strategically placed independents to control their operation was a common tactic in AT&T's fight against the larger independent movement.) The Missouri and Kansas Telephone Company reported in 1905 that out of twenty thousand rural telephones in its territory, only sixteen were connected to its lines but more than three thousand were connected to its sublicensees.[50] Southern New England encouraged farmers to organize as sublicensees, and a special agent for the Bell Telephone Company of Missouri openly promoted a mutual in 1905. Other companies worked in secret. The Northwestern Telephone Company clandestinely asked merchants to suggest that farmers, who were suspicious of the Bell system, organize and build their own lines to the Bell exchange. The Missouri and Kansas Telephone Company organized a mutual system behind the scenes in Hamilton, Kansas, to combat a countywide cooperative promoted by a local rancher. The rancher sold out and advised his patrons to connect with Bell.[51] These tactics drew the ire of A. B. Fishback in 1907. "This wolf in sheep's clothing, finding he cannot harm the large Independent companies in the city and towns, has turned his attention to the rural districts, spreading discord and strife, and is offering our farmer friends sublicensee contracts and renting them instruments at seven dollars per year and other schemes to break into our territory."[52]

Some Bell companies did have tremendous success in the countryside, especially after AT&T began selling a less expensive rural telephone set around 1906. By the end of 1907, the Pacific States Telephone and Telegraph Company, which had complained about farmer lines in 1903, had increased its number of farm customers to twenty-two thousand. Facing the prospect of losing its small exchanges to independents who hooked up farmers, Pacific States established a "fighting rate of $1.00 per year" for a switching fee and even began selling telephone sets. The company stopped leasing telephones, a wise decision, in its view, because of the rapid obsolescence of Western Electric's "farmer" sets (i.e., the magneto-generator telephones). By late 1908, customers owned almost 10 percent of the twenty-four thousand farm sets connected to Pacific States' lines, and the company was able to keep the opposition out of Salem, Eugene, and Cottage Grove, Oregon.[53] Southern Bell tripled its number of rural telephones between 1907 and 1909 by pushing AT&T's newly designed farmer sets, convincing businesspeople to finance rural lines, and encouraging farm people to organize cooperatives and build lines to a Bell exchange. Bell Telephone of Buffalo, however, spent so much money building rural lines that it got into financial trouble. Pickernell advised an AT&T vice president in 1907 that the company's investment of about $1 million in rural lines "must be disposed of" because it prevented more profitable investment in larger exchanges. Pickernell blamed the troubles on Buffalo's policy of resisting for years the "local ownership of farmer lines and equipment."[54]

As AT&T gained more control over its operating companies, it standardized its rural line polices. In 1912, Vice President H. B. Thayer explained that after a decade or so of experience, AT&T held an "exceedingly flexible" policy, consisting of setting flat rates, limiting lines to twelve parties (fifteen to eighteen in some cases), and permitting users to construct their own lines and own their telephones where they could not pay commercial rates, cutting rates by one-third in those cases. This resulted in operating losses, which were "borne as a matter of policy, and for the sake of the added value given to the urban service by the building of rural lines."[55] Thus, the aggressive style of access competition pushed by President Theodore Vail became AT&T's answer to the "farmers' line question" posed by his predecessor almost a decade before. AT&T, as well as the independents, took these actions because of the socioeconomic power exercised by farmer cooperatives in the hinterland.

Communicating in the Country

Although cooperatives exercised some power in the tangled telephonic web of farmer lines, small-town politics, and corporate power struggles, farm people built telephone systems primarily to communicate with rather distant neighbors and towns. As the first instantaneous and interactive communications technology to be adopted widely in the countryside, the telephone occupied a special place in American rural life.

Like all new technologies, the telephone met some resistance in its early years. Independent companies reported that farm people initially opposed the "new-fangled," "voice-in-the-box" contraption as a citified luxury, one they did not need because of the widening RFD network.[56] Others took action against telephone companies' erecting poles without obtaining a right of way from the landowner. New Jersey farm men cut down several miles of AT&T's long-distance line through the Moorestown area before coming to an agreement with the company in 1903. A farm man in Minnesota chopped down the poles of an independent company in 1905 but later had to pay for his apparently illegal act. In 1908, the *Colfax (Ind.) Messenger* repeated the classic story of an older farm man who found workers "putting up telephone poles through his best field. He ordered the men out, but they wouldn't go. They showed him a paper that gave them the authority to put up their poles wherever they wished. The old man looked at the paper, saw it was lawful, and walked silently away. He walked to the barn and turned a savage red bull into the field. The bull made for the men, the men fled at the top of their speed, and the farmer shouted 'Show *him* your paper.'"[57] Leaders of the more peaceful Old Order Amish in Lancaster County, Pennsylvania, simply decided around 1908 that member's telephones should be put away and that their farmsteads should not be hooked up to the telephone network, a practice they continue to this day. (The use of the telephone was never banned, however, and this most conservative of Amish sects has installed their own community telephones, in "telephone shanties," since the 1930s.)[58]

Some small-town merchants also objected to the telephone. "When we put in our system no merchant would take it," E. H. Collins, of central Indiana, wrote the *Rural New Yorker* in 1902. They were afraid that some farmers would order goods sent out by mail wagon or a neighbor, promising to pay on Saturday, but then neglect to pay. The telephone, however, greatly increased their business. "Another trouble our merchants feared was that it would start a delivery system which is hardly practical in small towns. The system was not in long before a new store was started with a telephone, and its orders were so many that everyone had to put in a telephone."[59] Many people thought that trees were excessively "trimmed" when lines were put up and that the lines drew lightning. James Rice, who helped organize a rural system in Westchester County, New York, disagreed. "Often the trees look better for trimming up, and to my mind the local telephone line adds strength and attractiveness to the landscape, giving the impression that the community is progressive and in touch with the world." Because of the telephone's ability to summon help in time of emergencies, "a house is safer from injury, with a telephone connection than without it."[60]

All of these forms of resistance were relatively short lived and passive, however, especially when compared with the longer and more violent crusade against the "devil wagon" automobile, a crusade described in the next chapter. Throughout the early part of the century, farm people, particularly those in the prosperous Midwest, enthusiastically took up the telephone and adapted it to their way of life.

Many of the early uses of the telephone were the same in the country as in the city. Rural and urban people developed similar calling patterns, used the telephone in innovative alternative ways, and created elaborate party-line etiquettes. But farm people tended to intensify these practices and develop them in ways that became distinctive of rural life. The socioeconomic geography of agriculture, the strong cooperative tradition in many regions, the long-standing customs of visiting and sharing labor, the established business and social reasons for communicating with the outside world, the gendered division of labor on the farm—all of these aspects of turn-of-the-century rural life helped shape the telephonic cultures forming on farms. The telephone did not reshape rural life in a deterministic manner. Farm people, like their urban and small-town cousins studied by recent scholars, wove the rather flexible device into existing communication patterns, transforming themselves and the telephone in the process.[61]

Like families in the city, farm people used the telephone mostly to make local calls, not toll calls between exchanges or long-distance calls. Reporting on conditions in 1899 and 1900, the Central Union (Bell) Telephone Company found that "farmers, as a rule, did not care especially for long-line facilities, their use of the telephone being confined to local service or nearby towns attached to their facilities." Independent companies and cooperatives noticed this phenomenon, as well. They turned it into a selling point for their cheaper system, which had more local connections than the more expensive Bell system, whose long-distance capabilities were of little use to most farm people.[62]

Statistics gathered by the Bell companies illustrate this usage pattern. The New England Telephone and Telegraph Company monitored traffic at eight exchanges in New Hampshire, Vermont, and Maine, having a total of eighteen farmer lines with 218 telephones, in the winter of 1904, using the typical method of a "peg count" (operators moved a peg along a row of jacks to record the number of calls). Operators recorded a total of 1,970 calls per day on these lines: 39 percent of the calls were between patrons on the same line; 32 percent were outbound calls to another line; and 29 percent were inbound calls from another line. The average number of calls per telephone was also high, ranging from 5.2 per day at the exchange in Burlington, Vermont, to 16.1 per day at St. Johnsbury, Vermont. When compiled with figures from a similar number of farm telephones in Ottawa, Illinois, the combined New England and Illinois phones averaged 10.4 calls per day. Peg counts for thirty-three farmer lines in Dayton, Ohio, averaged only 5.0 calls per day per telephone, but the local outbound and inbound figures, at 42 and 45 percent respectively, were higher than the intraline calls, which constituted only 10 percent of all calls. In all studies, the majority of calls were from farm to farm.[63] After collecting more data in 1905, AT&T's chief engineer assumed the following figures for doing cost analyses of the work of switchboard operators handling farmer lines: five calls made from each telephone per day (two calls between patrons on the same line and three outgoing calls) and three incoming calls to each telephone per day.[64]

Although these traffic studies did not distinguish between business and social uses of the telephone—a matter of some concern to the telephone industry at the time and to present-day historians and sociologists—telephone men thought social uses outweighed business ones on the farm. "The principal use of farm line telephones has been their social use; although among the reasons for their placing, the protective and emergency ideas have been prominent," G. R. Johnson told an Ohio independents' convention in 1909. "The business use for the farmer has been of slower growth. The telephones are more often and for longer times held for neighborly conversation than for any other purpose. The next greatest demand is for connection with that village with which the farmer does his ordinary trading, and next with the nearest largest city, or county seat. The demand for truly long-distance connections, as to Pittsburgh and Indianapolis . . . [is] very nearly negligible from the farms and villages." Another independent leader told a South Dakota conference in 1909 that farm people initially installed the telephone for the amusement of the youngsters, but they were beginning to recognize its economic importance.[65] Reprints of numerous local newspaper articles in *Telephony*'s humor section, "In the Rural Line Districts," leave little doubt that social uses were common on the farm, even though farm people and the telephone industry emphasized business use in the rhetoric of access competition (probably to combat the image of the telephone as an urban luxury).[66]

Farm people found all sorts of novel uses for the telephone, some of which telephone companies (including cooperatives) opposed, others of which they promoted. The most common resembled the urban "broadcasting" functions predicted by Edward Bellamy in his popular 1885 novel, *Looking Backward*. Rural churches broadcast sermons over the telephone, an Illinois political convention transmitted its speeches to farmers via the phone, and rural telephone companies provided news, weather, and market reports to their subscribers. Some of these uses persisted much longer in the less populous countryside than in urban areas.[67]

In 1899, A. R. Phillips, in Geauga County, Ohio, regularly "listened to the reading of the latest evening edition from the Cleveland papers [on the telephone] just before retiring, even though the reader was three miles away." He also obtained USDA weather forecasts over the phone. By 1902, subscribing to a telephone news bureau in the Midwest and Great Plains was not unusual for farm families. Every evening at a designated time, usually seven P.M., an operator would call all farms on a line and give the time, weather, and market reports, newspaper headlines, and local news, "with a spicing of gossip," according to an urban magazine. Newspapers opposed the service at first but found that the brief telephone report whetted appetites for more details in the paper (a phenomenon noted by Marshall McLuhan a half century later). Even with the coming of RFD, telephone news services still thrived in Missouri, Minnesota, Iowa, and South Dakota as late as 1906. A senator from Indiana introduced a resolution in 1903 to modify RFD so that postmasters could transmit the "contents of letters requiring quick delivery by telephone." This was a common practice in Michigan during the long shut-in winters.[68] Other alternative uses included placing a telephone receiver in a baby's cradle so the operator could babysit by phone, using the receiver as holder for darning socks, and swinging from cut telephone lines into the swimming hole.[69]

Weather reports were widely available, mainly because the USDA telegraphed this vital information to newspapers and telephone companies on a regular basis. In 1904 the agency began to reduce the number of farmers to whom it sent weather "forecast cards" by RFD and transferred "a large portion of this work to the free telephone service. The rural telephone lines are now the best and most economical means of distributing weather information." That year, more than one hundred thousand farms in Iowa, Ohio, Wisconsin, Oklahoma, Illinois, and Missouri received daily weather reports by telephone. The plan succeeded on a smaller scale in Indiana, Arkansas, Massachusetts, New York, and Nebraska. *Telephony* predicted that 1 million farms would receive the reports by 1906. An independent company in Kansas provided the service in 1909. When an Illinois company added it in 1911, the local newspaper advised farm families that "if you are figuring to put up hay, taking an auto trip to a neighboring town, or expecting a visit from your mother-in-law, just call up the weather prophet at Central and govern yourself accordingly." Radio came to supplant the service, but as late as 1925, Iowa farmers preferred to receive their weather reports by the familiar telephone than by the new, and often unreliable, radio.[70]

Broadcasting music over the line was popular in the early days of the telephone. "The opening of the new telephone line at Ten Mile was celebrated with graphophone, violin, banjo, French harp, guitar, and organ Friday night," reported the *Macon (Mo.) Democrat* in 1904. A Nebraska newspaper observed that "the telephone northeast of town is a most interesting line. Any time of the day, if a person takes down the receiver, someone can be heard playing the graphophone or singing." Country newspapers made similar reports in Indiana, Ohio, Illinois, and Michigan.[71] Large telephone parties were the rage in some areas. Farm families near Wadesboro, North Carolina, held these events regularly. "It begins in the early evening when the business calls are over," reported a local paper in 1910.

"Central" connects a number of subscribers on the switchboard and everybody talks at once. . . . If some of the subscribers have an organ or piano, members of the family play and sing for the delight of the other listeners. Sometimes it is a violin or banjo that furnishes the music, and if there happens to be a phonograph in the community, each new record is hailed with delight and its owner is a most popular person. Imagine an evening like this in the town or city. With an accommodating "Central," the telephone exchange is the source of countless pleasures to rural folks. Here "'Central" is not a person to be sworn at but serves as a social arbiter and distributes rare joys each evening to hundreds, for miles around.[72]

Commercial telephone companies were not so enamored of these "accommodating" centrals or of farmer lines broadcasting music to switchboards in town. Minnesota stockholders complained in 1904 about "patrons having their receivers down and holding concerts over the line." A North Dakota newspaper observed that the musical "fad has become such a nuisance to the manager that he threatens to take out the offending phones and never put them in again." The *Pulaski (N.Y.) Democrat* told the story of a hurried businessman who "was greeted with the strains of the latest rag-time march [when he took down a receiver in town]. The result was a letter to the management and the sending forth of the order that the farmers' chief amusement must cease."[73] The manager of Pulaski's telephone company enforced rules against playing music on the lines that were on the books of most telephone companies serving rural customers in this period, whether they were small cooperatives, large independents, or small-town businesses. The telephone company in Mound Valley, Kansas (population 1,250 in 1907), ruled shortly after it was formed in 1906 that "singing or playing of musical instruments over the lines will not be permitted." The rule was still on the books a decade later.[74]

Although telephone companies invariably considered playing music over the lines to be an abuse of their system, at least two manufacturing firms approved of the custom. The Tel-Musiki Company attempted to commercialize it in Chicago in 1909. In their system, telephone companies would distribute phonograph selections over their lines at rates of three cents for an ordinary piece and five cents for grand opera, each subscriber agreeing to spend at least eighteen dollars on the service annually. The company reported eighty residential subscribers, but this expensive urban service seems to have flopped. The powerful Western Electric Company accepted the popularity of music on rural lines and promoted it in their advertising. In a 1910 booklet showing farmers how to build telephone lines, the company told farm youth, "The telephone enables you and your friends to share pleasures. A new piece on the phonograph or piano, a good story, or an interesting bit of news may be passed along to a friend miles away as readily as to the one next door."[75] Bell operating companies, which had to deal with systems being tied up by music playing on the lines, probably did not welcome Western Electric's acquiescence to this rural custom.

By all accounts, then and now, the most popular rural institution was not listening to music over the telephone but listening to conversations on the party line (a custom that was popular in town as well). Plenty of opportunities existed for eavesdropping, or "rubbering," as it was called at the time (and well into the 1940s in some regions).[76] In 1903 an Indiana promoter explained that farmer systems, having ten to twenty parties on each line, used "telegraphic signalling—an arrangement of longs and shorts. On a busy day each house must be a diminutive stock exchange—a Babel of Longs and Shorts." A North Dakota newspaper described the party-line culture in 1907: "Usually when a country subscriber rings anyone up several of his neighbors immediately butt in—not to talk—just listen. . . . Then there are a number of persons gossiping by the way of the telephone, and the business of T. Roosevelt, even, would have to wait, once they get started, till the matters of the entire community have been wafted over the wires. And occasionally a real talkfest occurs when there isn't much difference in the cyclone of conversation and the flow of soul of a sewing circle."[77]

Following the patriarchal custom of publicly expressing gender bias in this period, men stereotyped women as the main gossipers and eavesdroppers. E. H. Collins of central Indiana wrote the *Rural New Yorker* in 1902 about a tenant who had sold "his last family cow to secure a telephone. . . . It robs his family of what they need much more, just because his wife probably delights in gossip and can't stand it to see others chatting over it and talking about their 'fun' without enjoying it herself." A writer in the *Lawton (Kans.) Herald* complained in 1911 that "when two old windy sisters on a party line once get astraddle of the wire, nothing short of re-enforced lightning will ever shake 'em loose under an hour." In 1914, *Literary Digest* published a photograph of a woman sitting at her sewing machine with a telephone receiver tied to her head by a piece of cloth so that she wouldn't miss a word while performing her daily tasks.[78]

Recent scholarship has challenged these views. In her study of women and the telephone in a rural community in Illinois, the anthropologist Lana Rakow argues that what looks like gossip to men, in the past as in the present, is actually work that holds together kith and kin. It is "both gendered work—work delegated to women—and gender work—work that confirms the community's beliefs about what are women's natural tendencies and abilities."[79] This work complemented that of ordering supplies and parts for machinery over the phone.

There is some evidence that men enjoyed rural party lines as much as women. "All day long there is the chance of friendly gossip that is dear to the hearts of all women, and of many men as well if the truth were confessed," *Telephony* admitted in 1905. An item in the *Great Bend (Kans.) Tribune* for 1911 reported that the "phone company is going to take out the phone of a man who lives on a party line. He is said to butt into all conversations on the line and to make things so unpleasant that if his phone isn't removed, the other patrons will have theirs taken out."[80]

Another reason telephone companies disliked extensive eavesdropping was that it tied up the lines and wore out batteries, which companies had to replace. They tried all sorts of measures to stop the practice, including passing rules against it, fining eavesdroppers, giving priority to business use, and limiting conversations to five minutes.[81] Manufacturers adopted several inventions to prevent eavesdropping; farm journals and local papers published editorials condemning what they viewed as a bad-mannered abuse of the new technology; and several states (including Ohio and Indiana) passed laws making it a crime to repeat the contents of a telephone conversation. *Telephony* printed poems, cartoons, and newspaper reports describing the neighborhood strife that could, and did, result when people "listened in."[82]

One farm community in Iowa held a meeting at the schoolhouse in 1908 to address complaints about eavesdropping and gossiping. "But this meeting," reported the local newspaper, "instead of helping matters had the reverse effect. Following heated arguments and acrimonious retorts the gathering ended in a free-for-all fight. . . . Lifelong friendships have been broken, relations have become estranged, and it is said that a dozen lawsuits will result from the feud." *Wallaces' Farmer* reminded its readers in 1910 of "an old maxim that eavesdroppers seldom hear anything good of themselves." A Michigan newspaper proclaimed that the "man who pilfers the private communication of another by pulling down the receiver at another's call, listens, then quietly hangs it up so that the click will not be heard, is no better than the thief who steams open a letter, reads its contents, and then seals it up so as to avoid discovery." The paper then added a sarcastic, gendered twist to the moral by saying, "We are speaking of the men, the ladies never do this."[83]

But many farm people viewed eavesdropping in a favorable light. When caught eavesdropping by her boarder in 1907, a farm woman replied, "We all listen. Why shouldn't you listen? I heard four receivers go up just when I stopped talking. There's lots of people [who] wanted to know about that [neighbor's] chimney [that caught fire]. It's all right of course." A telephone man could not understand the custom, telling a convention in 1909, "It may be a strange code of etiquette that would actually defend eavesdropping, but they defended eavesdropping in defending their beloved telephone. This explains why so many farmers can be found who, at first thought, say they do not want a strictly private service." When asked by a city woman if she didn't find country life monotonous, a farm woman supposedly replied, "Yes. . . . Storms sometimes blows down the telephone wires and it gets lonesome not bein' able to listen to anybody." A technical writer observed in 1914 that farm women "meet and talk in company on the rural lines in a way which should be regarded as perfectly legitimate." The opportunity to "listen in" was also appreciated on the Canadian prairies before World War II.[84]

Recent interviews of elderly farm people reinforce the view that eavesdropping was socially acceptable in some communities. A few women interviewed in the 1980s thought it was not polite to listen, as did several women and men interviewed in New York state in the 1990s. In New York, Eva Watson hung up if she heard a click, and George Woods could tell when a certain woman was on the line because he recognized her breathing. Lina Rossbach and her husband would speak German when they heard someone on the line in the mid-1940s; this infuriated one listener, who blurted out, "You cheat, speak English!"[85] Yet many women thought listening in was a friendly habit. Helen Musselman of Indiana remarked that it "wasn't really nosiness; it was just neighborliness. I know I missed it when we had our new phones put in." Another Indiana woman, Pearl Snider, recalled, "We had a party telephone— probably thirteen or fourteen on one line. When the bell would ring, why, if you wanted to listen, you could run in and hear the conversation of the neighborhood." Opal Cypert from Arkansas reported that "when you'd get a ring, why, everybody would take the receiver down and they'd listen. They was welcomed in on the conversation then, if they wanted to." Edna Dagnen, from Washington state, remembered that "a lot of times when you were in a conversation, somebody would come on the line and say, 'Is that you Mabel? Do you know your cows are out?' or, 'Are you going to be home?' or something like that. Pretty soon you'd have three parties on the line and sometimes four."[86]

Villages experienced the mixed blessings of eavesdropping, as well. In Harriet Prescott Spofford's 1909 short story, "The Rural Telephone," widow

Dacre has taken to her bed over the prospect that her daughter Nancy will marry beneath her station (the son of Mrs. Dacre's only real love). Accustomed to minding the affairs of the community, Mrs. Dacre keeps in touch by listening (and breaking in) to telephone conversations by using a receiver placed near her pillow. Nancy fears the telephone keeps her mother an invalid. A neighbor arrives to express the community's outrage at the breach of etiquette, then finds herself drawn into the excitement of eavesdropping. The day is saved when Mrs. Dacre overhears a newspaper story about a young woman contracting galloping, and incurable, consumption when she cannot marry her heart's desire. Thinking this could happen to Nancy, Mrs. Dacre rises from her bed and declares, "I'm a goin' to a weddin'. My heart, what a blessin' the telephone is!"[87]

(Re)designing a Rural Telephone

In the many interactions among producers, mediators, and users of the rural telephone, farm people's novel uses of the device resulted in several changes to the telephone system itself. Recognizing the difficulty of exerting social discipline over thousands of far-flung, rather independent-minded consumers, especially when farmers ran their own telephone companies, commercial firms redesigned the telephone network to fit the social practices of this "class" of customer. In this case, producers, rather than consumers, adapted the new technology to fit the social patterns of daily life.

Party-line practices were the main targets of the redesign efforts. Kempster Miller, a prominent engineer with an independent phone company and the author of *American Telephone Practice*, noted in the book's fourth edition (1905), "Probably no branch of telephone work has offered more inducements to the inventor and designer, and consequently received a greater share of ingenious application, than the party line problem."[88] Some independents attacked the problem of "listening in" by using the technique of battery testing. In 1903, A. E. Dobbs suggested adding an extra charge to phone bills if batteries ran down after ten months. A decade later, W. H. Barker, of Iowa, told a telephone convention that he had instituted the practice of testing local batteries twice a year and informing customers that the company could calculate from the remaining battery life, and the number of calls made, how much time they had spent eavesdropping. "Your knowing about how much each one listens materially cuts it down," Barker reported. He suggested another means of "disciplining" customers by telling them that eavesdropping reduced the volume of a receiver. Afterward, he overheard customers saying, "'We can't hear Brown's, they listen too much,' etc. Every subscriber is making a record for himself. Then when you have an occasion to speak to them about it, they don't hop up, get angry and say 'why, we never listen.' They are aware you know what you're talking about."[89]

An Oregon newspaper reported in 1906 that the local telephone company was going to install a device that would let operators know whose receivers were off the hook when they were not being called. But the device (probably a lamp signal on a switchboard) did not stop eavesdropping, by physical or social means. A Nebraska man devised a more elaborate invention in 1914. The device "sounds a warning when a third party breaks in on the wire, and also identifies the culprit to both the legitimate users of the telephone." The inventor supposedly got the idea from a farm woman who said it would make him a fortune. But the author of an article on the innovation predicted that farm women would not adopt an invention that broke up their community visiting on party lines.[90]

Many telephone companies wanted to do more than merely rely on customers to stop eavesdropping when told about it; they wanted hardware that would physically prevent eavesdropping. One way was to use selective signaling, a system than rang the desired telephone and not others on the party line. Inventors proposed three methods, which used step-by-step mechanisms, electrical currents of different polarity, and currents of different frequencies (the harmonic system). In 1896, Angus Hibbard of the Chicago (Bell) Telephone Company invented a four-party line with currents of different polarity. By 1905 it had become universal in the Bell system, especially when modified to work with common-battery switchboards. Hibbard's polarized ringing circuit replaced a more complex lock-out design and the troublesome and little-used harmonic system, both of which were difficult to maintain.[91]

Telephone companies placed a few of these systems on rural lines before World War I. In 1903, Hibbard's company offered a selective signal plan (only two telephones would ring at the same time) on its eight-party farmer lines for eighteen dollars a year. By 1905, Stromberg-Carlson was offering a step-by-step mechanism for twenty-party service. In the same year an Indiana newspaper suggested that a local man "who invented a device to detect and cut off eavesdropping on a telephone line has the thanks of all legitimate users of the telephone." A large independent company installed a "full selective or secret lock-out service" for three hundred rural telephones in 1909. The system supposedly stopped boys who monopolized the line by playing music. Only one person preferred the old system, but the firm claimed her eavesdropping "had been largely responsible for the adoption of the more modern secret service." An independent company installed a lock-out system on a rural line in Wisconsin in 1911, and *Telephony* publicized a new lock-out system for rural lines in 1914.[92]

In a striking example of the social construction of technology, some Bell engineers decided early in the century not to fight party-line practices but rather to modify Western Electric farm telephones so they would work under these social conditions. In late 1903, P. L. Spalding of the Pennsylvania Bell company wrote Joseph Davis, chief engineer of AT&T, that replacing the standard no. 13 induction coil with a no. 10 coil improved the volume on farmer lines, where "there is likely to be a great deal of listening in on the circuits by a third party." But since the no. 10 coil "was not designed to mount in the subscriber's set," Spalding wondered, was there a more suitable coil? Davis could "not recommend its use except in cases where it is found absolutely necessary to do so." Laboratory research showed that adding a noninductive resistance to the no. 13 coil improved matters considerably.[93] In 1905, C. E. Paxson, of the Chesapeake Bell company, wrote to AT&T's new chief engineer, Hammond Hayes, complaining that "when we ring on a [farmer] line, a sufficient number of subscribers along the line take the receiver off the hook to prevent us from ringing again"; would it be feasible to solve this problem by placing a condenser in series with the receiver? Hayes noted that several Bell companies were doing this, but he did "not recommend the use of condensers for this purpose unless absolutely necessary" because of the added expense and the slight drop in transmission volume.[94] Although AT&T resisted making these changes, this episode shows how the local practices of farmers and engineers could alter the technology of even the Bell system.

Weaving the Telephone into the Fabric of Rural Life

What social changes can we ascribe to the rural telephone? Reformers and telephone companies sold it as a means to increase the profits and productivity of farmers by improving communication with their markets.[95] Promoters and farm people also praised the telephone for relieving the alleged isolation of rural life, especially during northern winters (as long as the lines stayed up through ice and snow storms). "No one feature of modern life has done more to compensate for the isolation of the farm than the telephone," claimed an article in the *Cyclopedia of American Agriculture* for 1909. "The rural mutual lines have brought the farmers' wives in touch with each other as never before."[96] Recent interviews of elderly farm people support this optimism. A Nebraska farm woman, Nellie Yost, recalled that the telephone "was such a great help. . . . It was just wonderful to be able to talk to your neighbors." Mara Scheerer of Indiana said "telephones came in 1910 and we just couldn't believe receiving messages over the wire."[97]

We do not have much survey data to evaluate these claims. As mentioned earlier, agricultural economists included statistics on telephone cooperatives in their surveys of farmer cooperatives, but they did not investigate whether the phone increased profits and productivity. In the social arena, the USDA's Bureau of Home Economics gathered information on using the telephone when it began studying the daily activities of "homemakers" in the 1920s (see chap. 3). The bureau advised the investigators to place the time women said they spent on the phone into three subcategories to account for business, social, and other functions.[98]

Although this level of detail was not usually published in the time studies, home economist Maud Wilson listed the amount of time women talked on the phone. Her study of about six hundred farm, rural nonfarm, and urban women in Oregon, conducted from 1926 to 1927, showed they spent an average of one hour and twenty-two minutes (5.7 percent) of their weekly leisure time writing letters and using the telephone. The highest amount reported was seven hours and twenty minutes per week. The average figure seems very low, given the fact that AT&T in 1905 assumed eight telephones calls would be made each day on farmer lines. Men may have been the primary users in Oregon in the late 1920s; most of the women's calls may have been listed under the management of household and miscellaneous categories; and the women may have underreported their usage to the college-educated home economists because of cultural prejudices against women using the phone too much. In any event, the figure of one hour and twenty-two minutes per week slightly underrepresented the amount of leisure time these women spent writing letters and talking on the phone because it was averaged over the entire group, not all of whom had telephones (86 percent of the women reported spending time on these activities).[99]

A few rural sociologists included the telephone in their surveys of farm communities. In his extensive investigation of Dane County, Wisconsin, published in 1921, John Kolb located "primary rural groups" by asking farm people to name the neighborhood in which they lived. Kolb plotted the location of these neighborhoods in relation to historical neighborhoods, topography, native vegetation, trade areas, farmer organizations, school districts, high schools, parishes, social areas, highways, electric power lines, mail service, newspaper areas, village centers, and telephone exchanges. The school district was the only configuration that mapped well onto rural neighborhoods. All others, including telephone exchanges, did not respect neighborhood boundaries.[100] The Cornell sociologists Dwight Sanderson and Warren Thompson also mapped telephone exchanges in their study of Otsego County, New York, published in 1923. But their conclusion differed from Kolb's results.

> The areas served by telephone exchanges . . . approximate rather closely the larger community areas of the suggested school communities . . . and it is important that the telephone exchanges should serve the potential social areas. If it is necessary to pay a toll charge for telephoning to persons within a given community, or between closely associated communities, their association is impeded to that extent, and business and other relations thereby tend to be confined to the area served by the exchange. In this respect the exchange areas in Otsego County seem to be well located with regard to the relation of the centers of the areas served.[101]

There is little doubt that the social geography of the telephone network mattered to farm people. In 1901, more than a dozen families in a forty-square-mile area in Ohio were bound together by the phone and the "tie of blood relationship." The correspondent for "Mt. Zion Echoes" in the *Hastings (Nebr.) Democrat* reported in 1904, "Juanita is to have telephone service, so they say. Us fellows who live on the divide offer an amendment to the effect that the Juanita and Kenesaw companies combine, as [the new service] cuts our community in two." A Kansas telephone man described his promotional techniques a few years later: "In organizing a new [rural] line I always try to get enough subscribers in one neighborhood and run a line to them. We always considered it bad practice to get two different neighborhoods on one line, for there is sure to be trouble." An item in the *Cadillac (Mich.) News* in 1910 explained that "the 'Cold Water Line' is the name given the telephone line south of town, from the fact that every patron is a Prohibitionist."[102] As we shall see, these telephone communities were extremely resilient. When farm people, the government, and manufacturers rebuilt rural telephony after its decline in the Great Depression, they recognized that the remaining telephone networks had redefined rural communities as much as RFD had done.[103]

Although the mosaic of social history sketched in this chapter lacks the detailed data of social surveys, it reveals the prominent place of the telephone in farm life before 1930. The many uses farm people devised for the telephone and the responses by telephone companies and manufacturers show how these joint actions reinvented the telephone (both technically and socially) as they wove it into the fabric of rural life. Farm people used the telephone primarily to extend existing communication practices, so much so that social scientists may have been reluctant to study such a mundane artifact.[104] Farm men and women conducted a wide range of business deals and social events over the telephone instead of in person. They ordered supplies from town and checked on markets. They received news and weather reports over the phone rather than by going to town or relying on a newspaper delivered by RFD. They played music over the line in telephone parties rather than having the party in someone's home. They transplanted the venerated custom of rural visiting or neighboring to the party line. Because of the ease of eavesdropping and the consequent lack of privacy, the rural line was an ambivalent social space, a space where new communication cultures were created through neighborhood negotiations. Building telephone cooperatives gave farm people, especially in the Midwest, another arena in which to build a social organization that many considered to be the hallmark of the best that rural life had to offer in this period.

Yet others complained that the telephone destroyed established customs, an indication that weaving the new device into rural life could alter an admired pattern. "The only objection that comes to my ears is that of the lessened value of visits and social calls," an organizer for an Ohio line reported in 1902. "The telephone to some extent supplements calling, for little chats over the telephone with neighbors and friends near and far on the association lines, forestall much of the activity of the old-time visit."[105] Writing to the Country Life Commission in 1909, the social reformer Amelia Shaw MacDonald of Delhi, New York, lamented that

> guarded telephony with every other neighbor on the line does not furnish the same psychic relief to pent-up emotions that a genuine heart-to-heart gossip did. A glimpse of sun-bonneted women rocking violently to and fro on the front piazza and discussing still more violently some neighborhood affair has gone with the party telephone line. The out-of-doors walk and the enlivening criticism was surely more invigorating than the present custom of "listening on the line." As far as [the alleged] monotony [of rural life] is concerned, the conversations on the telephone indicate the perpetuation of monotony.[106]

The *Toledo (Ohio) Blade* repeated the criticism in 1911, saying, "Only in sorrow can we think of the farmwife taking down the receiver of the telephone upon a Sunday afternoon and having a chat with a neighbor, instead of dressin' up and goin' visiting in the family carriage." Two years later, a Kansas woman told the USDA, "The telephone soon cultivates reticence." An Iowa auctioneer recalled that farmers visited less with each other in town after they had the telephone. Magazines accused the telephone of contributing to the decline of literacy and letter writing.[107]

In criticizing the telephone, some rural reformers recommended reinvigorating social institutions. In 1910, Henry Wallace, the founder of *Wallaces' Farmer,* observed that "the rural mail delivery and the rural telephone have not had the effect of bringing the farmers together, but on the other hand, have perfected their communication with towns." Good roads, Wallace feared, would do the same. "What is really necessary is the socialization of farm life."[108] Farmers should establish a "social center in each neighborhood" and organize more Granges, cooperative creameries, cow-testing organizations, and larger schools. In her letter to the Country Life Commission, MacDonald concluded that

> when the schools and the church become a genuine stimulus to right living, the farmwife will be as broad minded as her husband, for then and only then will party telephone lines, trolleys, and rural mails translate better things into the lives of women. The mistake has been of supposing that rural improvements in communication generated ideals. No greater fallacy could exist. Telephony, visiting, and letters must reveal some interchange of ideas if they are to accomplish this much heralded successor to "Monotonous, humdrum lives."[109]

We need not agree with MacDonald's social prescriptions in order to endorse her belief that it was a "mistake" to invest new communication technologies with the power to reform rural life on their own. Indeed, we should go further and conclude that farm women and men had more say over the role of the telephone in their lives, in their business as well as in their social affairs, than Shaw and other reformers realized.

CHAPTER 2

Taming the Devil Wagon

In 1922, *Farm Life* published a cartoon praising the automobile's power to urbanize rural culture (see fig. 5). The top panel shows rural life before the automobile. A farm man and a person from town (perhaps a country doctor making a call) ask about the news in their respective townships when passing on the road in their horse-drawn vehicles. A farmhouse and barn standing close to the road symbolize the primacy of agriculture for both travelers. The bottom panel depicts the social changes supposedly wrought by the automobile: rural people enjoy wider horizons when a family from Vermont (apparently farmers, because of the driver's long beard) and a family from Colorado (marked as city dwellers, because of the driver's stylish cap) ask about the news in their respective states when passing on the highway in cars loaded down with camping gear. Farmhouses and barns are now in the background, as the car, a symbol and an agent of mass consumption, erases differences between geographical regions and the appearances of rural and urban folk.

The cartoon illustrates the ideology of technological progress that pervades twentieth-century American culture. The automobile, that always up-to-date icon of middle-class modernity, revolutionizes rural life by bringing city folk to the country and country folk to the city. Ironically, promoters early in the century asserted this belief from the moment autos entered the countryside, when farmers opposed them as a "devil wagon." Ignoring the often violent resistance to the car, the automobile industry predicted that the auto would join other consumer goods in the urbanization of rural life. "The Kansas farmer has paid off his mortgage, sent his son to college, bought his daughter a piano and his wife and himself fine raiment and jewelry, and he is now looking around to purchase something else which will make him more

Figure 5. Motor Camping Is Nationwide. A cartoon from *Farm Life* heralding one of the social changes supposedly wrought by the automobile in the 1920s

Source: National Automobile Chamber of Commerce, *Facts and Figures of the Automobile Industry,* 1922 (New York: National Automobile Chamber of Commerce, 1922), p. 17.

and more like city folks," an enthusiast wrote *Motor World* in 1902. "The automobile is the very thing; with it he cannot only make a splurge, but can run into town in half an hour, thus making, with the telephone and the delivery at his door of his daily paper, another step toward his securing all the conveniences heretofore regarded as the sole right of the habitant of the town." Two years later, the journal's editor claimed that the "telegraph and telephone, the rural free delivery, the bicycle and the trolley car, and now the automobile, have almost annihilated distance, and with it the evils which formerly made hypochondriacs of the farmer and his family and drove some of them to the slums of great cities out of sheer desperation."[1]

As we have seen with the telephone, such images and predictions usually tell us much more about the beliefs and methods of promoters than about the social changes occurring in the countryside. Farm families adopted the automobile in large numbers, especially in the Midwest, but only after a sustained period of resistance and outright opposition, which included attacks on the dangerous "devil wagon" and its drivers. When farm people began to buy cars, they used them in a variety of ways. They drove to more distant cities and went to town more often than they had with the horse and buggy. They turned the car into a general source of power around the farm. Taming the devil wagon on their own terms allowed them to weave the car into the fabric of rural life. Although the fabric expanded into a new social geography, farm men and women exercised a good deal of control over this process.

The Crusade against Cars

The earliest measure of control was to resist the incursion of the car into rural life. Farm people initially opposed the automobile, much more than they did the telephone, because it was very expensive and it roared into the countryside, disrupting rural culture. When cars first appeared on rural roads in the early years of this century, usually driven by rich city folk out for a spin, they often met a hostile reception. Indeed, farm people joined small-town residents, suburbanites, and even irate city dwellers in many parts of the country in hurling at the dangerous, speeding car such epithets as "red devil" and "devil wagon"—names that soon symbolized the rising clamor of rural protest.[2] Motorists and automobile journals countered with the traditional antirural insults of "hayseed" and "rube" but also coined such new terms as "autophobe" and "motorphobe" for all critics of the car—whether they lived in the city, town, or country. A group in St. Louis even defied the widespread opposition to "scorchers" in 1905 by calling themselves the Red Devil Automobile Club.[3]

Protest against the car was rife in small towns. A study of the diffusion of the automobile in Dexter, Michigan, found that from 1902 to 1905, ridicule and opposition dominated the tone of the references [to the car in the local newspaper]. Of the fifty editorial comments during these years, not one favored the automobile, only six were neutral, and forty-four were distinctly hostile. . . . Apparently there was a widespread fear in the Dexter region at this time that the automobile was escaping social control and mixed with this fear was not a little envy on the part of the farmers who sensed the automobile as a symbol of urban superiority.

With the passage of registration laws and speed ordinances in 1905, the report continues, the "fear element disappeared, but rural resentment was slower to die away for we still find evidences of it in 1909 and in 1910 despite the fact that young farmers near Dexter had begun to buy machines."[4]

The main antagonism between farm people and the early car and its drivers seems to have stemmed from the disastrous effects that cars had upon livestock. Horses reared at the car's noisy approach, often breaking away or upsetting buggies; chickens crossed the road for the last time. One confused ram even charged a car that had stopped to fix a flat, hitting a front tire head on. "The shock knocked the Ford off the jack and bounced the ram back on his haunches," recalled the surprised driver. "A look which combined amazement and increased animosity showed in [the ram's] eyes and he wasted no time in sighting for a second assault. His hurry made his aim even poorer and his head struck the front axle with terrific force," breaking the ram's neck. Its owner did not ask for compensation—after all, the ram hit the car!—but many drivers paid handsomely when they killed farm animals on country roads.[5]

Many farm women complained that recklessly driven autos prevented them from driving their horse-drawn buggies on country roads. A New York woman told a newspaper in 1904, "We farmers' wives and daughters think that the people who are able to own and run an automobile are able to build their own roads to run them on, and leave the public highways for the use of people who do not care to be sent from this mundane sphere by a horse maddened by one of those 'pesky' automobiles." A Maine woman lamented in 1909 that country roads were impassable for six months of the year; "the other half of the year [the farm woman] dares not drive [her buggy] over country roads . . . because of automobiles."[6]

The rural press generally supported this early criticism. The *Rural New Yorker* declared in 1904 that "reckless drivers of automobiles are laying up trouble for their class. The record of runaway horses and damage to life and property is a long one." Four years later, the editor was still adamant about the problem. "The auto hog or tiger has become a perfect curse on our public roads. Many car drivers are most accommodating. We have had them stop and help hold a frightened horse. Some of them, however, appear to be drunk. They rush along the road with little regard for human life. A load of buckshot fired into the tire of their wheel is about the only force they are willing to recognize."[7] When *Wallaces' Farmer* ran an article in 1909 suggesting how cars and horses could peacefully coexist on country roads, an irate reader in Northern Iowa complained that "the automobilist is classed as a reckless driver while the farmer you class as a road hog. As for the rights of the two, I think the man who is carrying on the business of the country and keeping up the road has the right of way." The editor, who agreed with this form of agricultural fundamentalism, replied that the journal had not called the farmer a road hog but was merely quoting a motorist.[8]

Even the goggles and dusters worn by early drivers while touring in open cars appeared monstrous to some farm people. *Motor Age* reported in 1904 that during a "century run" in Utah a motoring party saw a farm couple and their seven children picking berries along the road. "The motorists stopped and the driver and another from the party started towards the group of busy pickers. They heard him approach, and, as he wore goggles, they were so frightened they ran back to the farm house screaming. The party had to continue the trip without berries."[9]

The early car had other drawbacks. It was expensive, unreliable, and certainly not quiet. Beulah Mardis, an elderly Indiana farm woman, recalled recently that a boyfriend "had an old open Ford. He lived on a higher piece of ground about a fourth of a mile from here, and even in the winter when it was cold and with the doors closed, I could hear him start that Ford and I knew it was time for me to get my coat on and be ready to go."[10] Apart from the car's great speed, many country folk were unimpressed with it as a means of transportation. It was a common sight to see farmers with their horses towing a car that had broken down or pulling a car out of muddy country roads, a source of income for some farmers as well as moral satisfaction against the "devil wagon."[11] Townsfolk also criticized the car in this manner. "The Harris Telephone Company got swell and bought them an auto so as to look after the business of the line with dispatch," reported the *Burlington (Kans.) Jeffersonian* in 1905. "The maiden trip was taken Friday last. They 'phoned to Glenwood to look out for the devil wagon in thirty minutes, but instead at sundown they were seen to be stroking its mechanical intestines some miles away. The auto can be had at a bargain."[12]

Cars also created a dust problem because the vast majority of rural roads at this time were dirt roads. A New York farm man complained in 1906, "A new and serious menace to horticultural interests everywhere are the dust clouds so freely raised on country roads by swift-speeding automobiles." The dust was thought to harm hedges, trees, shrubs, lawns, hayfields, and even animals who ate it. "It needs no prophet to predict the days of fast running on ordinary roads are numbered. The dust evil, when widely recognized, will accomplish what the physical dangers and inconvenience to legitimate road users have failed to do. It is not a question of staying the hand of progress, but of utilizing a beneficent invention in a rational manner." Farmers in New Jersey said it was "impossible for them to have their windows open because the speeding automobiles raise dust which enters their homes." J. H. Haynes of Indiana wrote in 1909, "A trail of dust one-fourth of a mile long is no uncommon thing following in the wake of the fast-running car." Haynes thought the car was here to stay, but "farm papers should begin a crusade against the usurpation of the roads by auto owners."[13]

The early opposition toward the car was such that rural people resorted to both legal and illegal means to stop its influx. Counties in West Virginia and Pennsylvania passed laws that banned autos; Vermont required that someone carrying a red flag walk ahead of any car on the road. A flurry of legislation around 1908 required cars to slow down for horse-drawn vehicles or stop if the horse appeared frightened. Lucrative "speed traps" also date from this period. Support was withheld from road improvement schemes. The threat was perceived to be such that, as in the case of the bicycle, many farmers took the law into their own hands. Between 1902 and 1907, a period of widespread auto touring, the press reported numerous cases of farm men attacking motorists. Farmers shot a chauffeur in the back in Minnesota, stoned a motorist in Indiana, shot at a car passing a horse-drawn buggy in South Carolina, and assaulted a chauffeur in Wisconsin. New York farmers hit a motorist with a galvanized iron pail on Long Island, pushed a lawn mower into an auto's path, whipped a motorist for no apparent reason, and delayed a hill-climbing contest near Rochester by fighting with onlookers.[14]

Many farm people detested the "devil wagon" so much that they sabotaged their own roads to try to stop the growing menace. In 1905, Connecticut farmers spread a tire-cutting slag on roads (supposedly to fill in ruts!), and Minnesota farmers plowed up roads near Rochester. The *Dexter (Mich.) Leader* reported in 1907 that "automobilists near Ypsilanti have offered a reward of seventy-five dollars for information as to who placed three pieces of wood filled with nails and covered with straw on the south Ann Arbor road." As late as 1909, Indiana farmers, tired of being awakened by revelers return-

ing from a night of drinking in nearby roadhouses, weakened bridges and bar-
ricaded roads. In the same year, farmers near Sacramento, California, dug
ditches across several roads and caught thirteen autos in their traps. The organizers of a 1909 reliability run through the South routed the race around
Pinehurst, North Carolina, because of the resistance of local farmers. Rural
people booby-trapped other roads with an innovative assortment of rakes,
saws, glass, tacks, and ropes or barbed wire strung across the road. One
scholar has noted, however, that such tactics as plowing up roads were meant
to make them "impassable by automobiles *but not by horse and wagon*."[15]
Groups such as the Farmers' Anti-Automobile League near Evanston, Illinois, the Anti-Automobile Club of Grover, Missouri, and the Farmers' Protective Association in Harrison Township, Ohio, were formed to organize
rural opposition to the car. The Illinois league had a twenty-member vigilante
committee to mete out justice to reckless drivers.[16]

Farmers took these actions partly because they viewed country roads in a
proprietary manner. Building and maintaining roads was a local affair, usually
financed by road and poll taxes that many farmers paid by working on the
roads themselves (a system of corvée labor). When engineering reformers began to agitate for a centralized system in the 1870s and the League of American Wheelman headed a national "good-roads" movement during the bicycle
craze of the 1880s and 1890s, farm people opposed both efforts, often using
the rhetoric of populism in the Midwest. Opposition reappeared during the
good-roads movement of the early twentieth century—promoted by the U.S.
Department of Agriculture's Office of Public Roads and Progressive Era reformers—when the faster and more dangerous auto followed the path of the
bicycle into the countryside. According to recent studies, some Wisconsin
farmers "considered automobiling on the public highways as trespassing upon
their own private property" and thus resisted road reforms. Kansas farmers interpreted rural roads "as public in a limited sense; they were open to use for all
those who paid their upkeep" but not to urban motorists who transgressed
these boundaries. In the South, where corvée labor prevailed, farmers
"adamantly refused to pay additional taxes to have better roads." Farmers
voted down a bond issue in North Carolina in 1907 and chose to fund farm-
to-market roads over touring highways in the next decade.[17] The growth of
rural free delivery put more pressure on localities to improve roads by covering them with gravel or a macadamized surface (consisting of packed stones
bound together with earth).[18]

Yet the automobile harmed country roads far more than the horse and
the bicycle. A farm man wrote the *Rural New Yorker* in 1909 that the

siren song of the "good roads" propaganda is less enticing to the practical farmer than before the advent of high-power automobiles.
These speedy and enormously heavy motor vehicles rush through the
country, plowing the softened roads to their foundations during wet
weather and scattering the surface in the form of irritable dust over
the adjacent landscape during drought times. . . . The automobile has
come to stay, and will doubtless be greatly perfected in the future, but
there is little need in having our roads destroyed during its evolution.[19]

The "good roads" built by some states were not immune to this destruction.
Reformers had noted as early as 1907 that the auto's tires sucked up the binder
from between the compacted stones on macadamized surfaces, raising clouds
of unwelcomed dust and destroying these expensive roads built by public
funds. *Wallaces' Farmer*, which favored a cheaper system of constructing well-
drained dirt roads and maintaining them locally with an easy-to-build King
Road Drag, thought the proposed technical solutions (oiling the roads or using a coal-tar binder) were out of the question in the Corn Belt. The journal
was a little smug about the irony of the latest transportation technology making the modern macadam roads obsolete. "We have warned our readers in the
corn belt who have the so-called good roads fever that this was coming,"
crowed Henry Wallace in 1909. The *Progressive Farmer* instructed southern
farmers in how they could make their own road drag the next year.[20]

Adopting the Car

Motor World was a regular critic of this form of resistance, publishing such
editorials as "Why the Hayseeds [in Illinois] Oppose Good Roads" in early
1905. But the automobile industry also began to view farmers in a more positive light at this time, as an untapped market for cars. A few weeks after publishing the hayseed editorial, *Motor World* featured a wealthy, up-to-date
Minnesota farmer as a "pioneer who is showing his brother agriculturalists in
his locality and at the same time demonstrating to the world at large that the
automobile is the farmer's best friend and not his enemy." That summer, *Motor Age* reported that the "farmer is getting a little used to automobiles and the
country papers are beginning to realize that the farmer has been making mis-
takes about automobilists." The journal thought that local papers were following the lead of city papers, which themselves had only recently approved
of the automobile.[21] The urbane *New York Times* picked up the story, running

such headlines as "Farmers Are Buying Autos" and "Farmers Becoming Motor Enthusiasts." The latter story begins, "A hand of friendship, even of welcome, is being extended to the automobile by the farmer. The days of hostility to the horseless vehicle have passed. Instead of being looked upon with jealousy as a new luxury for the rich, a reckless speed annihilator, and a disturber of the old-fashioned rural peace," the car was being accepted on its "practical merits. . . . There is less hurling of invectives or something worse at the tourist as he journeys through the country. Indeed, he is more likely to receive gifts of fruit when it is ripe on the farm. Women in auto parties are presented with flowers. Directions regarding roads are given cheerfully. The golden age seems to be opening for the motorist who possesses good-nature and is a careful observer of the rights of others."[22]

Although the *Times* repeated the overly optimistic stories of the manufacturers and automobile clubs (in this case, the Kansas City Automobile Club), it correctly reported that prosperous farmers were beginning to buy cars. As we have seen, anticar sentiment did not end abruptly in 1906; it faded away around 1910 because of a combination of circumstances. Faced with the apparent saturation of the urban luxury-car market, manufacturers developed a large rural market by producing more affordable cars designed for rutted country roads. This came at a time of growing support for the car among farm leaders. The National Grange passed a resolution in the summer of 1908, stating that the "motor vehicle is a permanent feature of modern life" and had a right to use rural roads. The Grange followed the lead of such journals as *Wallaces' Farmer*, which had begun to promote the automobile in early 1908 using the same methods it employed for any new technology it favored: advertisements, editorials, articles, and requests for readers' experiences. The journal's editor, Henry Wallace, stated that "farmers have had their fun—and sometimes it was not fun, either—with the users of the automobile." Although farm people had justifiably "called it the rich man's plaything" and sworn at it for disrupting rural life, they had begun to value cars and to buy them for themselves. The *Rural New Yorker*, a former critic, started to promote the automobile in 1909. Another former critic, the *Dexter (Mich.) Leader*, noted the same year that a local farmer and a rural mail carrier in a nearby village had bought automobiles. In 1910, the paper printed a humorous poem suggesting that because he drove a car the farmer was no longer a rube.[23]

Wallaces' Farmer thought highly of two types of cars: the technologically out-of-date, but inexpensive, buggy car, whose high wheels cleared the hump in rutted country roads; and a touring car with a removable tonneau (backseat) that could be easily converted into a small truck. Manufacturers of both types flourished for a brief time, helping to introduce the automobile into the countryside. The inexpensive Model T, to take the most successful example of a car designed with rural roads in mind, sat high off the ground (also making repair easier) and had a high horsepower-to-weight ratio and a three-point suspension. When the Ford Motor Company introduced the long-lived Model T in late 1908, the firm advertised it successively as "the family car at an honest price," "the farmer's car," "the doctor's car," "the merchant's car," and, finally, the "universal car." Henry Ford's folksy ways and populist rhetoric also helped construct the austere Tin Lizzie as the "farmer's car."[24]

The National Grange backed the good-roads movement in 1909, and road improvement was supported by more localities as farmers began to buy cars. The *Atlanta Constitution* editorialized in 1912 that the "farmer and the automobile were [formerly] as incompatible as oil and water. But today farmers are among the foremost owners of cars." Wayne County, Iowa, even used an automobile to drag its roads in 1911. The possible income to be derived from wealthy city people did not go unnoticed. Tourism thrived, as did repair shops. Farm men, many of whom had operated steam engines and stationary gasoline engines, were well placed to become car users. As buggy cars, convertibles, and the Model T spread into rural areas—often driven at first by country doctors and mail carriers who showed their practicality on rural roads—the anticar movement vanished.[25]

In fact, the prosperity of the so-called golden age of American agriculture enabled rural people to become the main purchasers of automobiles. The Rambler company, which targeted the farm market, sold about 60 percent of its cars to country buyers in 1908. Automobile magazines reported big sales to farmers in 1909 and 1910, so much so that the *Wall Street Journal* criticized farmers for going into debt to buy cars. *Wallaces' Farmer* was indignant. The "western farmer has been the subject of much criticism at the hands of certain eastern financial interests, who say that to some extent the tightness of the money market has been due to his heavy purchases of automobiles. Stories have been circulated to the effect that hundreds of western farmers have mortgaged their farms to buy motor cars, and this will sooner or later mean financial trouble." Rural people, the journal reminded its readers, could take care of themselves. "The farmer has learned that the auto is a good investment" because it saved high-priced horses, time, and labor and provided recreation. In 1913, a New York farm woman objected to the "attitude most town people have toward the farmer. He is represented either as a 'Rube' with chin whiskers and his trousers in his boots, or as having several motor cars bought with his ill-gotten gains from farm products figured at the high prices."[26] Another buying spree accompanied rising farm prices in Wo I. *Motor Age* reported a huge demand for cars by farmers in Minnesota, the Southwest, Kansas, Wisconsin, and Illinois in 1915. *Wallaces' Farmer* reported the same for Iowa. At the end of 1917, the *New York Times* observed that "farmers are estimated to own nearly half of the automobiles in the United States and are the largest buyers at the present time."[27]

The number of autos on farms increased from an estimated eighty-five thousand in 1911 to about 2 million in 1920. As stated previously, the U.S. Census Bureau reported in 1920 that a larger percentage of farm households owned an automobile than nonfarm households (32 to 25 percent), probably because of the availability of trolley cars and other forms of mass transit in town. These figures, like those for the telephone, masked a wide variation by region. The prosperous Midwest topped the list, with 53 percent of farms owning cars. It was followed by the Far West at 42 percent and the Northeast at 33 percent. But cars were few and far between in the poorer South, where only 14 percent of farms owned automobiles in 1920 (see table A.3).

The automobile's chief motorized competitor was the interurban trolley, which promoters viewed as one of the new technologies urbanizing the countryside. Built by electric-streetcar promoters in a boom period from 1901 to 1908, the interurban reached its peak in 1916, when more than sixteen thousand miles of track connected towns and cities in almost every state. The most mileage was in prosperous farm states having a relatively high population density: Ohio, Indiana, Pennsylvania, Illinois, California, and New York. Often running parallel to the tracks of steam-powered railroads, the interurbans took passengers away from them with a combination of cheaper fares, more frequent service, and more stops, especially in the country. Interurban companies preferred the faster and more profitable express service, but the trolleys stopped at crossroads and farmhouses, especially where the farm-to-market runs were lucrative. The Grange favored the freight business of electrics because it came at the expense of their old enemies, the railroads. Other urban-rural tensions were evident with this new technology. In an article promoting the rural market to automobile dealers, a writer for *Motor Age* referred to the interurban as the "Great Inter-Reuben railway."[28]

Small towns promoted interurbans to build up their business and industry, a good deal of which related to agriculture. In 1909, the *Ogle County Republican*, in Oregon, Illinois, thought an interurban would allow the farm man "to take his produce to market in good or bad weather, in the busy season or the slack. . . . He can telephone his orders to our merchants for supplies and receive them at his door by the next car. . . . He can visit fifty miles from home and return in the evening with greater ease than to drive a dozen miles to town" in a horse-drawn vehicle. Although some farmers refused to grant rights-of-way and others saw no need for the new method of transportation, most farmers in this township favored the proposed interurban, primarily because property values were projected to increase along the line. As in the case of the telephone, however, townsfolk worried that other towns would build an interurban and capture the farmer's trade. Merchants also opposed the interurban, which never came to Oregon, Illinois.[29]

The extent of rural traffic on interurbans is difficult to measure. The estimates given in 1902, by eighteen companies operating fast, long-line systems, varied from 2.5 percent on the Wilkes-Barre to Hazelton line in Pennsylvania to 20 percent on two lines extending from Indianapolis into rich farmland. The Census Bureau attributed these relatively low percentages to the fact that "farm dwellings are so widely scattered. A system of interurban railways connecting all the towns and villages in a given section would be conveniently accessible to only a fraction of the agricultural community." The bureau noted in its 1907 report that rural traffic had increased.[30] But even at the peak of the industry in 1916, interurbans were accessible to probably fewer than 10 percent of the nation's farms, a sharp contrast to the wide diffusion of RFD, the telephone, and the automobile by that time. Unlike the telephone, which complemented rather than replaced RFD, by 1920 the auto had displaced the trolley in rural life, a process that occurred later in cities and towns.

Farmyard Mechanics

A major reason that by the end of World War I farm people were buying automobiles in large numbers was that they found all sorts of uses for the car. In this process of social construction, farm men and women interacted with automobile manufacturers in determining the meaning of the car in rural life. By 1909, most automobile manufacturers had settled on a standard design for the car—the "large, front-engine, rear-drive automobile" based on a European design.[31] Because they produced the car, the automobile manufacturers greatly influenced the form the technology initially took. But their position, although influential, was not overwhelmingly so. New manufacturers could (and did) produce new and different cars with different users in mind. Furthermore, although manufacturers may have inscribed a particular meaning to the car, they were not able to control how that artifact was used once it got into the hands of the users. Users, precisely as users, embedded new meanings into the automobile.

This happened with the adaptation of the car into rural life. At the turn of the century, farm families started to define the automobile as more than a transportation device; in particular, they saw it, like the horse, as a general source of power. Mrs. H. T. Hoskins of Vermont wrote the *Rural New Yorker* in 1901, "I hail with joy the prospect of this sure to come, blessing on the farm, the automobile. When it reaches here with its adaptability to do all kinds of work the church bells shall be rung in its honor. . . . I could have saved two hundred or three hundred dollars this year if I had owned or hired such a machine. With such a machine a woman with a large orchard to oversee can do most of her work, especially if some frame can be devised to load and unload barrels and boxes. . . . Are there automobiles for farms made yet in America?" The editors replied that "manufacturers have hardly thought of a machine strong enough to do all the work required" on a farm. One Kansas farmer, George Schmidt, wrote the journal two years later that he had had some success using the auto on the farm. Using language common to the horse culture the car eventually replaced, Schmidt advised readers "to block up the hind axle and run a belt over the one wheel of the automobile and around the wheel on a [corn] sheller, grinder, saw, pump, or any other machine that the engine is capable of running, and see how the farmer can save money and be in style with any city man." T. A. Pottinger, of Illinois, wrote *Wallaces' Farmer* in 1909 that the ideal farm car should have a detachable backseat, which could turn the vehicle into a small truck, and that it should be able to provide "light power, such as running a corn sheller, an ensilage cutter, or doing light grinding."[32]

The car was also used for domestic work, such as powering a washing machine, which seems to have been a source of some humor. One suburban commentator, in musing about the impact of the car on his family life in 1910, described his car falling off the jack and careering away across the backyard out of control, dragging the washing machine (and the luckless domestic servant, Mary) with it.[33] The photograph in figure 6, taken in the 1930s, shows a dramatic instance of this sort of use. Here a farm man has jacked up a Model T in the farmyard to provide power to operate a washing machine. Although the car was sometimes used to assist in traditional "women's work" (e.g., running the butter churn and cream separator), farm men—rather than farm women—more commonly used the car to provide stationary power, and mainly for what was considered to be "men's work"—that is, to run agricul-

Figure 6. Homemade Car-Powered Washer. Kansas farmer Bill Ott with his daughter Lizzie Ott and a car-powered washing machine he has devised for her to run, ca. 1930
Source: Photograph courtesy of Bill and Ruth Dick, Newton, Kansas, collection of the author.

tural machinery. Corn shellers, water pumps, hay balers, fodder and ensilage cutters, hay and grain hoists, cider presses, and corn grinders were all powered by the auto. One rancher even used a Cadillac to shear his sheep.[34] A Tennessee farmer said in 1920, "I make it run a cut-off saw for wood-sawing, and run the pumps." Five years earlier, a Maine farmer had put a car to so many multiple usages that tax assessors did not know whether they should classify it as a pleasure vehicle or a piece of agricultural machinery.[35]

In addition to providing a stationary source of power, cars found a wide variety of unexpected uses in their mobile form. Farmers used them as snowmobiles, tractors, and agricultural transport vehicles. Indeed, it seems from the earliest days of the car's introduction onto farms that farmers were acutely aware of its potential, whether simply to transport fodder or to power a feed chopper.[36] Adapting the auto to the myriad tasks of rural life was common enough practice that seven of the twenty-three farm families in New York who participated in a recent oral history project recalled that they or their neighbors had used the car as a hay rake, pickup truck, or power source. Eighty-eight-year-old Winfred Arnold remembered that neighbors used the car to power jobs around the farm, but he himself could afford to use stationary gasoline engines.[37] Others saw alternative uses as showing off. F. D. Brown wrote *Progressive Farmer* in 1920 that "we have heard of some folks hitching the hind wheel of a jitney to the churn, the buzz saw, the rubber-tired milking machine and other machines in and around the place, but we never tried it. We may some time. In the meantime, the 'jitney' pays for its upkeep and a little extra without the fancy frills."[38]

The remarkably different ways in which farm people interpreted the car had a strong tie to the flexible and historically variable gendered division of labor in the field, barn, dairy, poultry shed, and house.[39] Alongside the gender structure of the division of labor were gender identities among farm men and women that help explain the social construction of the rural automobile. Many farm men, especially in the Midwest, saw themselves as proficient mechanics who could operate, maintain, repair, and redesign most machines on the farm, from steam engines and threshers in the field to water pumps in the kitchen.[40] Although the characteristics attributed as "masculine" have varied historically, competence in the operation and repair of machinery formed a defining element of masculinity (and thus gender identity) for many male groups in this period, including linotype operators, other craftsmen, small entrepreneurs, and farm men.[41] Women might pump water, drive the horse and buggy to town, and occasionally operate field machinery, but men fixed a leaky pump, oiled and greased the buggy, and redesigned a hay binder to work over hilly ground. Technical competence helped to define their gender position as a form of masculinity and reinforced the rural gender system.

The gasoline automobile was symbolically inscribed for masculine use by Henry Ford and other manufacturers. Women purchasers of automobiles in Maryland and New Hampshire, for example, avoided the Model T, which had a reputation as a man's car because of the physical strength required to steer and shift it and because it lacked such amenities as a front door on the driver's side.[42] Consequently, when farm people purchased an automobile, they usually viewed it as the latest, highly sophisticated piece of farm machinery and put it in the hands of those defined as technically competent—farm men.

But male and female access to the driver's seat varied widely in farm families. At one extreme, some women drove the car to the exclusion of men. Alice Guyer, an Indiana farm woman, recalled that her father "had trouble with [cars], and he just gave up the driving to my older sister." Bertha Pampel remembered that "my dad never did drive. My mother did all the driving." At the other extreme, some women who had been proficient with the horse and buggy never mastered the car and thus became more dependent on men and less technically competent. Laura Drake, another Indiana farm woman, recalled that her family had a car when she was growing up, "but we weren't allowed to touch it. Nobody touched that [car] but him [i.e, her father]." Contemporary studies show a wide variation, as well. *Farmer's Wife* reported in 1922 that one-fourth of the farm women and girls surveyed in Steele County, Minnesota, drove the family car. Surveys in New York in the late 1920s found that from one-eighth to one-third of farm women knew how to drive. The figures improved with time. The home economist Jean Warren noted that almost one-half of the New York farm women she studied in the late 1930s had a driver's license, but only 15 percent of those drove to town regularly. At least two of the twenty-three farm families interviewed recently in New York said that a mother or daughter had not learned to drive.[43]

Age was probably a factor, since many accounts say that teenagers—daughters as well as sons—took readily to the car. A Kansas newspaper editor explained in 1910 that

> farmers' sons and daughters are taught early to be venturesome. They have managed the farm machinery, much of which is far more difficult to handle than is the motor car. Soon after a car goes to the farm does the daughter take lessons, and it is a common sight on the streets of the interior towns to see cars driven by girls who have come in from the farms, or to see the farmer's children taken to the country school by the motor. Their handling of the machines would be a credit to the professional chauffeurs, for they have to encounter conditions on the roads of which the city driver knows little.[44]

The *Rural New Yorker* marveled about this trend in 1916. "We find that our own children know more about the different kinds of cars and the ways of handling them than we ever did about the different breeds of livestock." When a farm family buys a car, the "girls are ready to drive, in fact it is probable that when father and mother start out they will go as passengers on the back seat with some of the young folks at the wheel. We all begin to realize now what the modern car has done for farm life. It is stirring up the people as nothing else has ever done, mixing them together, and giving every member of the family a new sense of power." Yet a study of farm youth in Missouri in the 1920s showed that about three-fourths of the boys drove cars, while only about a third of the girls did.[45]

A motor-wise farm woman was rare enough to be news. A New York woman told a reporter in 1915 that she was "thoroughly familiar with the machine" and then proceeded to fix a flat tire by vulcanizing it. More typically, farm men seem to have adopted the gender ideologies expressed by manufacturers, that women needed conveniences like a self-starter in order to become motorists. An ad for the Case 40 in the *Rural New Yorker* read, "Big Value in a Big Car. An Auto Your Wife Can Run. Your wife can manage this big, powerful car with the utmost ease and safety. The self-starting motor, the easy control, and the reliability of the machine in general make it *safe* in the hands of a *novice*."[46]

In general, farm journals and oral histories indicate that farm men, rather than farm women, maintained, repaired, and tinkered with the new addition to the farmstead, especially because repair facilities were few and far between in this period. The technical competence with autos varied considerably, of course. Some writers for automobile journals said farmers needed a lot of education in the intricacies of auto mechanics.[47] Many more authors and auto manufacturers thought farm men had the necessary technical background to handle cars. In 1908 the *Ford Times* reprinted a statement by the National Grange to this effect: "As the farmer is accustomed to handle machinery, he is able to 'use' and not 'abuse' his machine, and as he is his own machinist, he keeps it from getting out of 'kelter' by always having it in the best of running order."[48] Farm men also sent in numerous tips to agricultural journals about the care of cars, and the journals printed regular columns about maintaining the car.[49] The consensus seems to have been that, although most farm men were not expert auto mechanics, they could probably maintain and repair cars better than city men.

The farm man's technical competence, rooted in his masculine gender identity, enabled him to reinterpret the car's function, jack up its rear wheels, and power all kinds of "men's" work on the farm and, less frequently, the "woman's" cream separator, water pump, or washing machine. This version of the gendered division of labor on the farm—in which men maintained agricultural machinery (including cars) and women performed household tasks—could not be more strikingly illustrated than in the photograph in figure 6, where the man has jacked up the car but his daughter still operates the washing machine. The evidence overwhelmingly shows that farm men, not farm women, reconfigured the car in order to use it in an alternative manner. One exception was that of an independent "woman farmer" who used her car to pull a hay rake during the "manpower" shortage of World War I.[50] Farm men also converted the car from a passenger vehicle to a produce truck. Showing off further, they returned the car to its original configuration, as defined by the manufacturer, and either drove family members to town and church or handed it over, in this more symbolically feminine form of usage, to women to operate—often to go to town to get parts to repair field machinery.[51]

The mutual interactions between the artifact, social groups, and intergroup power relations are clearly evident in this case. The gender identity of farm men, defined in contrast to the constructed femininity of farm women, enabled men to interpret the car flexibly and to socially construct it as a stationary power source. This, in turn, reinforced technical competence as masculine, thus reinforcing farm men's gender identity vis-à-vis farm women. Gender not only shaped the motor car, but gender identities were, in turn, shaped by using the motor car.

Capitalizing on Farm Uses

Four groups responded to farm men's and women's reshaping the automobile to their own ends. Automobile manufacturers, farm equipment manufacturers, gasoline engine firms, and the newly emergent accessories companies designed, built, and sold numerous artifacts that either assisted or replaced the work of the barnyard mechanics. The timing of their efforts indicates that these commercial groups responded to the flexible interpretations of the rural auto developed by farm men and women during the first decades of the century. The phenomenon is similar to the situation in rural telephony, where AT&T redesigned their farmer systems to accommodate a party-line culture.

Many automobile manufacturers countered, rather than supported, the flexible interpretation of the rural auto in the early days of the industry. Although before World War I many companies made cars with removable tonneaus, they usually discouraged using the car as a stationary power source by jacking up its rear wheels, just as phone companies had disapproved of listening in on the party line. When asked by the *Rural New Yorker* in 1901 if autos had been used to turn farm machinery, one manufacturer replied, "We have heard of a few cases where automobiles are put to such usage. They are not built for this usage and [we] would not advise the use of an automobile for other work than that for which it is constructed." Another manufacturer said the "attempt to make use of the automobile, independent of the road service, for which it is built, we think entirely wrong. It does injury to the vehicle, and it is shortly unfit for either service. . . . You are using probably a thousand-dollar investment in a splendid wagon to do ordinary farm work, when nothing is required but a simple engine and boiler." In response to a later survey on this question by the same journal in 1906, six out of seven auto manufacturers adamantly opposed the common practice of using the rear wheels of an auto to provide stationary power, mainly because it could damage the engine or differential gear. The representative of the Reo company was more equivocal. Although he "did not approve of using the automobile for a traction engine, to drive a sawmill, or the different purposes which I have named," he thought that farmers could "save a considerable amount of labor" if they harnessed the auto's power properly.[52]

Based on these responses, the *Rural New Yorker* advised farmers over the next decade to purchase a stationary gasoline engine, which was regularly advertised in the journal, instead of using the car as a stationary power source, even though several technically competent farm men wrote that they had good luck with the practice. The journal also opposed using the car to pull implements, advising readers in 1914 that,

> as a rule, it requires a man of considerable mechanical skill to fit up an outfit of this sort [a converted car pulling a mowing machine] so that it will work properly, and as a rule we do not advise the average man to tinker very much with an expensive car in his efforts to make it do more than carry matter swiftly about the country. Where a man has the skill, however, and enough knowledge of mechanics to fix the attachment properly, there is no doubt that the car can often be used for the lighter jobs of farmwork.[53]

The journal modified its position in 1919 and recommended kits that safely took power from the crankshaft or rear axle—the type of kits that were now being advertised in its pages.[54] Agricultural engineers supported the position of auto manufacturers when they wrote for farm journals on this topic in 1907, 1917, and the early 1920s. They advised against using the auto for general farmwork and warned that homemade kits could damage a car's engine and differential gear.[55]

The Ford Motor Company took a more positive view of the alternative uses of the automobile. One of the earliest published photographs of a car providing a stationary source of power, in fact, shows Henry Ford sawing wood with a new Model A Ford in 1903. Power was taken from a pulley connected to a long shaft inserted into a crankshaft connection in the side of the car.[56] The *Ford Times* published a photograph of a Model S sawing wood in 1908. Shortly after the Model T came out later that year, H. B. Harper, the editor of *Ford Times*, wrote an article for the magazine, in which he matched the Model T's technical characteristics point for point with those he thought farmers required, concluding that "with a little ingenuity the engine [of any automobile] can be made to run the cream separator, saw the wood, or pull a trailer loaded with farm produce or housing supplies." For much of the Model T's long life, Ford magazines and sales bulletins published numerous stories of how farm men had harnessed the Tin Lizzie to do their chores, including plowing, in support of the advertising slogan that the Model T was the "universal car."[57]

In 1912, at the height of this publicity, *Ford Times* reprinted a poem from a Peoria, Illinois, newspaper, which began thus: "The auto on the farm arose / Before the dawn at four. / It milked the cows and washed the clothes / And finished every chore." After reaping, threshing, plowing, pumping water, grinding corn, and hauling the baby "around the block" to put it to sleep, the tireless "patient auto stood outside / And ran the dynamo" so the up-to-date farmer could read by electricity. The magazine changed the title of the poem from "The Auto on the Farm" to "Farming à la Ford." *Ford Owner Magazine* published a similar poem by "rural bard" Wheeler Croy in 1916.[58] In a similar vein, a humorous postcard series, "Let Lizzie Do It," apparently produced independently during World War I, showed the Model T doing many farm chores, including running a washing machine and plowing a field. Interestingly, women operated the cars for both applications. One card depicts the traditional sexual division of labor, with the woman using the car to launder clothes, wash dishes, churn butter, and rock a baby cradle while the man urges her to hurry up and finish so he can plow the fields. The other card shows Maud Muller, a poetic representation of women working in the field, now up-to-date because she plows with a Ford. As with the popular books of Ford jokes, the company saw no reason to discourage the free publicity.[59]

Several accessory manufacturers took advantage of the car's flexible interpretation and began to commercialize it. Although as early as 1912 firms had brought out kits to allow the car to be used as a stationary source of power, advertisements for these kits—and those used to convert the car into a tractor—did not appear in large numbers until 1917, during wartime shortages of farm labor and horses. Some companies simply sold a pulley to be attached to a jacked-up wheel, but most kit manufacturers realized that jacking up one wheel puts an undue strain on the differential gear, as one wheel is turning while the other is stationary on the ground.[60] Most kits, therefore, were designed to overcome the problem with the differential. *Scientific American* described a kit in 1917 that avoided wearing out the differential gear by jacking up both sides of the car and taking power directly from the axle. The Lawrence Auto Power Company in St. Paul, Minnesota, advertised a thirty-five-dollar kit that would take power directly from the crankshaft in the front of a car without having to jack up the car. The company claimed that the device—consisting of a tie-rod, two pulleys, and a metal stand—could operate a feed grinder, corn sheller, silo filler, wood saw, and cream separator. From 1917 to 1919 many other companies sold a variety of kits designed to take power off the crankshaft without jacking up the wheels. These included the E. F. Elmberg Company of Parkersburg, Iowa, the Ward Work-a-Ford company of Lincoln, Nebraska, the Auto Power and Malleable Manufacturing Company of Omaha, and the Knight Metal Products Company of Detroit. The advertisements for the latter company stressed that its conversion device and a Ford car cost less than a fourteen-horsepower stationary gasoline engine. The Maxim Silencer Company sold a kit for sawing wood as late as 1948.[61]

Firms also introduced more elaborate kits that allowed the car to act as an agricultural tractor during the high wartime demand for farm products in 1917. Food shortages led the federal government to encourage farmers to "plow to the fences," which provided an added incentive to buy tractors or kits. Before the war, at least four men had yoked the automobile to the plow. A prosperous Ohioan got his picture in *Scientific American* in 1903 by plowing a field with a plow attached to his touring car driven by a chauffeur; a Montana barber-farmer accomplished the same feat with a Model T in 1916. John Mason, a Rhode Island "back-to-the-lander" with twelve years experience working with gasoline motors, made a tractor in 1915 from "an old discarded automobile given to me to experiment with" at a cost of ninety dollars. That same year, A. W. Bell, a Canadian farmer, converted his three-year-old Overland into a robust tractor by replacing the car's rear wheels with larger steel reaper wheels attached to a heavier axle.[62]

The conversion kits, which came out in a flurry in 1917, did not deviate that much from Bell's design. Typically consisting of tractorlike drive wheels, a heavy axle, reduction gears to lower the speed to about three miles an hour, and a large radiator, a force-fed lubrication system, and other means to reduce overheating problems, these kits sold for $97.50 to $350.00. Advertisements claimed they could quickly convert the popular Model T (and other cars) into a tractor that would pull plows, harrows, mowers, binders, and other implements in the field or, with a different set of wheels, into a traction engine that would pull road graders, wagons, and other heavy loads on country roads. A four-page ad for the "Any Auto" also boasted a pulley attachment so farmers would not have to jack up a car to take power from the rear wheel. At least twenty-two companies made these kits, including the Smith-Form-a-Tractor and Uni-Ford Tractor in Illinois; the American Ford-a-Tractor and the

Hardy Hank Company in Minnesota; the Geneva Adapto-Tractor in Ohio; the F. R. Corcoran Company in New York; and the L. A. Tractor Company in California.[63]

Most of these products seem to have led a relatively short life, but the Pullford Company of Quincy, Illinois, brought out a kit for $135 in 1917 and advertised it continuously from that year until at least 1940.[64] Pullford and the Shaw Manufacturing Company of Galesburg, Kansas, which also made garden tractors, responded to the economic crisis of the Great Depression by targeting their ads toward farm men who were prosperous enough to have an old car to convert permanently into a tractor but not wealthy enough to buy a tractor. Pullford followed this strategy as late as 1940, when it made kits that would fit a Model T, a Model A, a 1926–31 Chevrolet, or a powerful 1932 V-8 Ford. Beginning in 1930, the company changed its standard advertisement picture from that of a young farm man cheerfully plowing with a "Flivver," which he could later reconvert and drive to town, to that of a middle-age, no-nonsense, professional farmer plowing vigorously ahead with a kit that had turned a now-unidentifiable car into a permanent workhorse that emulated the gasoline tractor.[65] Pullford followed its practice, and that of other kit manufacturers, of supporting the flexible interpretation of the car by commercializing what farm men were doing in the field. Ford, for example, received letters and photographs from farmers in the late 1930s and early 1940s describing how they had converted an old Model T or Model A into a tractor.[66] The fond recollections in recent oral history interviews of nicknames for these cars—"Skeeter" in the South, "Puddle Jumper" in the Midwest, and "Doodle-Bug" in the Northeast—indicate the prevalence of this practice.[67]

How did auto manufacturers respond to these kits? The Ford Motor Company seems to have been ambivalent about them before developing its own line of tractors and trucks. When a Canadian man asked Henry Ford in 1908 to comment on his idea of designing cars that could provide stationary power, Ford's secretary replied that even "ordinary automobiles" had been used in this manner, and Ford was designing a tractor that would fully meet these needs. Ford gave a similar answer to a proposal for an auto-tractor combination vehicle that year. But in 1912, *Ford Times* published a photograph of a Model T using a Home Auto Kit to saw wood. Replying to a proposed car-tractor-truck vehicle in 1919, Ford's secretary said that "this is a combination we do not believe would work satisfactorily."[68] By that time, the Ford company had introduced a complete automotive ensemble for the farm—car, truck, and tractor—and thus had little interest in multiple-use vehicles or conversion kits. The firm told its dealers in 1916 that it did not want them to convert "Ford cars into trucks and other makeshifts not recommended or sanctioned by us." Making such alterations would cost them their dealerships. The company advised dealers the next year not to sell truck kits because it had just put a Ford truck on the market. Despite these tactics, the sales of truck kits climbed, and the company warned dealers in 1918 that owners would void their warranty by altering their cars in this manner.[69] One limit to Ford's power, of course, was that Ford dealers could go into another line of business or become dealers for another manufacturer. Dealers who remained with Ford assisted in the reinterpretation of the automobile on the farm as simply a passenger vehicle by not selling conversion kits.

In the long run, however, the redesigning of tractors and trucks during World War I was probably the most effective means by which manufacturers of automobiles and farm implements responded to the flexibile interpretation of the car. Before this time, gasoline tractors had many of the drawbacks of the heavy, expensive steam-engine tractors on which their design was based. After the decade-old gasoline tractor business nearly collapsed in 1912, farm equipment firms and some automobile manufacturers like Ford designed smaller, less expensive tractors. By 1940, large numbers of farm families began to buy mass-produced tractors having belt-power capabilities (a popular item), rubber tires (instead of steel wheels), and a three-point hitch (to prevent the light tractor from rearing) to replace their horses, steam engines, the stationary-power auto kits, and the auto-tractor conversion kits.[70]

Although tractor manufacturers dismissed the tractor kits as impractical, they seem to have posed something of a threat to these firms. The Pullford Company advertised for several years that its kit had competed successfully with gasoline tractors in a national plowing contest in Fremont, Nebraska, in 1917. Companies entered four conversion kits in the 1918 plowing contest at Salina, Kansas, and demonstrated them at national and state tractor shows as late as November 1919. In a retaliatory move, the American Tractor Association, a powerful trade group, requested in late 1918 that the War Industries Board modify its order reducing the amount of iron and steel allotted to the manufacture of tractors by adding a provision that would "prohibit entirely the manufacture of attachments for converting automobile and motor trucks into tractors for farm use."[71] Henry Ford, who ironically conducted most of his tractor experiments with automobile engines, obviously hoped that farmers would buy the newly introduced Fordson tractor instead of the conversion kit, which the company discouraged in general, as we have seen. The advertising manager for the La Cross Tractor Company told tractor dealers in 1918 that he advised farmers to use a stationary gasoline engine, rather than a tractor, to drive small devices like washing machines. He favored saving the tractor's belt power for larger jobs. But the "jacking up of an automobile and attaching a belt to one of the hind wheels to drive a grindstone or a cream separator, is simply ridiculous and should not be given serious consideration by anybody."[72] *Ford News,* in the early 1920s, is filled with stories about the Fordson's power-takeoff option being used to power all types of farm chores and industrial processes.[73]

This reaction, in which manufacturers criticized alternative uses of the car and introduced new products to eliminate its flexible interpretation, continued with the motor truck. The truck became a serious product during World War I, when demand for troop and equipment movement led to a sturdier vehicle and a huge manufacturing capacity. Recognizing these trends, Ford introduced a one-ton truck in 1917 and sold a Model TT truck chassis in 1917, to which owners could add custom bodies. These products, in combination with the company's warranty policy on alterations, helped put an end to kits that converted a Model T car into a truck.[74]

Gasoline engine manufacturers also reacted to the new source of power on the farm. These firms had, since the turn of the century, sold stationary engines for belt-powered work on the farm that were powerful but rather expensive and difficult to start; they now faced stiff competition from the power-takeoff capabilities of automobiles and tractors. The Maytag company responded in 1915 by adding a kick-start gasoline engine to power its washing machines. One historian has said that only then did farm women "begin to feel at home with [gasoline] engines." Yet several oral histories relate that farm women still had problems and that many relied on a male to start the balky gasoline-engine washer.[75] Maytag and many other companies sold these washers into the 1940s, when rural electrification became more common, often with the option of using a gasoline engine or an electric motor (see chap. 6). Farm people who could afford the device no doubt preferred it to jacking up the car to do the wash.

Despite the proliferation of new power sources, many more farm men and women owned automobiles than tractors, trucks, and gasoline engines before World War II. Census data shows that from 1920 until the war automobiles were far and away the most popular form of inanimate power on the farm. A major reason was that during the agricultural economic crisis of the 1920s and 1930s, farm men and women used their autos, often purchased during the boom times of World War I, for multiple purposes like going to town, hauling produce, powering farm equipment, and even fieldwork (for those who bought conversion kits or made their own). Large numbers of prosperous farms did, however, buy tractors, trucks, and gasoline engines. A survey of 538 prosperous Minnesota farms in 1929 showed that more than 90 percent of them had autos, two-thirds had stationary gasoline engines, nearly one-half had tractors, more than one-third had electricity, and about one-third had trucks. The families used their autos almost equally for farm and family purposes, but the study did not mention any belt-powered use of the car. The families made heavy use of tractors, gas engines, electric motors, and trucks to pull agricultural implements, provide belt power, and haul farm products. Southern peanut farmers in the 1930s used trucks to obtain belt power in the traditional way by connecting a large belt from a jacked-up rear wheel to a threshing machine.[76] More and more farms bought trucks and tractors after the federal government established a New Deal program in the 1930s that provided low-cost loans for the purchase of farm equipment. The program led to a large increase in the number of tractors on farms, thus helping to displace the horse and the rural auto as an all-purpose power source.[77]

Restructuring Rural Life

The fact that it could be driven faster and thus much farther than the horse in a given amount of time has led many commentators to view the automobile as a major agent of social change. When farmers started buying cars by the trainload during World War I, a writer for the *New York Times* suggested that previous technologies had not solved the problem of rural isolation. "The trouble with the farm was three miles an hour—three miles by horse and buggy or two by team and wagon." This limited the size of farm communities and encouraged serenity but also monotony. "What was needed for the farmer's family was twenty miles an hour. This the automobile supplied and this is the real meaning of the remarkable increase in the number of automobiles now being bought by farmers."[78] Although, as we saw in the introduction, the claim that farmers were "isolated" is highly problematic, being mostly a construction of urban reformers, a wealth of social-survey data documents that the structure of rural communities changed considerably with the adoption of the automobile.[79] Adoption of the car displaced a whole network of institutions such as open-country schools, churches, and small trading centers that had been created based upon the shorter range of the horse and buggy. Children could go to consolidated schools further away, and churches other than the local one came within range. In this view, the "automobile was the most forceful machine for creating a new social and economic geography." Autos "killed those distinctive institutions supporting conservatism and traditional values: the general store, the open-country church, and the one-room school."[80]

In the 1930s, the sociologist Arthur Raper thought cars would improve race relations in the South because whites and blacks would meet on the road on a basis of equality. But cars were often associated with racial tensions. The legend of the welfare Cadillac began during the New Deal, when many blacks bought cars while whites bought mules for the farm. In the Cotton Belt area of Georgia studied by Raper, planters complained about the increased mobility the car gave to their sharecroppers and renters. According to a recent study, the

> effects of the auto on racial harmony were ambiguous. At best, the proliferation of the machines built upon a few fair, color-blind customs that already existed, then began the obliteration of the oppressive "community mores" that ruled the countryside. . . . The impersonal, colorless democracy of the car was most effective when the machines were in motion. Every driver, barely visible behind glass, frame, and wheel, had his or her right to the right side of the road, by law, no matter what his or her shade or tenure.

By being able to shop in chain stores in towns, sharecroppers found lower prices and more variety, which led to the abandonment of the southern credit system and general stores after World War I.[81]

On the other side of the issue, several scholars have argued that automobile owners wove the car into established social patterns before World War II. The social scientists Malcom Willey and Stuart Rice conclude in their 1933 study of communication and transportation systems that using the automobile did not decrease localism, as they expected, but actually increased it. Traffic surveys showed that "an even greater intensification of mobility takes place within circumscribed local areas . . . [and] contacts in a community are multiplied out of proportion to contacts at a distance." This seemed especially true for rural people. Surveys exploring usage on state highways showed that farmers drove their cars much shorter distances than did people from towns and cities, who did more touring. The average point-to-point mileage per trip was 13 miles for farmers and 8–9 miles for nonfarmers in New Hampshire, 12 miles and 102 miles for these groups in Vermont, and 12 miles and 88 miles for those in Ohio (which questions the veracity of the *Farm Life* cartoon in fig. 5).[82] Other contemporary studies indicate that auto travel decimated the number of crossroad hamlets but did not significantly increase the hinterland of villages and small towns. One historian asserts that in the Midwest, "rural space was not so much enlarged with automobile use as it was reshaped into a more centralized and hierarchical form." This was part of a larger social process that had been occurring since the late nineteenth century, with the growth of commercialized farming and consumerism.[83]

A study of travel on the Saskatchewan prairie reaches a similar conclusion about automobiles and localism. By the 1920s, most autos had the ability to travel on rural roads for a long period of time at an average speed of twenty miles per hour, which was much faster than the horse's speed of four to seven miles per hour for this type of travel. But the poor roads, harsh winters, and the persistence of "slow travel systems," like the peddler, country doctor, mail-order delivery, and open-country schools on the Canadian prairies, delayed the transformation that occurred in the social geography of the midwestern United States until after World War II. Before then, Saskatchewan farmers could have traveled further with their cars, even under poor road and weather conditions, but they did not. "If anything, people might frequent the same destinations more often; but rarely would they make a regular habit out of visiting further destinations." The main reasons were the already great distances between towns on the prairies and the fact that "travel behavior was not narrowly instrumental. People's loyalty to particular destinations was not reducible to reasons of economic utility." Saturday night in the nearest trading center remained a festive social occasion until the 1950s. "For sociability a familiar location served better than a more distant one."[84]

Northern communities in the United States shared many of the elements of "slow travel systems" before World War II. According to the 1929 survey of well-to-do farm families in Minnesota, mentioned earlier, the car was driven an average of 108 hours per year for "family use" and 99 hours for "farm use." The total comes to an average of only 4 hours per week, indicating that these families had not altered their travel patterns that much. Harsh winters and bad roads were probably factors, as they were in Canada, because during this period northern farmers and townsfolk regularly drained the radiators of their cars and put their cars on blocks for the winter. A survey of twenty-six thousand rural mail carriers in 1920 showed that more than eighteen thousand of them drove an automobile on their routes about eight months of the year. They used a horse-drawn wagon for the remaining four months of the year because of bad roads and temperamental autos.[85]

How did farm women fare with the automobile? Some studies maintain that farm women gained independence by using the car to extend their spheres of influence and redefine their gender roles. By marketing their prod-

ucts more widely, they gained more economic power on the farm, and by using the car to visit friends and relatives, they expanded their horizons beyond
the farmstead.[86] Many contemporaries professed this view, especially such
modernizers as editors of farm journals, auto manufacturers, and home economists, who preached the Country Life gospel of saving the supposedly overworked farm woman. In 1910, defending the farmers' purchase of cars against
"eastern" criticism of rural extravagance, Henry Wallace, who was also a
member of the Country Life Commission, said farm men did not use autos
"as the townsman uses his, mainly for pleasure, but as time-savers and for the
purpose of breaking up the monotony of farm life for the women folks." Wallace was more patronizing the next year when he said that farm men often
wrote the journal about how helpful the automobile was on farms, "how the
women folks use it for marketing and making their little social calls." The
Ford Times claimed in 1913 that "automobiles have done more, perhaps, in un-
fettering the farmer's wife than any other one thing. What to her previously
were prohibitory distances now are but matters of a few minutes." Home
economist Florence Ward came to the same conclusion in her 1920 study of
farm women. "The telephone and the automobile free the farm family from
isolation."[87]

Many farm women liked the automobile. A Kansas woman said in 1913
that the "only drawback [to farm life] is isolation, which is slowly being reme-
died by rural free delivery, telephone, and automobile." An Iowa woman
longed "for an automobile cheap enough to be available and cheap in construction." The rural reformer Mary Atkeson claimed in 1924 that chores were
speeded up if a ride in the automobile was promised. In a different measure of
acceptance, Charles Lindberg recalled that his mother liked the Model T the
family had purchased in 1912 so much that she named it "Maria." Mrs. Arthur
Hewins of Massachusetts responded to muckraking concerns about the overworked farm woman in 1920 by saying the car reduced her workload. "In our
'Lizzie' I carry the milk three miles to the creamery every morning, Sundays
included. . . . I have time to go for pleasure rides, and once or twice a week we
go to the 'movies' in the nearest town, which is nine miles away." Writing the
Rural New Yorker in 1925, Francis Ingersoll included access to the auto as one
measure of the increased liberty of the modern farm woman: "We farm
women drive the family auto. Many of us use a joint checking account with
our husbands; we have a personal income for chickens, butter, etc."[88]

Other studies argue that farm women traveled farther but stayed within
their traditional, supportive gender roles when they shopped for domestic
goods or went to town in an emergency to buy parts to fix the tractor.[89] Hold-

ing a position similar to that which some scholars have taken for the automobile and the telephone in urban life—that they widened and deepened existing transportation and communication patterns rather than revolutionizing them—historian Mary Neth has shown, on the basis of diaries and oral histories, that farm people in the Midwest were not isolated before the automobile, because most lived in established open-country neighborhoods. They "did not necessarily use the automobile to become more urban or to abandon their local communities. They often used the automobile to enhance the social ties that shaped their lives, rather than to alter them." For women in these families, the "automobile became a new tool for building rural neighborhoods in traditional ways." In this argument, the car reinforced rural gender roles, as it had for suburban women.[90]

My account of the alternative uses of the car supports this position. Rather than upsetting the gender relations and associated meanings involved with the automobile, the flexible interpretation of the early auto reinforced them. By turning the auto into a general-purpose power source, farm men reinforced their gender identity as technically competent males and the gender structure of what counted as men's and women's work. The auto's replacements for farmwork, the truck and tractor, did not upset the flexible patriarchal system either, even though women demonstrated during World War I that they could drive a tractor, just as their sisters proved they could do factory work during the war. For instance, some urban women learned to drive and maintain tractors and trucks in the American Woman's Land Army—a voluntary organization that hired out "farmerette" squads to farms during the war—and farm women drove tractors at home to meet the "manpower" shortage.[91] Yet when the war ended, farming by horse, car, or tractor was still considered to be primarily men's work.

A related question concerns whether the automobile led to "more work for mother," whether it saved the work of the farm woman's helpers, promoted a higher standard of living, and restructured work patterns, as it had for her sisters in the city and suburbs. The time-use studies mentioned in the previous chapter provide some data because they focus on the "problem" of the overworked farm women. But these studies provide much more information about time spent on household work than on automobile use, and the farm women surveyed were probably atypically well-to-do and had adopted the middle-class, urban domestic ideal to a great extent. Nevertheless, these two thousand women, the vast majority of whose families owned automobiles, still worked a full week in the house, dairy, garden, and poultry pen.[92]

One home economist included time spent with the automobile in her published report. Maud Wilson's study of farm, rural nonfarm, and urban women in Oregon in the late 1920s shows that about 20 percent of the farm women reported spending some time using the automobile. This averaged thirty-six minutes a week (Wilson did not say whether the time was spent riding in a car or driving one). It is noteworthy that the percentage of farm women who reported spending time with the car was double that of the other women in the study and that farm women spent slightly more time with the car. About 10 percent of rural nonfarm women averaged thirty-two minutes per week with the car, and about 10 percent of urban women averaged thirty minutes with the car. Wilson classified this time under "miscellaneous" use rather than under "leisure," "farm work," or "homemaking." It seems reasonable to assume that the time spent "going and returning" was also with the automobile for these middle-class women, which adds forty-four minutes per week for farm women. The average time they spent with the auto per week would then be one hour and twenty minutes.[93] The "miscellaneous" figure, which seems low, may refer to time women spent driving the car.

An Ohio man's story of a farm woman who used her car on a regular basis unwittingly provides one explanation of why the auto did not lead to more leisure. L. B. Pierce wrote the *Rural New Yorker* in 1919 that one morning, a farm woman cooked that night's dinner in a "fireless cooker" (an insulated box in which a boiled dinner could cook all day), drove forty-one miles to visit her daughter in Cleveland, shopped in the city in the afternoon, then drove home in time to put a late supper on the table from the fireless cooker. Before the family had a car, which the woman also used to run a butter-and-egg route, she would have had to skimp on her after-breakfast work, and her husband would have had to get his own dinner. "After the car was bought she could wash the breakfast dishes, sweep the kitchen, and then get to her customers as early as before, and generally get home in time to serve the dinner which the fireless cooker had been preparing in the basement."[94] The car thus enabled this farm woman to do more work—to expand her egg business and still perform the gendered tasks expected of her within the expanded sphere of "woman's work" on the farm, including shopping for bargains in the city and maintaining kinship ties.

The car could also provide recreation and promote mutuality between husband and wife. L. C. Greene wrote to *Wallaces' Farmer* in 1909, "One time in haying the mower broke, and it was ten miles to the supply of repairs. The next morning I invited the partner of my joys and sorrows and pocketbook to lay aside her after-breakfast work and we would go after mower repairs." They were in town by seven in the morning and back home in an hour. He was out in the field as soon as the clover was ready to cut. "That was business with a whole lot of pleasure attached," Greene concluded.[95] But his partner might not have felt the same way, because doing the dishes was probably delayed whether she went to town for repairs alone or with her husband.

The increased diffusion of the car into rural life did not stop farm people from criticizing the automobile. A survey of the wives of more than two thousand crop correspondents for the USDA in 1915 recorded many complaints. "According to several communications the increase in automobiles has made it hazardous or dangerous for women to do much driving on many country roads." The report published criticisms of automobile drivers by rural women in Vermont, New York, Illinois, Wisconsin, South Carolina, West Virginia, and Mississippi. A Kansas woman countered the technological optimism of the Country Lifers by writing, "We farm women are so alone. The automobile is really a separator instead of a link in the chain" of community life.[96] When auto touring began in the early part of the century, it seemed a boon to rural people. Farm men and women became entrepreneurs, sold produce to motorists, and established autocamps and tourist cabins. According to a student of the subject, "Here was a new market for home produce as well as an opportunity for outside contact. By the early 1920s, however, many rural areas had had enough. Roadsides were strewn with garbage, especially with tin cans, the autocamper's emblem. . . . Farmers took matters into their own hands, posting no trespassing signs and standing guard with shotguns."[97]

Farmers also resisted moving mailboxes to the "gee" side of the road because of increased auto traffic. Socializing at the mailbox was cut down when mail carriers switched to the car. During his 1921 study of rural Wisconsin neighborhoods, rural sociologist John Kolb asked farmers about the social changes that had occurred during the past fifty years. "Complaint was often heard that now with good roads and [the] automobile, less 'neighboring' was done. 'The young people go miles away,' someone said, 'but fail to get well acquainted with those nearby.'"[98]

Kolb also disliked so much mobility. He described a "well-educated couple just starting to farm," who had trouble identifying the boundaries of their neighborhood: They received their mail and attended meetings of farm organizations in Dane Village, traded in Waunakee, went to church in Martinsville, sent their milk to a creamery at Springfield Corners, and lived within the school district of Elm Grove. "The question is not raised as to which of these bonds should be cut or that any loyalties should be weakened, but it is contended that this scattering of interests is obviously impairing the efficiency of this family unit. It was a unit by itself looking for a larger group relationship." Kolb lamented further that the "early acquisition of the machine has excited a rather exaggerated sense of freedom, for where five miles was formerly a limit, now it is twenty-five or more. The automobile must be 'domesticated' and made to do service for the neighborhood and community as well." Like his fellow rural sociologists, Kolb generally praised the RFD, the telephone, and other up-to-date technologies. But he wanted them to uphold rather than tear down the type of village-centered "rurban" life he favored.[99] Of course, for those who wanted to leave farm life, especially the young, the auto provided a convenient means of escaping to town or to the city.[100]

Much has been written about how the automobile changed urban and rural culture. Although rural life was restructured and many local institutions were destroyed by farm people driving a car rather than a horse and buggy, they were not passive consumers in the early years of this process. Rather, farm men, women, and youth helped to determine what role the automobile would play in their lives. Their use of the car in alternative ways not only prompted manufacturers to introduce new products like the tractor, the pickup truck, and the gasoline-powered washer, it also enabled farm people to "tame the devil wagon" and weave it into rural culture, often as a general-purpose power source for the farm. The substantial social changes typically associated with the car in rural life generally followed this pattern—traveling a "domesticated" path of expanding traditional transportation practices rather than obliterating them beyond recognition. The experience with the telephone was similar, but telephone cooperatives allowed farm people to have more control over telephony than they had over rural roads, which were taken over by centralized highway departments in the 1920s. By that time, farm people had managed to domesticate the automobile and make it a defining element of rural modernity—quite a change from the turn of the century, when farmers attacked the devil wagon as an alien technology.

CHAPTER 3

Defining Modernity in the Home

In the eyes of urban dwellers, the telephone and the automobile connected farmers more closely to the town and city, but the lack of "modern" conveniences in the home still marked them as rural. Increasingly, most urban and many rural folk came to view the material culture and work practices in the farmhouse as rustic holdovers from the nation's agrarian past. Reformers in the early part of the century thought a wide range of technologies were necessary to modernize a house. *Wallaces' Farmer* elaborated on the point in 1910. "Modern conveniences, as understood in the city, mean furnace heat, hot air or hot water, sewage, hot and cold water, bath room, and some system of lighting other than the old-fashioned kerosene lamp." The journal advised well-to-do farmers to modernize the house in order to lessen the drudgery for farm women and to enjoy well-earned city comforts.[1]

Other writers disagreed about what conveniences were required to make a farmhouse modern. Rural sociologist Ellis Kirkpatrick summed up the debate in 1929: "It is difficult to determine the degree of 'modernness' in farmhouses owing to confusion as to the items of equipment which should be included in a so-called modern home. Some workers place the emphasis on the plumbing facilities, including running hot and cold water for laundry, kitchen, bath-room and toilet uses. Others are more concerned about the heating facilities. Still others are inclined to stress the lighting system." Kirkpatrick advocated a three-part classification. "Homes with central heating and central lighting systems and with running hot and cold water, including sewage disposal, for kitchen, bath, and toilet uses, are regarded as completely modern. Homes with one or more of these facilities are regarded as partly modern and those with none of them are regarded as not modern." Kirkpatrick noted that some researchers used other measures of modernity, such as window screens and laborsaving devices. Pianos, phonographs, and radios were usually seen as luxuries.[2]

Kirkpatrick developed his taxonomy at the end of three decades of efforts by Country Life reformers to modernize the farmhouse. The initial goal was to save the allegedly overworked farm woman, an ideology that informed the urbanization programs of reformers, manufacturers, and government agencies. Most farm people agreed with these aims, but they defined modernity on their own terms and exerted a fair amount of control over this process. They modernized farmhouses with an innovative assortment of new technologies, usually without the benefit of high-line electricity. Although the household technologies were "modern," they held mixed blessings for farm women.

Saving the Farm Woman

The issue of the overworked farm woman, which had come to the fore in the nineteenth century, gained a new sense of urgency when the Country Life Commission brought the topic to national attention in 1909. "Realizing that the success of countrylife depends in very large degree on the woman's part, the Commission has made special effort to ascertain the condition of women on the farm." The commission concluded that because the farm woman's work was not as seasonal as the farm man's, "whatever general hardships, such as poverty, isolation, lack of labor-saving devices, may exist on any given farm, the burden of these hardships falls more heavily on the farmer's wife than on the farmer himself. In general her life is more monotonous and the more isolated, no matter what the wealth or the poverty of the family may be."[3]

Sparked by the commission's report, reformers attacked this problem in traditional Progressive Era fashion, collecting data and promoting legislation to fund experts to make "objective" scientific surveys. Mattie Corson, a farm woman whose mother had supposedly died from overwork, conducted one of the earliest surveys. With the help of the "Bachelor Girls' Club," which she organized, Corson solicited letters from "girls and women of marriageable age and over, single and married, in the country." She asked them several questions about farm life, including whether they would marry a farmer or encourage their daughters to do so. In 1909, the *Ladies Home Journal* published excerpts from more than nine hundred letters received by Corson, many of which indicted farm men for buying laborsaving technologies for their work in the field and barn before they would buy such things for "woman's work" in the house. The *Journal* ignored the fact that most farm women had charge of the garden and poultry pen and that many worked in the dairy and "helped out" in the field on a seasonal basis. Yet one women considered the conditions so bad that she "would rather take my chances with my girl on Broadway . . . than to have her walk the sure road to the county asylum that I am heading for."[4]

A series of five articles in *Harper's Bazaar* by Country Life reformers Martha Bensley Bruère and Robert Bruère, entitled "The Revolt of the Farmer's Wife," took up the crusade in 1912. Addressing the issues of overwork, forestalled education, the "social significance of a bumper crop" in purchasing new transportation and communication technologies, unhealthy sanitary conditions, and the debilitating effects of growing old on the farm, the Bruères helped to raise the problem of the farm woman to the level of such turn-of-the-century muckraking issues as urban poverty. They also voiced an urban concern that rural migration was leading to food shortages and rising food prices in the face of a rising urban population. "Shall the nation go hungry because the farmers' wives don't like their jobs? For, after all, a man will not live on the farm without a wife."[5]

Government funding led to more extensive studies of farm women. As mentioned earlier, the U.S. Department of Agriculture (USDA) asked the wives of its volunteer crop correspondents in 1913 how the agency could help farm women. Most of the two thousand women who replied seemed dissatisfied with farm life. Extracts of letters describing the lack of modern household technology and the drudgery of farm life documented in great detail the complaints raised by earlier surveys. Women from all parts of the country bemoaned their loneliness and geographical isolation; but the USDA noted that many were content with farm life. "A study of the counties whence the letters complaining of isolation seems to establish the fact that the condition is peculiar to individual farms rather than attributable to any general state of farming," the condition of roads, or a low population density. The *New York Times* covered the report in an article entitled, "Farm Women Find Life Hard, Many Criticisms in Hundreds of Letters to Department of Agriculture in Novel Poll." A study of the long, toilsome workday of fourteen hundred farm women by the New York State College of Agriculture at Cornell University prompted the *Literary Digest* to title its 1919 article on the subject, "Some Solid Reasons for a Strike of Farm-Wives." The original story in Cornell's *Extension Service News* reported that the monotonous character of the closely confined work of farm women "is believed to offer an explanation of the prevalence of insanity among rural women."[6]

This widely held belief was not seriously challenged until the USDA published its 1913 survey. In response to a letter denouncing the insanity charge, the agency reported that

> this statement is borne out by the department's investigations. For the past twenty years a statistician has investigated every published statement to the effect that farm women contribute largely to the inmates of asylums. Similarly a number of such statements in these replies were investigated. In no case have the facts established any such condition, and on the contrary seem to indicate that the percentage of insane women who came from the farms is below rather than above the average. For these reasons statements to the effect that life on farms drives women insane have been omitted [from the report].[7]

A survey of rural mental health in the 1920 census and the publication of related studies from the 1920s supported these conclusions.[8]

To address the issues of the farm woman, in 1915 the USDA established an Office of Home Economics, which in 1923 became a full-fledged research Bureau of Home Economics.[9] In 1920, Florence Ward, in charge of the USDA's Extension Work with Women in the northern and western states, published the results of a survey of ten thousand farm women by home demonstration agents in thirty-three states (including the 1919 New York survey). Although Ward implicitly denied the insanity charge by saying that the average farm woman enjoyed better living conditions than her city sister, the survey found "five outstanding problems" of the farm woman: a long workday (more than eleven hours), regular performance of heavy manual labor, low standards of beauty and comfort in the house, perilous health of the mother and child, and few income-producing home industries. To overcome these problems, Ward recommended that farm women adopt measures that had long been advocated by home economists: "improved home equipment," "more efficient methods of household management," and education in nutrition and child care. Ward thought farm women should be taught that cultivating the "comfort, beauty, health, and efficiency of the farm home" was "perhaps the only means of stopping the drift of young people to the city."[10] Offering respite, a farm bureau rented the YMCA's Camp Brewster near Council Bluffs, Iowa, in the summer of 1925, as a vacation camp for overworked farm women.[11]

What was behind the great interest in the farm woman? A recent study has argued that it was part of broader concerns about both the quantity and quality of rural-to-urban migration in this period. Although Ward and others feared that depopulation would weaken rural institutions, most agrarian reformers (including rural leaders) thought some migration from the country was economically necessary and socially desirable because of the increased productivity caused by farm mechanization. But many Country Life reformers (especially urban elites, probably a majority of the movement) worried that the farm men and women left behind would be of inferior quality (e.g., less enterprising and intelligent) and would thus create an inferior rural life and send inferior, surplus youth to the city (a fear compounded by the reports of insanity among farm women). Many reformers also worried that immigrants from southern and southeastern Europe would migrate to the farm in the place of those who had left, thus creating a peasant (agri)culture in the United States. Thus, "to many, the Country Life problem was in no small part an eugenics problem, and the need for the 'best' people [i.e., native-born whites from northern European stock] to stay on the farm and reproduce freely served more than strictly agricultural ends."[12]

Although only a few authors, such as Assistant Secretary of Agriculture Willet Hays in 1911, referred explicitly to the science of eugenics, many reformers shared Hays's view that "folks are the best crop of the farm [and] the farm is the best place to raise folks." The Country Life Commission had said that "upon the development of this distinctively rural civilization rests ultimately our ability, by methods of farming requiring the highest intelligence, to continue to feed and clothe the hungry nations; to supply the city and metropolis with fresh blood, clean bodies, and clear brains that can endure the strain of modern urban life; and to preserve a race of men in the open country that, in the future as in the past, will be the stay and strength of the nation in the time of war and its guiding and controlling spirit in time of peace."[13] Farm journal editors, rural sociologists, farm women, and other commentators on rural life expressed this ideology after the commission published its report, using similar metaphors about "child crops" that drew on the long-standing agrarian myth of the primacy of rural life.[14] The reformer Mary Meek Atkeson argued in 1924 that urbanites should support the Country Life movement because the countryside continued to provide most of the nation's leaders and suitable (i.e., Americanized) workers for its factories. "In payment for this interest shown by the city people the farm will return not only its corn and hogs and cattle, but also a steady stream of bright-eyed young people to carry the best American traditions into every city in the land. As the farmer will tell you jokingly, his young folks are indeed the 'best crop' of his farm."[15]

Leading home economists adopted a similar rhetoric. Speaking at an agricultural conference in 1926, Louise Stanley, chief of the USDA's Bureau of Home Economics, justified her agency's efforts to improve the farm home by quoting her former boss, the recently deceased Henry Cantwell Wallace, secretary of agriculture and son of Henry Wallace, an original member of the Country Life Commission. Stanley quoted from the younger Wallace's *Our Debt and Duty to the Farmer* (1925) that the "surplus of this crop [of children] makes the steady stream of fresh, virile, potent blood which flows into the cities and is such an important contribution to the vital human force of the nation."[16] Because healthy women and healthy rural institutions (such as the home) were needed to produce both a healthy surplus youth crop and the important institution of rural life, the allegedly overworked farm woman stood at the center of the Country Life problem in this period.

Yet many rural leaders objected to the stark portrayals of farm life. Henry Wallace said the *Ladies Home Journal*'s allegations against farm men in 1909 were slanderous. Wallace blamed farm women for preferring old-fashioned ways of doing housework to buying laborsaving technology. The Country Life Commission heard these reports only from poor, white southerners. "If farmers' wives really want labor-saving devices they can in most cases have them" by exercising their womanly influence in the home. Wallace also thought "this story that the farm contributes more than the average number to the insane asylums is not founded on fact. . . . The farm with its abundant opportunity for pure air, good health, independence, comparative freedom from temptation, offers the best place for rearing a family to virtuous womanhood and clean, sterling manhood." Clarence Poe, editor of the southern *Progressive Farmer*, agreed that the "descriptions of the farm woman frequently given in the magazines seem to us unrecognizable caricatures." But as one of the rural leaders who instigated the USDA's 1913 study of farm women, he also suggested how farm men could address the problems documented by the survey.[17]

The farm woman Clara Williams criticized Wallace's view that women only needed to ask their husbands for household technology. "I wish you would ask Mrs. Wallace if she was not disappointed to learn that her husband expected her to demand her rights from him." A Missouri woman supported Wallace's position in 1913, but probably not in a way he would have liked. "It is a poor excuse of a woman who can not get help from her husband. I read somewhere of a woman who asked her husband for a wringer and sewing machine the first summer of their marriage. On being refused she hired out in harvest to make money. After that lesson she had every labor-saving appliance she saw fit to ask for."[18]

More commonly, farm women defended their way of life. Mrs. T. R. wrote *Wallaces' Farmer* in 1909 to complain that "some of our magazine writers are fairly waxing eloquent over the tragedy of the farmer's wife, portraying the farmer as a moneyed miser whose only ambition is to acquire more land, in the meantime sacrificing his wife's health and happiness to satisfy his ambition." Her response to this common view of rural patriarchy was to retort that "God created man first as head of the house, then he created women to manage him."[19] Adeline Goessling, the home editor of *Farm and Home,* asked her readers in 1920 what they thought about recent newspaper and magazine descriptions of the hardships of farm women. Nearly ten thousand women wrote in to deny the charges. Extracts from these letters, published in the magazine and in *Literary Digest,* extol the clean air, healthy food, neighborliness, the satisfaction of honest hard work, and the urban-style technologies available in the country.[20] In 1922, farm women on a committee of a national agricultural conference "resented keenly the present fashion of magazines and newspapers to belittle the country woman, in stories representing her as having few home conveniences and, apparently, fewer brains." The women conceded that many farm women were "overworked and inefficient," but these did not represent "normal or usual conditions" in their view. As late as 1925 a farm woman wrote the *Rural New Yorker* to complain about magazine writers who held up "that old rag-baby scarecrow of the preponderance of insanity among farm women. And this in the face of the fact that reliable statistics have proved it time and again to be utterly false."[21] Responding to this outpouring of protest from farm women, Emily Hoag, an economist for the USDA, interviewed farm women around the country and studied the mostly upbeat letters solicited by *Farmer's Wife* in 1922. Although many women complained about overwork and the lack of laborsaving technology, Hoag's report deemphasized the "tragedy" of the farm woman in favor of the virtues of rural life praised by this group of women.[22]

Modernization off the Grid

All of these commentators—irrespective of their position on the status of farm women—viewed modern household technology as a progressive social force, a solution to the problem of farm women, and its presence on farms as an indication that they were better off than the muckrakers suggested. A flood of advertisements for consumer products—bright appeals to adopt urban forms of modernity—reinforced the message and marked the widening gulf between the material culture of rural and urban life. Even farm journals carried urban images in ads for consumer goods, but historian Hal Barron has shown that advertisements directed to farm people and the concerns of rural life in the 1920s were more successful than those based on urban images.[23] An earlier example of this approach was the Fels-Naptha Soap campaign that ran in the *Rural New Yorker* in 1912. In these one-frame cartoons, Anty Drudge, a woman from the city, calls on farm women to tell them about the laborsaving properties of Fels-Naptha Soap. Anty claims that the soap saves labor because washing can be done in cool or lukewarm water and because its powerful action removes dirt easily. The ads used many themes common to the campaigns to urbanize rural life: modernity versus the conservatism of farm women; saving the overworked farm woman; urban-rural tensions over dress styles; and—the perennial favorite—keeping youth on the farm.[24] In response to these and other appeals, thousands of farm people modernized their homes by purchasing new consumer products, from soap to washing machines.

Reformers and home economists recommended many nonelectrical technologies to ease the lot of farm women, ways for them to modernize "off the grid," to use a recent phrase. In 1912, Martha and Robert Bruère were especially keen for farmers to use windmills and stationary gasoline engines as power sources to provide running water in the house. "The power engine—usually gasoline—is the heavy artillery of the farmers' wives' war on drudgery." The Bruères thought the gasoline-powered washing machine "is to the farmer's wife what the self-binder and the thresher are to her husband." They also praised cooperative laundries, cooperatively owned vacuum cleaners, and fireless cookers (the insulated box for cooking a boiled dinner mentioned in the previous chapter). Although the Bruères cited instances of cooperative laundries being established in northern states and the use of cooperative vacuum cleaners in Virginia, they regretted that "farmer's wives seem strangely reluctant to adopt" the fireless cooker. They tried to show farm women how to make them out of old trunks, newspapers, and hay, but farm women typically told them, "Oh, yes, I've heard of them, but I guess I won't bother today. I haven't much faith in them to tell you the truth."[25]

The USDA concluded from its survey of the wives of crop correspondents that this group had more faith in new technology. "Among the appliances for which there seems to be the most demand are vacuum cleaners, washers, wringers, fireless cookers, cream separators, power for lighting and various other purposes, better systems of heating, oil stoves, gasoline or electric irons, etc." Letters proposing a cooperative laundry came from all parts of the country. Yet there was some resistance to these technologies. As a Wisconsin woman said, "It isn't labor-saving machines that the farm woman wants nearly so much as warm, good-roofed houses with more rooms, sufficient light, easy-working pump in the well, good rain barrel, a garden fence that will keep out pigs and hens; above all, a woodshed with dry wood." An Illinois woman wrote, "We have our domestic science society, but as one woman said, 'We can't come in from the garden at half-past eleven and cook a domestic science dinner by twelve o'clock.'" A Wisconsin woman revealed that "our club has recently launched the project of a cooperative laundry. In canvassing the territory for stockholders it was the women who were discouraging—taking a personal view instead of a community view."[26] A recent study of women on the Canadian prairies supports the view that running water and good floors were more important than household appliances at this time. In response to the USDA survey, Secretary of Agriculture David Houston announced that his agency would help establish combined cooperative creameries and laundries.[27]

Well-to-do farmers built impressive facilities for the farmhouse, without the benefit of electricity, during the agricultural prosperity in the early part of the century. E. S. Sweet of Sangamon County, Illinois, wrote *Wallaces' Farmer* in 1909 that he built a ten-room house with a laundry room in the basement. "It has pipes with hot and cold soft water supplied from a six-hundred-gallon air-pressure tank, also a pump for supplying well water, a laundry stove, a washing machine, and a little gasoline engine." The engine pumped water from the cistern to the tank and also could be hooked up to the washer. "My boy, thirteen years old, does the running of the [washing] machine and the engine," an indication of gender flexibility with this type of technology. Sweet's house also had a "hot water heating plant, acetylene lighting plant, and the six-hundred-gallon air-pressure tank supplies the kitchen sink, lavatory, bath tub, and water closet with water all the time. A thirty-gallon range boiler supplies hot water for sink and bath room, also to laundry room. A six-inch sewer carries off all the waste from the bath room, kitchen, and the gas engine." The Sweets also had a telephone, RFD service, and grocery delivery twice a week, but no electricity. "We have no notion of moving to town," he wrote.[28]

A Kansas woman reported a veritable cornucopia of consumer goods in the 1913 survey of crop correspondents.

In our township about one-third of the landowners have water in the house pumped either by windmill or gasoline engine—one-third have a modern lighting plant of some kind. All the women have improved washing machines, some of them run by gasoline power; several have gasoline irons, so they can do a large ironing without a fire in the stove, and I think the best thing to help the farmer's wife yet is the coal-oil stove; within the last four years I think one-half of the homes have been supplied with them. We hatch our chickens in incubators. All that care for them have horses and buggies; some few have automobiles that they drive. We have our church society, a library association, a grange, and the farmers' institute and other social functions that the women have a part in, and they are able to do it well.

She thought one reason for the favorable situation in her township was that most families had homesteaded in the community.[29]

How representative were this township and Sweet's house? Although the U.S. Census Bureau did not enumerate household appliances to this extent, rural social scientists conducted numerous surveys of consumer goods on the farm in this period. To meet the Country Life movement's goal of gathering statistics on farm life, rural sociologist Charles Galpin of the USDA convinced the Census Bureau to start counting the number of households having automobiles, telephones, running water, and electric lighting in the 1920 census (the data to 1960 are compiled in fig. 2). Galpin also initiated, supervised, and helped coordinate many farm-life studies conducted by rural sociologists and home economists, almost all of which enumerated durable consumer goods.[30] Other agencies, such as the General Federation of Women's Clubs, conducted their own surveys of farm homes.

The results of three national and six local surveys are shown in tables A.6 and A.7. The higher percentages of telephones, automobiles, running water, and radios in these surveys, as compared with the figures from the censuses of 1920 and 1930 shown in figure 2, are a result of the fact that the Census Bureau attempted to be inclusive, whereas rural sociologists, home economists, and women's clubs were usually exclusive. The latter groups tended to select more prosperous farms through contacts with county agents, home demonstration agents, and (white) women's clubs. Imbued with the ideology of uplifting rural life, as well as the ideals of farm management, they usually focused on the "family farm," that is, on well-educated, middle- and upper-class, native-born white men and women, the traditional clientele of the USDA Extension Service. Yet, precisely because the surveys are biased in this manner, they provide good evidence about how prosperous farm people modernized their homes in the 1920s and early 1930s.

This modernization looked much different from that in prosperous urban areas at this time and in rural settings after World War II. The tables show that most well-to-do farm families in the interwar years owned a telephone, an automobile, and, later, a radio. But many did not have a refrigerator (gas or electric), running water, a stationary bathtub, an indoor toilet, or a lighting system—all of which were more common in the electrified city. The modernization of housekeeping took a back seat to possession of new communication and transportation technologies.

Rural gender relations and the small-business character of the family farm help explain this priority. The USDA reported as much in its 1913 survey of farm women. "That any device which will lessen the labor in the fields is purchased without hesitation, but that no labor-saving devices are introduced into the house to lighten the woman's work is an almost universal grievance."[31] Old Order Amish families also preferred to mechanize operations in the barn and field rather than those in the house. Cornell's 1919 study of farm homes struck a similar chord. "In one quarter-section of high-grade farm land, there were but two homes having running water in the house, and but one of these had a furnace and a bathroom, whereas there were many agricultural implements and sometimes even a Ford in the barn." The attraction of the city seems to have been a factor in deciding how to spend hard-earned cash. When asked by a researcher why her family had purchased a Model T but did not have indoor plumbing, one farm woman replied, "You can't go to town in a bathtub!"[32] Many of the wives of crop correspondents praised indoor plumbing. A California woman, who probably had an automobile, said that the "luxury of having the water right in the house, sinks, and a really, truly bathtub are some of the principal things we long for."[33]

Some reformers recognized that there were trade-offs involved. Mary Atkeson wrote in 1924 that farm women sometimes asked her whether they should buy an automobile or a water system. "It is a question which I never answer, because no one can answer it for her." Farm women had to weigh the advantages and cost of each for their family. Atkeson optimistically thought a water system might save enough time to allow women to do more work for cash, which the family could use to buy an auto. She also noted that "some farm families that I know gave up the automobiles they had bought and spent the money on water systems and electric light plants."[34] Figure 2, however, indicates that this number must have been very small.

One reason farmers did not buy more household appliances was that many were not designed for farm use, which was also the case with early telephones and automobiles. "Labor-saving articles especially designed for the needs of farmers' wives" were required, a Michigan woman told the USDA in its 1913 survey. "For instance, the washing machines on the market are all right for the slightly soiled clothes of city people, but we need something that will work effectually with clothes grimed with mud and water."[35]

As they had with the telephone and automobile, farm men and women found innovative ways to modernize the house, especially before rural electrification became more common in the late 1930s. A survey conducted by the General Federation of Women's Clubs in 1925 shows the wide variety of ways farm people lit their houses, cooked and stored food, washed their clothes, and brought water into the house without the benefit of electricity. Natural gas, acetylene, and gas mantle lamps lit houses; wood, coal, gas, and kerosene stoves cooked meals; cellars, caves, iceboxes, and mechanical refrigerators preserved food; machines powered by hand and gasoline washed the clothes; and a variety of pumps, powered by hand, gasoline, the wind, and electricity, supplied the house with water. Many farms in the Midwest had natural gas on the premises; farmers in other areas made their own acetylene gas, using generators that combined carbide and water to produce the gas. Reports claimed that more than two hundred thousand farms had installed acetylene systems by 1912.[36]

Although lighting and cooking with gas was a symbol of modernity, running water and sanitation was the main focus of the Country Life campaigns to save the overworked farm woman. Running water would lift the daily burden of hauling water from a creek, spring, or well; indoor plumbing facilities and sewage systems would remove the outhouse—that much-joked-about symbol of rural backwardness—from the landscape. Although the country was far behind the city in this area, prosperous urban and suburban families were able to install "all the modern conveniences" in the late nineteenth century, before most cities installed centralized water and sewage plants.[37]

Farm people were just as innovative in the early part of the century. Despite the very low percentage of southern farmhouses reporting running water (see table A.8), the *Progressive Farmer* published several letters from prosperous farm men in 1910, recounting how they used windmills, hydraulic rams, and gasoline engines to pump water from lakes, wells, and springs in Mississippi, Georgia, and the Carolinas. Costs varied widely, from $34 for a pump, bathtub, and piping to $310 for a windmill and plumbing fixtures and $1,000 for a gasoline engine, fixtures, and drilling a deep well. A South Carolina farmer installed a simple gravity-feed system, supplied by a spring at a point higher than the house. In 1915, a North Carolina man installed a water wheel on a creek to power a water pump at a nearby spring.[38] *Wallaces' Farmer* described farmer-built compressed-air, gravity, and hand-pump systems in 1915. Junius Dixon of Ohio wrote,

A tank holding fifty-six barrels of water is placed on the second floor of the barn about a hundred feet from the house. . . . A pipe from the tank goes down through the barn and is conducted underground to the house. . . . The bathroom is on the first floor, and there is plenty of force to raise the water to it and to the pantry taps in the kitchen. There is a slate roof on the barn, and the water is clear as crystal. . . . We have never supplied a closet because the water supply is not sufficient, but for any family this arrangement will save a wonderful lot of carrying of water, and does not necessitate the gasoline engine for pumping.

Dixon's decision to install running water in the house, but not a "closet" (a flush toilet), was a common one. In fact, a USDA bulletin on sewage systems in 1922 gave instructions on how to build sanitary outhouses before describing how to construct septic tanks.[39]

Early Efforts to Electrify the Farm

While many of the nation's farmers found all sorts of innovative ways to modernize the house without electricity, a small percentage of them managed to electrify the farmstead before the Rural Electrification Administration started its program in the mid-1930s. The main methods were hooking up directly to a utility company's high-voltage line (high-line), forming an electrical distribution cooperative, and generating electricity on the premises.

The building of interurban trolleys at the turn of the century provided an early source of central-station power. A Wisconsin farmer tapped an interurban's line as early as 1897; the Aurora, Elgin, and Chicago Railway began selling electricity to farmers on its right-of-way on a regular basis in 1906. Distribution costs were low because no new lines had to be built, and the increased usage during off-peak hours enabled utility companies to sell more electricity with less capital investment. Municipal systems also began extending their lines into rural districts in this period. Following the construction in 1906 of a five-mile line serving five farms in Oregon, by 1910 utilities had built rural lines in California, Colorado, Indiana, Illinois, and Washington. The use of cheap hydropower to run electric water pumps for an irrigation-based agriculture made California the foremost state in rural electrification. One-fourth of its farms had central-station service in 1924, more than one-half in 1932.[40]

The unfavorable economics of rural electrification in the rest of the country—as interpreted by the utility companies—prevented most states from reaching these figures. The Detroit Edison Company calculated that the average revenue from a farm would be only twenty-one dollars a year. At a density of four farms per mile, the revenue would not pay the interest on the two thousand dollars needed to build a mile of line.[41] Acting on figures like these, most utility companies would serve farmers only if they paid for the lines, which nonetheless became the property of the electric company. High rates thus created a Catch-22 situation by keeping usage (and revenue) low. Without a detailed understanding of distribution costs, utilities usually charged farmers the household rate in town plus a surcharge, which totaled anywhere from nine cents to eighteen cents per kilowatt-hour. Because the average net income from farming plummeted from an all-time high of $1,395 per farm in 1919 to $517 in 1921, during the first stage of the long agricultural depression (see table A.1), it is no surprise that less than 3 percent of the nation's farms were connected to the high-line in the early 1920s.[42]

Faced with high prices from utility companies, many farm men and women drew upon the antimonopoly tradition of American agriculture and formed electric-power cooperatives. The scheme appealed to many farmers because of the success of telephone co-ops earlier in the century and the similarities between the two technologies. Telecommunications and electric light and power were both delivered by electricity running along wires strung from pole to pole in a network connecting farms to a central office. But there were important differences. Farmers could build private telephone systems without having to connect them to each other or to the urban world. The low-voltage and low-current equipment was relatively inexpensive, and local people operated the switchboard and ran the co-op. Far more complex technology and expertise were associated with running large engines and dynamos to generate thousands of kilowatts at a central source, stepping it up to several thousand volts, transmitting it safely to farms, then stepping it down to 110 volts for lights and appliances. The complexity of the task undoubtedly convinced light and power co-ops to buy electricity wholesale from utility companies, which already had the generating equipment, and distribute it to the countryside themselves (the part of the job utility companies thought was unprofitable). Besides, distribution lines did not seem that much different from telephone lines.

Several farm groups made decisions like these. Aware of electric power co-ops in Canada and Europe, farmers built systems during the energy and labor shortages of World War I and many more thereafter. At least thirty-four rural electrical cooperatives were formed in the decade following World War I, most of which were in localities in the Midwest and Northwest that had a nearby source of public power. Iowa topped the list at eleven, all of which were served by the municipal plant at Webster City. Five co-ops were established near a municipal plant in Tacoma, Washington, four on Bureau of Reclamation projects in Idaho and Washington, and eight in Minnesota and Wisconsin. The remaining co-ops were scattered across Michigan, Missouri, Virginia, and North Carolina. Members in at least twenty of these organizations had been involved with other agricultural cooperatives. One report, reflecting the eugenic theories common at this time, noted that the "racial characteristics" of electric co-op members were those often associated with cooperatives in the Midwest, that is, the "Nordic stock" of Scandinavians and Germans. The co-ops were financially healthy, although most of them were rather small; eighteen co-ops had less than fifteen miles of lines, and only nine had more than forty-five miles of lines. Rates varied considerably. About one-half of the co-ops charged from three to six cents per kilowatt-hour for using up to fifty kilowatt-hours, while the other half charged six to ten cents; three co-ops charged ten to fourteen cents for this amount of power.[43]

Several of these early cooperatives enjoyed a long life. A ten-year-old mutual association, the Peninsula Light Company, served nearly a thousand farms, villagers, and rural industries in western Washington in 1937, thanks to favorable public-power laws in that state. The Berwick Transmission Line Company of Nemaha County, Kansas, was organized in 1919 with seventy-eight customers (sixty-nine of whom were farmers) on thirty miles of line. Shares cost $250 per member. The company bought electricity from the City of Sabetha and built a distribution line along a public highway. Farmers built lines from the highway to their houses. The co-op grew to more than one hundred members in 1944 and was still in operation in 1960.[44]

Other cooperatives failed miserably, for a variety of reasons. Farmers near Prairie du Sac, Wisconsin, established thirteen co-ops to buy power from a utility company hydroelectric power plant in 1915, then saw their experiment fail when several of the co-ops had to sell their assets at half price a decade later because of overdesigned lines and high wholesale rates charged by the utility. Farmers organized an electric co-op near Hope, Arkansas, in 1929 and bought power from a municipal system, but the enterprise dissolved in 1938.[45]

Profitable cooperatives were also bought out by utility companies. In the mid-1920s, a local Grange in Pennsylvania refused to pay a bonus of thirty thousand dollars to a utility company to obtain power for its members, then organized the Morrisons Cove Electric Light, Heat, and Power Company. The company obtained charters from the state legislature to deliver power to primary rural line associations (groups of farmers living a certain distance from power lines). Much like the earlier telephone cooperatives, these power associations procured rights-of-way, furnished the poles and all other equipment for the line, erected the line, and maintained it. Morrisons Cove thrived. Within eighteen months it formed three associations serving about fifteen miles of customers, returned 6 percent on its capital stock, and cut the cost of building rural lines to five hundred dollars, one-fourth of the typical price charged by power companies. The co-op then sold out, at a profit, to a utility company. But the utility attempted to discontinue service to the associations, claiming, much like AT&T had in the case of telephone cooperatives, that the farmer-built lines were substandard.[46]

Tom Martin, president of the Mutual Light and Power Association in Spanaway, Washington, described his battle with the utility industry in 1930:

> Our hick lines down our way are a black mark on the vaunted efficiency of the power trust. They are trying to do anything to discredit us. Not long ago a circular letter was sent out by one of the private companies to all our users. This letter attempted to discredit our system and purported to show our people that they were losing money by having their own system. It was not very flattering to the intelligence of our members. The net result has been to solidify our mutuals and to make them realize all the more the fate that awaits us should the District Power Bill lose.

Utility companies had attempted to purchase the mutuals but wanted to raise the rates because they thought the co-ops received "bootlegged juice" from the municipal plant in Tacoma.[47]

As we shall see in chapter 5, the situation changed dramatically in favor of electrical distribution cooperatives in the mid-1930s. When the REA was established in 1935, it made co-ops the mainstay of a highly successful program that eventually electrified more than 90 percent of the nation's farms. Yet long before that happened, farmers managed to gain control over the new technology by making their own electricity. The schemes for generating home-grown electricity were as innovative as those for using the automobile as a workhorse and bringing water to the farmhouse. Before World War I, prosperous farmers bought dynamos and yoked them to turbines powered by water flowing over homemade dams, to windmills that normally pumped water for the livestock, and to gasoline engines.[48] As in the case of automobile kits, manufacturers soon capitalized on this inventiveness and started selling complete home-plant sets.

Many inventors designed wind-powered electrical generators for the farm after World War I. An early firm, the Fritchie Wind-power Electric

Company of Colorado, sold about fifty plants between 1919 and 1922 before going out of business during the agricultural depression of the early 1920s. The Perkins Corporation of Indiana, a windmill firm that got into the electrical generating business in 1922, was more successful with its Aeroelectric. Farmers could choose between 32-volt and 110-volt sets, costing from one thousand to twelve hundred dollars—a price outside the budget of most rural families. Known as the Cadillac of the wind-generating business, the Perkins unit held up well (especially in high winds) against cheaper sets made by a variety of firms, including HEBCO (named after engineer Herbert E. Bucklen, the company's founder), Wincharger (the most popular set), Air Electric, and Wind King.[49] In the same period, two firms designed wind turbines with propeller blades (the design presently used) that were more aerodynamically efficient than sets with the older multibladed wind wheel. HEBCO sold a small number of turbine sets to isolated air fields and large farms; the Jacobs Wind Electric Company of Montana, and later Minneapolis, produced more than thirty thousand units between 1927 and 1957.

Farmers wishing to build their own wind generators thus had plenty of examples to emulate, as well as a 1918 book on the topic to consult. J. F. Forrest of Poynette, Wisconsin, was so proud of the generator he built in 1910 that he called his place "the Electric Farm." In 1922, Forrest said he was much better off than "some farmers who have the high power lines, they built the lines, put in the transformers and turned the whole line over to the company who charges one dollar per month for lights and eight dollars for power." A North Dakotan used his wind plant to generate electricity for powering a washing machine and cream separator, as well as to charge batteries to power lights on his sleigh, which replaced the automobile during the high winter snows in his part of the country.[50]

Although farmers made and companies sold a large number of wind energy plants, the less expensive gasoline engine set, which did not depend on the availability of wind, dominated the field. Fairbanks-Morse, makers of stationary gasoline engines, advertised an engine-dynamo set in 1907 that not only provided electricity for lights, fans, pumps, and motors but also permitted the farmer to "shift the belt to the other flywheel and drive any kind of machine."[51] The number of manufacturers skyrocketed during the prosperous war years, from at least five in 1911 to about one hundred in 1916. General Electric and Western Electric joined scores of lesser-known firms in creating a sixty-million-dollar-a-year industry by 1921. The automotive engineer Charles Kettering, who established the research laboratory at General Motors, invented the durable Delco plant in 1913, which became a household name. Delco had sold forty thousand plants by 1918 and claimed one-third of the market in 1923.[52]

These sets constituted one of the most complicated technologies on the farm. Consisting of an engine, dynamo, switchboard, and instruments, sets were usually installed in the basement, where they charged a bank of storage batteries by manual, semiautomatic, or fully automatic control. The batteries supplied all the electricity—usually at 32 volts, instead of the high-line's 110 volts—for lights, small appliances, and motors in the house, barn, and shop. Although available in sizes up to twenty-five kilowatts, the average set produced a little more than one kilowatt, just enough for lights and a small motor. The purchase price for most sets ran from $350 to $800, while estimates of operating costs varied widely from seven to sixty-nine cents per kilowatt-hour. Maintenance was apparently not a trivial matter, especially with the troublesome lead-acid batteries. Here is one instance where the industrialization of the home meant more work for father, instead of for mother, since the gendered female tasks of cleaning and filling kerosene lamps were transformed into the gendered male tasks of tinkering with engines and batteries.[53]

Some manufacturers promoted home plants on economic grounds as a way to replace hired help with "wired help"—as in the case of the telephone—but most advertising centered on the themes of modernity. One ad in 1916 claimed that "Delco Light brings city advantages to the farm . . . takes much of the drudgery out of farm work—adds comfort and convenience to farm life." Western Electric provided its dealers with stock stories to send to local newspapers, announcing, for example, the "progressiveness of . . . county farmers shown in many ways." Dealers were instructed to fill in the blanks with the name of the county and farmers who availed themselves of "every city convenience" by purchasing a "Western Electric Power and Light outfit." Western Electric told its dealers, "These writeups will prove to be the most valuable advertising you can run." A writer in *Electrical Contractor–Dealer* advised his audience in 1919 that when advertising home plants to the farmer, "talk saving labor and increasing profit, not luxury," the same approach taken by promoters of the telephone and automobile.[54]

Several surveys in the 1920s and early 1930s show that farm people used home power plants primarily for household purposes: lighting and running washing machines, irons, and cream separators. Motors and water pumps were far down on the list.[55] But we should not conclude from these figures that farm men necessarily agreed to buy home plants because they were persuaded by the Country Life rhetoric to do something to save the overworked farm woman. Many of the farm men who responded to a New York survey of Delco owners in 1926 criticized the plants for not providing enough power for the larger motors required for *their* work. In response to whether he preferred gasoline engines or electric motors, one man replied, "Engines of course. How could you carry lines all over the farm with current for a tractor or a saw outfit? Think it over."[56]

Despite being generally satisfied with their home plants, many of the Delco owners in New York said they would rather be hooked up to the highline but could not because of the expense. The ideal of self-sufficiency had apparently lost some of its appeal after their experience running what amounted to a small utility plant at home. Some respondents also noted that having a little electricity, even on a part-time basis, whetted their appetites for more and made them long to be near a high-line.[57]

Yet home plants were successful by one measure. An estimated 200,000 had been sold by 1919, 600,000 by 1929. Although not all were sold to farmers and many went out of service, the number of farms with home plants probably outnumbered the estimated 180,000 farms receiving central-station electricity at the end of 1923. By 1930, the figures were reversed. Six hundred thousand farms obtained electricity from central stations, 250,000 from home plants.[58] Still, these numbers represented very small percentages: 10 percent of farms had central-station service in 1930; less than 4 percent used home plants (see table A.5).

Less Work for the Farm Woman?

A wide range of social groups thought that modernizing homes with electric lights, running water, washing machines, and vacuum cleaners would eliminate drudgery, save labor, and increase leisure time. Ironically, however, time studies conducted by home economists in the mid-1920s challenged this interpretation. These studies provided data that was later used to support the argument that although new household technologies might have reduced the energy required to perform specific tasks, ownership of these appliances did not correlate with less time spent on housework by full-time home workers. In surveying two centuries of this technology in the United States, Ruth Cowan argues that an "industrialization" of the home often resulted in "more work for mother" because the use of artifacts such as coal stoves, water pumps, and vacuum cleaners tended to save the work of the helpers of married women (husbands, sons, daughters, and servants) and to promote a higher standard of housework. The full-time home worker's patterns also shifted from production to consumption, which included household management and child care."[59]

A few commentators on rural life recognized some of these ironies before they were documented by the time-use studies. In 1910, Ruth Stevens, a home economist and the first woman's-page editor for the *Progressive Farmer*, told her readers, "Household appliances have been invented in many instances by men to sell and many are more than useless. . . . Any household appliance should save more time and labor than it takes to adjust it and clean it after use." A South Carolina woman wrote the USDA in 1913, "You may read bright, sunshiny letters telling how this or that saves labor, but these labor-saving devices need hands to make them go." The agricultural economist Thomas Carver said in 1914, "It must be remembered that these labor-saving improvements [in the home] seldom reduce the amount of work. They merely enable people to accomplish more with the same effort." Mary Atkeson echoed this view in 1924 when she said that older women had expected that the sewing machine would "liberate them from toil." Yet styles changed accordingly, and "women simply elaborated their garments until they were as much work as before. Such a tyrant does even the best labor-saving invention sometimes become!" A New York farm man replied in the survey of Delco home plants that electricity "only makes the work easier and less tiresome with the exception of the vacuum cleaner. There the work is cut nearly ½."[60] But if Cowan's thesis holds in the countryside, we should expect that it was the man's work and that of his sons in removing carpets from the house and beating them on a clothes line that was cut in half, not necessarily that of the farm woman.

This interpretation is supported by the time-use studies. Funded by the Purnell Act (1925) and conducted at state agricultural colleges under the guidance of the USDA's Bureau of Home Economics, the studies were carried out by home economists imbued with a progress ideology of technology, the middle-class urban domestic ideal, the Country Life goal of saving the overworked farm woman, and the ideals of objective social science research. The head of the project, Hildegarde Kneeland, had been trained in sociology and stressed the survey as a method of social reform. She and her colleagues sought to discover how much time farm women spent on homemaking (generally defined as housekeeping, household management, and child care), farmwork (e.g., work in the garden, poultry pen, dairy, and field), leisure, and sleep by asking them to fill out the bureau's standard forms or time charts, graded in five-minute intervals, at the end of each day for a "typical" week (see

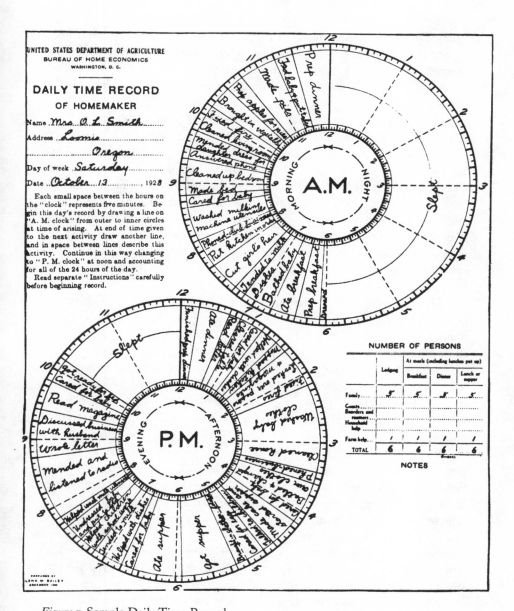

Figure 7. Sample Daily Time Record

Source: Maud Wilson, "Use of Time by Oregon Farm Homemakers," Oregon State Agricultural College Agricultural Experiment Station, *Bulletin*, no. 256, Nov. 1929, p. 41.

fig. 7). The taxing requirements of keeping these charts and the selection of women through contacts with the Farm Bureau and women's clubs biased the samples toward well-educated, middle- and upper-class, native-born, white women (the female half of the traditional clientele of the USDA Extension Service).

Home economists studied about two thousand farm women over a forty-year period. Researchers used the bureau's time charts and guidelines to survey women in Idaho, Washington, Rhode Island, Oregon, South Dakota, and Montana from the mid-1920s to the early 1930s. Most researchers then switched to interview methods and studied the time habits of farm women in New York, Vermont, Wisconsin, and Indiana from the mid-1930s to the mid-1960s. Urban women were surveyed in five of these studies and in four separate nonfarm investigations. A second study of Rhode Island investigated married rural and urban women employed outside the home. An agricultural economist performed a time-use survey of more than three hundred Nebraska farm homes in 1924, using his own methodology.[61]

The reformist ideals of the researchers often clashed with the objectivist goal of the social survey. Louise Stanley, head of the Bureau of Home Economics, said in 1926, "We need . . . to know what equipment is going to help the homemaker most. Time studies will show us this."[62] It came as something of a surprise, then, when all of the studies revealed that prosperous farm women still spent at least fifty hours per week on housework. Many of the studies showed that "laborsaving" household technologies—which home economists had been recommending for some time—did not live up to their billing.

The home economist Maud Wilson conducted one of the most perceptive studies on this score. Her survey of about six hundred farm, rural non-farm, and urban households in Oregon, from 1926 to 1927, shows that women in "well-equipped" houses (those which had both electricity and plumbing, representing 19 percent of the sample) spent an average of three hours and twenty minutes less per week on meal preparation, routine cleaning, and washing clothes than did women in "poorly equipped" houses (those which had neither electricity nor plumbing, representing 57 percent of the sample). But the women with more advanced technology spent an average of two hours and ten minutes more per week on ironing, sewing, child care, and care of the grounds. Both groups spent nearly the same amount of time on specific tasks, and the town women, who had a much higher percentage of "laborsaving" devices than farm women, spent almost exactly the same amount of time on housework as their country sisters (51.5 hours per week versus 51.6).

Why? Wilson, who was working on her doctorate in home economics from the University of Chicago, gave four possible explanations: (1) "No time reduction is possible, and the equipment is of value because it makes the job more pleasant or because it reduces [the] energy requirement"; (2) "Time habits tend to persist, with the result that the family living standard is raised"; (3) The "homemaker spends more time on the parts of the task which she most enjoys doing"; and (4) "Time given by other members of the family or by hired help is reduced rather than her own time." Although Wilson thus articulated what Cowan would later call the "ironies of household technology" and suggested that many people had "an exaggerated idea of the time reduction possible" from using this technology, she promoted it as a means to improve housework. The caption of a photograph showing a collection of household appliances in her report reads, "Good equipment means better housekeeping at the *same* time cost."[63] The caption expresses an interpretative shift made by many time-use researchers. When confronted by the intractability of apparently paradoxical data, they turned from advocating technology as a means to reduce labor time to promoting it as a way to increase the quality and productivity of housework.

At least four other researchers identified these ironies of household technology before World War II, and their explanations for them were somewhat different from Wilson's. Inez Arnquist and Margaret Whittemore elaborated on Wilson's second point by speculating that women in Washington and Rhode Island allotted a specific amount of time to a task and thus either slowed down when an appliance allowed them to be ahead of schedule or raised their standard of living by doing more work in the same amount of time. Whittemore also advised the farm woman to "be careful to see that her [washing] machine does not cause her to work harder" as had happened when the sewing machine was first introduced.[64] Jessie Richardson published data in 1933 showing that slightly more time was spent on washing clothes and preparing food by Montana women with electric or gasoline-powered washing machines, running water, and refrigerators (three minutes, twenty minutes, and forty-five minutes per week, respectively). Rather than criticizing the baneful effects of technology—a common practice in the debates over technological unemployment during the Great Depression—Richardson asserted that this equipment made household tasks more enjoyable and "thus induce[d] the worker to spend more time than formerly."[65] The data published by Jean Warren in New York in 1940—which she collected using a different methodology, the interview—showed a more complex picture. About five hundred farm women spent less time in individual tasks by using a washing machine, electric iron, and vacuum cleaner, but women with running water and those having both running water and electricity spent more time on housework than those having neither technology (7 percent more time in both cases). Warren explained the paradoxical data by noting that technology was not the only factor (others included family size and composition, number and size of rooms, kitchen arrangement, etc.). Like Wilson, she implicitly advocated the use of up-to-date technology by arguing that women with running water and electricity did a better job of homemaking than those without these conveniences.[66]

Three other time-use researchers in the 1920s and 1930s, using the same methodology and time charts as these home economists, did not find a negative correlation between household technology and time spent on housework (even where some data seemed to indicate this phenomenon). A study in South Dakota did not present data to compare ownership of technology with time spent. One Idaho study attributed large amounts of time spent on housework to lack of conveniences, but its data also indicated that the two women who spent the most time on housework both had kitchen sinks. The third study, of married women in Rhode Island who worked outside the house, did not comment on data showing that women without running water tended to spend more time on housework (although this was not a linear relationship).[67]

A survey of ten test farms in Illinois before and after they received electricity, conducted by the agricultural engineer Emil Lehmann at the University of Illinois, even claimed in 1929 that electricity reduced the amount of time women on five of these farms spent on laundry, house care, and leisure but increased the time they spent on farmwork. However, Lehmann's research was not comparable to the Bureau of Home Economics studies because he used an extremely small sample, excluded the increased time spent on meal preparation, and was funded by the electric industry. The Illinois State Electric Association, a group of utility companies, financed the study, which Lehmann conducted during a sabbatical leave in which he worked for the Illinois Power and Light Company. (Lehmann was so optimistic about the benefits of electricity that he explained the decreased leisure time of the five farm women by saying that since electricity eliminated the drudgery of household tasks, they did not have to rest so much!)[68]

The Bureau of Home Economics took an ambivalent view of the time-use research. In a 1928 report on records received from seven hundred farm women in five states, Kneeland lamented the large amount of time still being spent on housework. "Are we, then, to conclude that the farm home has not been touched by the industrial revolution? . . . Such a conclusion would, of course, be unjustified. For though the working hours of the farm woman are still long, they were undoubtedly even longer fifty years ago."[69] But as researchers began to prepare reports for publication that explicitly questioned the timesaving capabilities of household technology, Kneeland softened her position. In a 1929 magazine article, she repeated her argument that women had worked much longer in the past but added other reasons to explain the continued long hours, including the fact that "much of the gain which the Industrial Revolution has so far brought has gone into reducing the work of the household to a one-worker job."[70] When Lehmann wrote the Bureau of Home Economics in 1930, asking for estimates on time saved by electrical appliances for a pamphlet he was preparing for a utility group, Kneeland wrote a memo to her boss that she could not "give Mr. Lehmann the information which he wants, as our [summary] study shows a negative correlation between time expenditure and equipment, due undoubtedly to the fact that the housewife with better equipment raises her standards of housekeeping or slows up in her time expenditure."[71] However, neither this analysis nor that of the ironies of household technology, noted by Wilson and others, made their way into the bureau's summary report of the prewar time studies.[72]

By 1940, then, researchers were about evenly split on the issue of whether ownership of modern household technology correlated with reduced time spent on housework for farm women. Five studies showed that a direct correlation did not exist between technology, especially electrical appliances, and reduced time. The coordinator of the time-use studies at the Bureau of Home Economics was ambivalent, admitting in a private, internal memo (though not in print) that a negative correlation existed. Three other studies assumed that the technologies examined were laborsaving without analyzing the correlation; and one study done with a very small sample and from the point of view of the electrical industry presented data showing that electricity reduced time spent on housework.

We should also note that the time-use data was drawn from farm women who seem to have accepted to a large extent the urban domestic ideal promoted by home economists. Many others did not. A study of Nebraska farm women shows in detail the time spent on the nonhousehold work of gardening, taking care of poultry, milking, and farming in the field.[73] Indeed, the historian Katherine Jellison argues that the time studies neglected to consider the fact that women could use the time saved by household technology to do a good deal of farmwork. Even the time-study subjects spent some time on these tasks in a typical week. Although the number of hours was much less than that devoted to housework, it was not insignificant, ranging from 4.5 hours per week in Rhode Island to 11.6 hours per week in South Dakota before World War II. Time studies conducted after the widespread electrification of the farm following the war support the position that electrical appliances did not lead to less work for the full-time home worker on the farm.

As had been the case with the telephone and automobile, new household technologies did not revolutionize the farm. Instead, middle- and upper-class farm women brought these technologies, many of which had been designed— and were then redesigned—by farm men, into the traditional work patterns of rural life. The work of farm women expanded in the process, as they got more done and raised their standards of living. But they did not become urban or suburban women of leisure, the stereotypical view many farm women had of the unproductive city and town woman.

CHAPTER 4

Tuning In the Country

Of all the new technologies used in the home in this period, radio was the most popular—on the farm and in the city. The advent of radio broadcasting in the early 1920s took the nation by storm. In 1923, three years after Westinghouse's station KDKA began regular broadcasting in Pittsburgh, nearly six hundred stations were on the air throughout the country, broadcasting church services and live music, playing phonograph records, reading the news, and covering sporting events. It was an unforeseen "revolution." As late as 1919, when the United States Navy, acting on national security interests, encouraged General Electric to buy out the British-owned American Marconi company and form the Radio Corporation of America (a patent cartel and operating company whose owners soon included Westinghouse and AT&T), the principal players in this twenty-year-old field interpreted the medium as point-to-point communication (wireless telegraphy or radio telephony), not as broadcasting to the public. Radio amateurs played a major role in defining broadcasting as the dominant meaning of radio, much like farm users of the telephone and automobile helped to define the meanings of those artifacts and systems in their lives. Both groups of users were thus agents of technological change.[1]

The history of radio in rural life has many similarities with that of the telephone, the automobile, and household technology: promoters used the shibboleths of Country Life rhetoric to predict that the new technology would revitalize farm life; farm listeners helped shape the meaning of the radio; and broadcasters and manufacturers responded by designing programs and radio sets aimed at the farm market. But there are important differences, as well. Partly because of their experience with the telephone, farm people welcomed the radio enthusiastically and did not resist it to any significant degree. The federal government became a much more active player, through the efforts of the Federal Radio Commission, established in 1927, and the U.S. Department of Agriculture's Extension Service, which quickly used the new medium to reach farm people. Reformers thus viewed radio as a powerful agent of modernity: it would eliminate differences between urban and rural life by connecting farm people directly to the "civilizing" aspects of the city and transmit modernizing messages directly to farm people in an efficient and effective manner. As had been the case with the telephone and the automobile, farm men and women had other ideas about how to use the new technology.

The Radio on the Farm

The diffusion of the radio in the countryside proceeded apace during the 1920s and 1930s, despite the prolonged agricultural depression. County agents counted 145,000 radios on the nation's farms in 1923. The percentage of farms reporting radios grew rapidly thereafter, from 4.5 percent in 1925 to 21 percent in 1930 to 60 percent in 1940. These aggregate figures masked wide variations by region, with poor southern states far behind the rest of the country (see table A.4). The adoption of the radio proceeded differently from that of the telephone and automobile, in that prosperous midwestern states did not dominate the picture. Instead, the Northeast region, which had large cities with high-powered radio stations, led the way. Although the Midwest and Far West had fewer stations, high-powered stations like WLS in Chicago were popular in those areas.[2] The ranking of these regions did not vary from 1925 to 1930, further evidence of the importance of these economic and geographic factors.

Another important consideration was the availability of electricity on the farm. This was not much of a factor in the early days, because radios were either "cat-whisker" crystal sets, which did not need a power source, or vacuum-tube sets, which ran on direct current from batteries: an acid-filled "A" battery; a dry-cell "B" battery; and, for some sets, another dry-cell "C" battery. Some consumers who lived in electrified houses used costly battery eliminators (which converted household alternating current to direct current) to save the expense and inconvenience of buying and maintaining batteries. The situation changed considerably with the introduction of all-electric sets in 1927, made possible by a new type of vacuum tube that removed the "hum" produced by connecting alternating current (AC) directly to the filament. The AC sets also included a built-in battery eliminator to provide "B" and "C" voltages.[3] By the

1930s, urbanites and others had begun to view battery radios a
much like the wall magneto telephones became marked as "rural'
velopment of the central battery system in town.[4]

The question of the importance of electricity in farm ownership of radios
before the New Deal can be seen by comparing table A.4 with table A.5.[5]
There is a rough correlation between regions with high percentages of electrified farms and those with high percentages of radios in 1930. But the figures
show that a large number of midwesterners listened to radios without being
hooked up to the grid. On the other hand, many farm people in the Far West
had electricity but no radio.

By all accounts, then and now, the radio was one of the most popular of
the allegedly urbanizing technologies on the farm. As can be seen in figure 2,
by 1940 the percentage of farm households owning radios surpassed that of
those having a telephone (whose numbers actually declined during the agricultural depressions of the 1920s and 1930s), those with electricity and running
water (which were not widely adopted until the Rural Electrification Administration was formed in 1935), and—just barely—those owning an automobile
(whose diffusion came to an abrupt halt during the Great Depression). Although the radio spread rapidly to the farm, it did not keep pace with urban
figures. In 1930, 50 percent of urban households owned radios, compared with
21 percent of farms. The fact that the latter figure had risen to only about 33
percent in 1935 was of some concern to Morse Salisbury, a radio expert in the
USDA. Salisbury blamed the poor showing on the neglect of the farm market by radio manufacturers, who concentrated on AC sets, the high price and
costly upkeep of batteries, and the depression of the 1930s.[6]

Economic factors were surely important at this time, but radios came in
many types, sizes, and prices. In 1925, before the introduction of AC sets,
manufacturers advertised complete radio sets in the midwestern *Wallaces'
Farmer* that ranged in price from $15.00, for a Rowe two-tube set, to $35.00
for a Radiola and $142.50 for a top-of-the line Kennedy Model XV.[7] In 1936,
an appliance trade group listed prices ranging from $13.25 for a Simplex Com-
pact AC-DC set to $64.95 for a Philco AC-DC set and $750.00 for a Zenith
Console AC set. Battery-only receivers ran from $19.95 for a Crosley Table ra-
dio to $69.95 for an Emerson Console to $165.00 for a Pilot Phono-Combi-
nation.[8] Ivan Bloch, a staff member of the Rural Electrification Administration, reported a similar range in 1936. "Complete battery-operated rural
receivers" listed from $30.00 to $100.00, 32-volt "rural radio receivers" (for
home power plants) cost from $30.00 to $50.00, receivers "operated from
storage batteries charged by wind-driven generators" started at $55.00, and

AC sets ranged from "$7 to several hundreds of dollars each." Of course, sound quality, reliability, sensitivity, and range varied greatly with these prices. Bloch estimated that the range of a typical radio was a hundred miles and that the operating cost for battery sets was one dollar per month (assuming a six-month battery life) and thirty-six to sixty-three cents for 32-volt sets, while the windcharger sets incurred no monthly cost (as long as there was enough wind). The AC sets on REA lines averaged about seven kilowatt-hours per month (at about thirty-five cents in 1936).[9]

Small windcharger units, usually mounted on the roof, seem to have been fairly popular before widespread rural electrification. *Wallaces' Farmer* carried advertisements for a half dozen varieties in 1935, ranging in price from $5.00 for a unit sold with a Coronado radio to $15.00 for one made by the Wincharger Company. Wincharger claimed an operating cost of only fifty cents per year for its radio unit. A popular windcharger made by the Zenith radio company stood about seven feet tall on a tripod mount.[10] Two of the twenty-three farm families interviewed recently in New York state remembered windchargers. Gerald Cornell recalled that "my neighbor up here had a Zenith. They came out with a windcharger on top of the house and when your battery got weak you'd release the break on that and the propeller would really shake the house. It did charge up the battery [though] so you could hear the radio."[11]

There were other ways to charge the "A" battery. An Atwater-Kent sales-man in West Virginia recalled in an oral-history interview that "we would al-ways sell the farmers battery sets. If they had a Ford car, why, we'd put a Ford battery on their radio. If they had a Chevrolet, we put a Chevrolet battery, so that when the battery on the radio ran down, they'd switch it with the one in the car and get it recharged."[12] New York farmer, Leroy Harris, recalled re-cently that when his family had a car in the 1920s, "we drove the car up to the window and ran the cable out [to the car's battery] and we had music . . . that way."[13]

Agricultural leaders seized upon the new technology's growing popular-ity as a way to reform rural life. Modernizers drew on the rhetoric of the Country Life movement to proclaim the radio's ability to relieve isolation, bring the city's culture of highbrow music and higher education to the coun-try, and keep youth on the farm. (Promoters of radio, including advertisers, thought this uplift function applied to urban audiences, as well.)[14] Secretary of Agriculture William Jardine, who had pioneered in radio extension work while president of Kansas State College in the early 1920s, spoke for many re-formers in 1925 when he said that "radio's greatest contribution to civilization may lie in its influence upon the life and action of the farm population. . . . It is to become a vital necessity for their economic, spiritual, and intellectual life . . . by delivering the farmer and his family from the sense of isolation, by coping with class and sectional differences, by keeping boys and girls on the farm, and by making possible a system of agricultural education through the radio-extension courses of the agricultural colleges."[15] Many modernizers praised the radio over the telephone and automobile. The manufacturers of the Kennedy radio and a professor of radio engineering at the University of Minnesota thought the radio's advantage was that it brought the city to the country, rather than requiring farm people to go to the city for urban culture. I. W. Dickerson, a frequent writer on radio for farm journals, thought radio excelled at relieving isolation because bad winter weather did not shut down radio as often as it did the telephone, the automobile, and rural free delivery by knocking down wires and making roads impassable. The radio correspondent for the *Rural New Yorker* wrote in 1924 that the radio "connects you with the whole world and breaks down the invisible but real barriers that separate the country from the city."[16]

Many commentators writing in the 1930s agreed that the radio's ability to bring the cultural benefits of the city directly to the country, without farm people traveling to town, was a distinct advantage. Republican senator Arthur Capper from Kansas, founder of the influential farm paper *Capper's Weekly* and owner of a radio station, told a radio conference in 1932 that the "radio on the farm is providing service [through market and weather reports], culture, and entertainment without the necessity of any of the young people leaving the family hearth. . . . The greatest service radio has given and can give, as I have said, is to dynamite the barriers of space out of the way of the farm family's daily contact with the life of our whole civilization." After quoting from a letter from a farm man who said the radio had convinced him to retire on the farm, instead of in town as many had done in the past, Capper concluded, "The skies of that man and his family had been lifted. Their horizons had broadened until they encompassed, not a prairie farm neighborhood, but great cities and far places."[17]

In 1935, the Radio Institute of the Audible Arts (founded by the Philco Radio and Television Corporation) solicited letters from agricultural leaders (eighty-five men and five women from at least thirty-six states) and published them as a "symposium on the relation of radio to rural life." In his opening remarks, the rural sociologist Edmund de S. Brunner proclaimed that "radio has become one of the most potent of these agencies such as the telephone, the automobile, and rural free delivery, that are rapidly banishing the physical and cultural isolation to which rural people of a few decades ago were inevitably subjected." The published excerpts concurred with the views of de S. Brunner, who organized the comments under such headings as "Breaks Down Rural Isolation," "Farmers Same as Other People," "Makes Farmer Understand Place in the World," and "Keeps Young People on the Farm."[18] Farm leaders quickly spread the report's message in farm journals as the latest social science gospel.[19]

One of the symposium's respondents, Floyd H. Lynn, secretary of the Farmers' Union, wrote that "automobiles and good roads have tended to take farmers away from home. The radio, on the other hand, tends to keep these same folks at home. It is, therefore, a counter influence in its relationship to those influences which have come with mechanical and scientific development and which have had the tendency to eliminate or stifle the social life and identity of rural communities."[20] Indeed, many farm people might have felt more comfortable listening to a famous city preacher's sermon at home, for example, than venturing into a culture that regularly caricatured ruralites as "hayseeds."[21]

To agricultural reformers, most of whom spoke in terms of technological determinism, the radio thus seemed to be the long-sought solution to the Country Life paradox of how to uplift and urbanize farm life without destroying its benefits. Here was a technology that promised to bring the city to the country while keeping people on the farm; it could thus maintain rural institutions while modernizing them. The fact that these often conflicting goals assumed an urban superiority that was far removed from the farmer's historic distrust of the city did not seem to worry Capper and other heirs of the populist tradition.

Enamored of radio's potential, agricultural agencies soon developed programs for farm people. The USDA joined the radio boom from the start. In December 1920, one month after station KDKA went on the air, the department used a naval station to broadcast its market quotations by radio telegraph. The USDA also experimented that year with Daily Radio Marketgrams, which were "wirelessed at 5 P.M. each business day" from the radio station of the U.S. Bureau of Standards to "hundreds of amateur wireless operators within a two-hundred-mile radius of Washington." The amateurs relayed the information to farmsteads, farm organizations, and newspapers, thus relieving the skepticism of farmers who thought they would have to become wireless experts to receive the reports. The USDA switched to the radio telephone in 1921 and relied less on amateur operators, who were relatively few in number. By early 1923, the department had broadened the service by using more government and commercial stations. United States Post Office stations in Washington, D.C., Cincinnati, Ohio, St. Louis, Missouri, Omaha and North Platte, Nebraska, Rock Springs, Wyoming, and Elko and Reno, Nevada, each with a radius of about three hundred miles, broadcast the market reports, as did naval stations in Virginia and Illinois and fifty-three stations operated by agricultural colleges and commercial agencies. Station KDKA began broadcasting the reports regularly in the summer of 1921. The number of stations carrying the popular market quotations rose to 85 in 1924 and to 107 in 1928. The USDA also sent out its weather reports by radio; 121 stations in forty states were broadcasting them in 1925.[22]

The radio, however, did not completely replace other methods of communicating the USDA's reports to farm people in the 1920s. The agency relied on its extensive telegraph system to receive market reports from observers located near market centers and to send these and its weather reports out to newspapers, banks, post offices, country stores, and radio stations that subscribed to the service. Since many more farmsteads had rural free delivery and telephones than radios in the early 1920s, most farm people received the weather reports by newspaper or telephone. Secretary of Agriculture Howard M. Gore admitted in his report for 1924 that the "most direct and successful means of furnishing timely weather information to farmers has been through rural telephone systems. Weather forecasts and warnings issued about 9:30 A.M. daily are now made promptly available to over 7 million rural telephone subscribers in the United States." Gore did, however, predict that radio "bids fair to outstrip all other means of communicating weather information to farmers," a view shared by radio expert I. W. Dickerson in 1925.[23]

These "broadcast" usages of the rural telephone played an important role in the development of rural radio. As we have seen, government and private agencies sent the correct time, market and weather reports, and news to farm people over the telephone, while telephone users broadcast music to party-line members, much to the dismay of telephone companies. Promoters and farm people had thus jointly created a pattern of rural telephone "programming" that was later adopted by radio stations.

The connection between the telephone and radio in rural life was literally made by a cooperative arrangement reported by *Wallaces' Farmer* in 1925. "A rather new departure in the radio broadcasting line is that followed by a few of the smaller telephone companies of setting up a good receiving set and loud speaker in the telephone exchange and then tuning in on a broadcasting and sending out the music over as many telephone lines as care to listen in. The listeners say it is just like being in the room with the music, and often three or four people will get close enough to the telephone receiver to listen in."[24]

Agricultural reformers soon expanded their radio programming to include more explicit modernizing messages to farm people than market and weather reports. In 1926 the new secretary of agriculture, William Jardine, applied his enthusiasm for radio on a national scale by establishing a Radio Service in the USDA, headed by his Extension Service editor at Kansas. Ninety broadcast stations throughout the country loaned their facilities for an average of thirty minutes a day to broadcast the service's programs the first year. The USDA started four programs, which reflected its long-standing gender bias: the United States Radio Farm School, taught by "schoolmasters" at each station; Noonday Flashes, a daily conversation between a county agent and a farm man; Housekeepers' Chats, "a fifteen-minute period devoted five days a week exclusively to up-to-date information on subjects of interest to women"; and Aunt Sammy, a cooking and recipe program for women, named after the fictitious wife of Uncle Sam. The expensive-to-maintain farm school closed in 1928, but versions of Noonday Flashes and Household Chats continued well into the 1950s.[25] Aunt Sammy, who was played by fifty women in as many radio stations during her first broadcast in 1926, was apparently a big hit. By 1927, the USDA claimed the program reached 1 million city, town, and farm women. Thousands wrote in for advice and the free "foolproof" recipes prepared by the Bureau of Home Economics.[26]

Other agricultural groups and companies interested in the farm market followed the USDA's lead. State agricultural colleges enthusiastically adopted the new technology for their extension work, with thirty-one colleges broadcasting programs by 1927; nineteen state colleges owned and operated their own radio stations by 1935. The Chicago Board of Trade bought a radio station (WDAP) in 1923 to broadcast its grain prices. The Ford Motor Company also had a station on the air that year. The Sears-Roebuck Agricultural Foundation started its powerful station WLS in Chicago in the spring of 1924 and then began a farm program on WSB in Atlanta in 1926. General Electric's main station (WGY in Schenectady, New York) started its Farm Paper of the Air in late 1925, and the National Broadcasting Company (NBC) approached the USDA to develop the National Farm and Home Hour in October 1928. Music and entertainment (the Homesteaders band and comediennes Mirandy of Persimmon Holler and Aunt Fanny) was provided by NBC; Sears' rival, Montgomery Ward, sponsored three quarters of the hour, and the USDA, which would not participate in a sponsored program, provided information for the remainder of the hour for five days a week for the 11:30 A.M. show. Other farm organizations, such as the Farm Bureau, the National Grange, and 4-H Clubs, supplied material on Saturdays. Broadcast six days a week until the 1950s (it was resurrected in 1945 after a year's hiatus), the National Farm and Home Hour was probably the most popular program of its type during this period.[27]

Listening Patterns

One perennial difficulty for radio stations was determining who, if anyone, was listening to the programs sent out on invisible waves into the ether with so much effort. Here was one technology users could control by simply turning it off! Country Lifers praised the radio for bringing the city to the farm, but this was only a capability. How often did farm people listen to their radios, and what did they listen to? Promoters of the new technology tried several ways to answer this perennial question for rural (and urban) listeners. Farm journals solicited experience letters, as they had done for the telephone and automobile when these technologies were new; radio stations counted and read listener letters; and advertising agencies began conducting listener surveys in the mid-1920s, when it became clear that the answer to the question of who was to pay for radio in the United States was the advertiser.[28]

The letters chosen for publication reflected the Country Life ideology of the editors of farm journals (and the editors of urban magazines who reprinted them). The *Kansas Farmer* published an enthusiastic letter in 1922 from farm boy Hugh Scott, who had used a "radiotelephone set" for about a year to pick up market reports, concerts, news items, police reports, church services, and the correct time. The urbane *Literary Digest* prefaced its account of the story by saying the "youth who is speaking lives way out in Kansas, but in his daily experiences he is a cosmopolite." An Illinois farm man also thought the radio was a "device to make home more attractive to his two boys, seventeen and nineteen." An Iowa farmer wrote in 1923 that he had received a wide variety of programs, including sporting events and political speeches, that were clearer than what he heard over the telephone. Ashley Dixon, a city man who had moved his family to a ranch in the Rocky Mountains, noted in 1924 that they had spent eleven winters with books and the phonograph. Radio brought back the city life they missed.[29]

Although farm people and modernizers had debated whether the telephone should be used for business or entertainment purposes, this was a minor chord in the discourse about radio. Promoters (and many farm people) emphasized the reception of market and weather reports, as well as other business uses.[30] These included making barn work more pleasant in tobacco country and sheep herding less lonely in Wyoming, where ranchers often put radio sets in sheep wagons. An Iowa farmer used radio announcements to sell his peaches, and a Minnesota cow tester took a portable radio along on his rounds. A radio station in Pierre, South Dakota, performed a community service by broadcasting messages instructing ranchers to pick up relatives at the hospital or asking neighbors to water stock when a rancher was kept in town to wait out a blizzard.[31] On the other side of the question, farm woman Mrs. F. H. Unger told readers of the *Rural New Yorker* in 1925 that "no farm family uses a receiving set solely to obtain market reports," and I. W. Dickerson admitted that the "splendid entertainment features provided by radio broadcasting stations perhaps is doing more to encourage increased installations than the utilitarian portions of the [farm] programs." Mary Puncke of the Sears-Roebuck Agricultural Foundation stated in 1926, "Although the silver-tongued salesman may sell the farmer his radio set as a business investment, once it is installed in the home, its chief function is entertainment as far as the man of the house is concerned."[32]

Puncke based her statement on the results of a national survey she had directed for the Sears foundation. Conducted by twelve hundred women acting as neighborhood observers of eighteen thousand farm homes, the survey found that 25 percent of the men preferred music to anything else, the largest preference category. Yet 24 percent of the men also received daily market and weather reports. The latter figure rose to 42 percent in the Corn Belt and dropped below 3 percent in Mississippi. The National Farm Radio Council conducted its own survey in 1926 in conjunction with cooperating agencies and radio stations. More than forty thousand replies largely corroborated the Sears findings that farm people preferred to listen to music. Reporting on both surveys, *Radio Broadcast* poked fun at the Country Lifers by observing, "The average urban listener doubtless has the impression, as had we, that the farmer is most interested in having himself uplifted and educated—for the reason, no doubt, that every program we hear announced as a 'special feature for farms' is of such uplifting or educating nature. But lo and behold, it seems the tired farmer, just as the tired businessman, is more eager to be entertained by his radio than taught."[33]

Radio groups undertook most of the later surveys. In 1928, NBC conducted a survey of thousands of urban, rural nonfarm, and farm families east of the Rockies. Reflecting the reformers' hopes that radio would eliminate distinctions between urban and rural life (a homogenization that many feared, as well), NBC concluded that all groups held the same "preferences for most of the different types of programs." But significant differences did exist in regard to classical music and grand opera (favored more by city folk) and religious services, crop and market reports, and children's programs (favored more by farm people).[34] According to a 1937 survey, the well-known and urbane CBS (Columbia Broadcasting System) shows featuring Major Bowles, Kate Smith, and Eddie Cantor were the three most popular shows among rural audiences.[35]

As with the telephone and household appliances, reformers thought the radio would solve the "problem" of the isolated and overworked farm woman. The Radio Press Service remarked in 1923 that leisured city women enjoyed the radio, "but radio is even more popular in the homes of women who reside on isolated farms and in small towns. . . . To women thus situated, radio is not merely a joy, it is rapidly becoming a necessity." The service claimed that radios were becoming easier to tune and many women were making their own sets (one of the few references to women as anything more than passive listeners).[36] The Sears-Roebuck study found that 41 percent of the farm women surveyed preferred homemakers' programs, 31 percent music programs, and only 3 percent church services. Southern women liked music and church services much more than northern and midwestern women, according to the survey. Like most studies of farm women in this period, the results were biased in favor of women who had accepted the middle-class, urban domestic ideal fostered by home economists. In this case, the Sears foundation probably drew its voluntary observers from the ranks of local women's clubs. In the Symposium on the Relation of Radio to Rural Life, in 1935, several home economists in the USDA Extension Service reported similar wonders wrought by radio for farm women, especially the lessening of the drudgery of housework.[37]

Experience letters from farm women published in farm journals usually followed this line as well. "A farmer's wife" wrote *Prairie Farmer* in 1925 that radio "takes the dullness and monotony out of living on the farm, and the greatest thing is it keeps our children happy and contented at home" and thus kept them from getting in with a bad crowd. Marian Jane Parker wrote that "one may also lessen the fatigue and monotony of many tasks by doing them to the tune of radio music. . . . Dishwashing does not seem like drudgery and goes much faster accompanied by sprightly jazz, while brushing up the living room is more fun if done to a gay two-step." Mrs. Unger, the New York farm woman quoted earlier, praised the radio for keeping farmers informed of market reports, raising the musical appreciation of farm men, and relieving isolation for house-bound folk. She did not think the radio interfered with farm-work, because daytime reception was so poor. Reflecting common gendered divisions of labor, she painted a picture of father relaxing in front of the radio after a hard day's work, while "mother is there with her sewing, for there is nothing better than a little hand work along with the listening." Unger also appreciated hearing the dishes rattle in the background during a cooking program. "We hear her sifting the flour and when she breathes heavily, poor dear, we smile in remembering that all good cooks are stout." Unger parted somewhat from the Country Life ideology by describing a mutual relationship between the radio, youth, and the city. Youth who left the farm for the city welcomed radio there because it created a sense of community they had left behind. "A bit of the country in the city and a bit of the city in the country—this is the new movement to be brought about by broadcasting."[38]

Some women resisted the new technology. *Wallaces' Farmer* said in 1925 that a "farm woman from Illinois reports that she was at first skeptical about the radio and refused to listen in for some time after one had been purchased, but she finally weakened" when she heard good housekeeping advice, and she became an avid radio listener to home economics shows. A 1925 ad for the Atwater Kent company said that a doctor's wife in a small town had "fought radio for three years" before becoming a fan of it.[39]

A more common type of consumer resistance was that farm people did not listen to the sort of programs favored by the reformers. Elizabeth Wherry, a Wyoming farm leader and a columnist on radio for *Wallaces' Farmer*, complained to a conference of educational broadcasters in 1936 that "no matter what we do about it, people like hillbilly and old-time music; at the other extreme we find them listening to the jazziest of modern tunes. That to me is a strange development." Her hired hands tuned in a broadcast from Jack Yager's Cut-Rate Clothing Store in Cedar Rapids, Iowa, a "low comedy" show with plenty of hillbilly music, then listened to a show that interviewed people on the street in Cedar Rapids. She thought some of the allure of Jack's show came from the "same instinct that keeps people listening at the rural 'phone." She was also skeptical about the appeal of homemaking programs. "Another thing I have never heard discussed is housekeeping hours and cooking schools. Somebody must listen to them, but I never hear anyone mention a recipe they got over the air. They're more of a joke than to be taken seriously, it seems." Gerald Ferris of the Extension Service in Ohio supported Wherry's point by noting that surveys of Indiana and Kansas farm people showed that they preferred "more reports of actual farmer and homemaker experience, delivered by the farmers and homemakers themselves."[40] This reflects a little-studied preference for local programming noted by Arthur Capper in the mid-1930s, a desire that did not disappear after World War II.[41]

How much time farm people spent listening to their radios is difficult to determine. The 1928 NBC survey reported that farmsteads listened more from noon until two and less after nine in the evening than households in small towns and cities, as would be expected from agricultural lifestyles.[42] But the survey does not seem to have asked how much time farm people spent listening to the radio, a concern in the countryside in this period because of the expense and bother of maintaining at least two batteries for a radio. Farm people found innovative ways to charge the "A" battery, such as using the automobile or a windcharger. But oral histories indicate that concerns about batteries limited radio use far more than the NBC survey implies. Grace Hawkins, an Indiana farm woman, recalled that "you'd get interested in something, and about the time you wanted to find out something, the blamed battery would go down." Four other farm women and one farm man in Indiana and New York recalled that because of the battery problem, their radio listening was restricted to "news and important things," "main programs like Amos and Andy," and some music in the evenings. Problems with batteries were so common that a woman "radio poet" in South Dakota wrote a poem about the frustrations of trying to listen to a baseball game "when the juice runs out."[43]

These impressions are borne out by social surveys conducted at the local level before World War II. A survey of almost three hundred Wisconsin farm families in the late 1920s found an average listening time of only 3.7 hours per week. "Homemakers" listened the most (an average of 4.1 hours per week), and farm "operators," usually designated as their husbands, the least (3.3 hours). The Wisconsin sociologists studied nine hundred farm families in 1930 and found that the average listening time for owners and tenants had increased to 5.4 hours per week per person (a little less than the time they spent reading). The sociologists revisited three hundred of these families in 1933, during the depths of the depression, and found that time spent listening to the radio had dropped in half: from 8.2 to 3.9 hours per week in Green County; from 4.2 to 1.9 hours per week in Portage County; and from 3.3 to 1.2 hours per week in Sawyer County. The decline was "largely because of the fact that although many families had radios, these were battery sets which needed re-charging. Since money was not available for the purpose, the radios remained unused." In contrast, a study of the village of Marathon, New York, and its surrounding territory in the late 1920s found that one hundred open-country families with radios listened 14 hours per week, almost three times as long as their Wisconsin counterparts.[44]

The lower figures were supported by several time-use studies of the "typical" workweeks of middle- and upper-middle class, mostly rural women conducted by home economists at agricultural experiment stations. Maud Wilson's survey of farm, rural nonfarm, and urban women in Oregon found that women who had radios listened an average of 4.1 hours per week. A survey of ninety-one rural women in Montana in 1929–31 reported that those with radios (about one-half of the group) spent almost seven hours per week more on leisure than those who did not own a set. But a 1935 study of sixty-nine married city and rural women in Rhode Island who worked outside of the house found that the thirty-eight women who reported listening to the radio listened an average of only 1.75 hours per week.[45]

Less than one hour a day seems to have been about all that most farm families could afford to listen with battery-operated radios. Most families before 1930 seem to have had this type of set, even if they had electricity, because AC sets did not appear on the market until 1927. The situation changed dramatically in the countryside with the introduction of AC sets and the founding of the Rural Electrification Administration in 1935. For example, a national survey conducted in 1937 and sponsored by the two major networks at the time, CBS and NBC, concluded that listening hours in about twenty thousand rural homes (about roughly divided between farms and village homes) averaged four hours and forty-seven minutes a day, compared with four hours and nine minutes a day for urban families.[46]

We know quite a bit about one listening habit: the huge appetite for "hillbilly music," what is now called "country music." As noted by historian Susan Smulyan, these programs were "rooted in rural white and African-American folk traditions . . . and proved to be a hit with both urban and rural dwellers." Listeners flooded radio stations with letters applauding old-time fiddlers. Surprised at the result, big-city stations like WLS in Chicago, WSB in Atlanta, and WSM in Nashville introduced the extremely popular barn-dance musical variety shows. The one in Nashville became the Grand Ole Opry. Smulyan sees the mass migration from the country to city in this period as a partial explanation for the popularity of these shows. Although urbanites seem to have had an ambiguous view of rural life and industrialization, Smulyan concludes that "early listeners used the barn dance broadcasts to help sustain a rural culture in the middle of the city," a conclusion supported by rural commentators like Unger.[47]

These unexpected listening habits indicate that the effects of the radio on urban and rural life are more difficult to discern than the modernizing rhetoric suggests. Well into the 1930s recognized experts on the relation between radio and rural life continued to use what the sociologist Claude Fischer calls an "impact-imprint model" of technology and society. In this model, which Fischer rightly criticizes for being nonempirical, the physical capability of a technology leads to a predictable social change. According to this view, because radio could broadcast from the city to the farm, it must have brought them closer together. Arthur Capper noted in 1932 that "radio has revolutionized the marketing system in agricultural sections," a point echoed by Morse Salisbury in 1935, but neither seems to have studied the matter empirically.[48] The influential sociologist William Ogburn and his students did more in-depth studies of the relationship between invention and social change in the 1930s, but their list of 150 "effects of the radio telegraph and telephone and of radio broadcasting" is problematic. Many of the items, like "homogeneity of peoples increased because of like stimuli," are so general as to limit the ability to study them, and others, like "isolated regions are brought into contact with world events," describe specific matter-of-fact changes that supposedly led to the more general "effects." Ogburn also seems to have assumed an impact-imprint model in some cases, such as "the penetration of musical and artistic city culture into villages and country" and "much more frequent opportunity for good music in rural areas."[49] Although Ogburn translated a technocultural ability into a social reality in these cases, we have seen that—contrary to his prediction—country music became popular in the city.

This unpredictable effect challenges the tenet of technological determinism and the assumption that cultural influences flow in one direction only, from the city to the country. Just as many city dwellers chose to listen to barn-dance music rather than to Beethoven, farm people could tune in the programs they wanted to hear and ignore the uplifting messages of the reformers. Although farm men and women did not create novel social organizations for the radio as they did for the telephone nor modify it for alternative uses as they did with the automobile, they were still able to weave it into established patterns of rural life—as they had with household technology—by tuning out the city and tuning in the country.

Part Two

A NEW DEAL IN
RURAL ELECTRIFICATION

CHAPTER 5

Creating the REA

The increasing presence of the telephone, the automobile, and nonelectrical household appliances on American farms in the 1920s did not overturn urban opinions about the inferiority of rural life. The media presented, instead, images of antimodern farmers demanding restrictions on immigration and opposing the teaching of Darwinian evolution. The American Farm Bureau Federation was furious and resolved at its 1923 convention "to compel the motion-picture industry promoters to quit caricaturing the agriculturalist on the silver screen." It recommended showing the "modern farmer, a type of business man with a capital of $25,000 to $100,000, an owner of automobiles and the latest farm machinery."[1]

Although telephones and automobiles were proportionately more common on farms than in nonfarm households in 1920, farmers lagged far behind in electricity and indoor plumbing (2 percent of farms had central-station service in 1920, compared with 47 percent of nonfarm households). Many farmers gave up their telephones during the Great Depression (see fig. 2). As electricity became widely available in urban areas, it began to mark the up-to-date city and town from an old-fashioned countryside dotted with kerosene lamps and the ubiquitous outhouse. To paraphrase the title of a recent book on technology and colonialism, electrification was seen as the measure of farm men and women, tangible evidence of how far they had advanced on the road to modernity.[2]

As we saw in chapter 3, electrifying the farm before the New Deal was no mean task. Central-station electricity—more powerful and usually more reliable than the home-grown kind—was a public utility, like the telephone, automobile, and radio, but one whose network was slow to develop. It took four decades after Thomas Edison and his co-workers built a central station in

New York City in 1882 to electrify more than one-half of nonfarm homes in the United States. Utility companies were slow to develop the domestic market because they, like most social commentators, initially viewed electricity in the home as a luxury. Early in the century, utility companies and manufacturers thought it was much more profitable to build networks of generating stations, interconnected transmission lines, and distribution systems to light city streets, power streetcars, and run factories. When electrical companies turned their attention to the domestic market in the 1920s, they focused on the densely populated and wealthier urban centers. Like AT&T, electrical firms were reluctant to serve farmers because of the high cost of building lines into sparsely populated areas not known for high disposable incomes.[3] Prosperous farmers responded by finding innovative ways to make their own electricity or to obtain it from a high-voltage line, or high-line.

The fight for public power was more successful. The creation of the Rural Electrification Administration (REA) by New Dealers in 1935—after a decade-long struggle between public and private power interests—provided the means to eventually electrify the nation's farms. The REA coupled its own mediating network to the vast network of agricultural modernization, shown in figure 3, to sell electricity to farm people. But that was not enough. Ironically, the urbane New Dealers established a government agency that achieved their goal by relying on the rural institution of the agricultural cooperative.

Public Power and the REA

The debates between the proponents of public and private power in the 1920s and early 1930s, between those who saw government action as the only way to bring affordable electricity to farmers and workers and those who thought laissez-faire capitalism was more efficient, led to the establishment of the REA in 1935. A major force behind the struggle in the 1920s was the threat of public ownership posed by a successful hydroelectric project in Ontario, Canada. Although the growing number of municipal plants in the United States worried the electrical industry—there were three thousand in 1922—private companies still generated well over 90 percent of the country's electricity.[4] The Canadian system was a different beast. Established in 1906, the Hydro-Electric Power Commission of Ontario became the largest public-power system in North America in 1910, when it began to buy power from a plant at Niagara Falls (a plant it later purchased), generate its own electricity, and transmit it to municipalities, which distributed "power at cost" to urban and rural areas. The U.S. electrical industry lost little time in using its powerful trade group, the National Electric Light Association (NELA), to attack the specter of socialism north of the border with a broadside of newspaper propaganda and secretly sponsored engineering studies of the "true" cost of "Hydro." The NELA paid the chief engineer of Superpower, an industry scheme that would have interconnected power systems along the Boston-Washington corridor, to criticize Hydro in 1922.[5]

Advocates of government involvement responded by trying to emulate Ontario Hydro at home. Governor Gifford Pinchot of Pennsylvania proposed a Giant Power interconnection plan in 1925, based on generating electricity at coal mines. Senator George Norris of Nebraska began pushing bills through Congress to establish a public-power system at Muscle Shoals on the Tennessee River, where the federal government was completing a dam it had started at the end of World War I to make electricity for the production of nitrates for explosives and fertilizer. Although Norris succeeded in keeping the dam in government hands—by blocking its sale to Henry Ford, for example—three presidents vetoed bills that would have created an Ontario-like "yardstick" for private utilities in the United States. Frustrated at every turn by opposition from the electrical industry, Norris and Senator Thomas Walsh of Montana initiated, in the mid-1920s, a decade-long investigation by the Federal Trade Commission (FTC) of a suspected "power trust." The FTC found little evidence of a disciplined cartel, but it compiled eighty volumes of damaging evidence about the NELA's underhanded propaganda methods and serious financial misdealings by the pyramid of holding companies that had come to dominate the electric utility industry.[6]

Rural electrification occupied a prominent place in these debates. It hardly seems coincidental that the NELA first took up the matter of rural lines in 1911—the year Ontario Hydro began to plan a system for rural service —nor that it established a Rural Lines Committee in 1921—just after Ontario Hydro made major changes to its rural program. Under the new plan, the Ontario government funded one-half of the cost of constructing rural lines erected in Hydro's rural power districts. Hydro would provide electricity to rural neighborhoods that signed a twenty-year contract at the rate of three farms to the mile. Progress in rural electrification had been slow in Ontario before this, but now more than six thousand contracts were signed within the first two years of the plan at a cost of two to seven cents per kilowatt-hour, well below rural rates in the United States. The rural program energized advocates of private and public power alike. The NELA created a national organization, the Committee on the Relation of Electricity to Agriculture (CREA), in 1923 to do research; Norris preached the gospel of rural electrification through public power; and progressive engineer Morris Cooke, the architect of both Giant Power and the REA, made rural electrification a central part of his plans.[7]

Giant Power was a government-assisted solution to the problem of rural electrification; CREA was the darling of private enterprise. Although worldly and urbane, the promoters of Giant Power certainly had rural interests. Governor Pinchot, the former head of the Forestry Service, had served on the Country Life Commission at the turn of the century and worked with the Pennsylvania Grange on rural electrification. Although Cooke was an independently wealthy engineer, scientific-management expert, and municipal reformer, he came to share Pinchot's concern for rural uplift. Having fought a successful court battle to lower the rates of the Philadelphia Electric Company before World War I, Cooke thought an Ontario-like plan could provide low rates for rural Pennsylvania. In the Giant Power scheme, locating large plants at coal mines would save fuel transportation costs and enable cheap electricity to be transmitted at high voltages in an interconnected network throughout the state. In this privately owned but publicly regulated and planned system, proposed legislation would permit the formation of rural power districts (similar to Ontario's) in unserved areas if a majority of farmers voted for them. New laws would also allow farmers to form cooperatives, which would receive engineering and business advice from the state agricultural college. In exploring the feasibility of stringing electric lines on existing telephone poles, Cooke told independent telephone companies that electric-power co-ops "would have much in common with the methods adopted in promoting telephone service in the less densely populated parts of the State."[8]

Rural electrification made up a substantial part of the Giant Power report, published in early 1925. One of its six technical sections was devoted to rural electrification, as were two of the five appendixes, one of which was authored by Cooke. In addition to articles about rural wiring and rate setting, the report contained detailed information on rural electrification, which Cooke and his board had collected in the Giant Power survey. This included an in-depth study of electrical consumption on a Pennsylvania farm, a survey of electrical usage on about 150 farms in Pennsylvania and Wisconsin, and much material about rural costs, rates, and usage from utility companies in Pennsylvania, Missouri, and California. Cooke also cooperated with various farm groups to establish a demonstration farm powered by an interurban line in 1925.[9] Although the Pennsylvania legislature did not approve Giant Power, the project distinguished Cooke as one of the country's foremost experts on rural electrification.

Cooke was also influenced by the goals of the Country Life movement. In addition to his collaboration with Pinchot, he worked with Country Life reformers Robert and Martha Bensley Bruère. Cooke's biographer notes that in 1923 he "enjoyed a pilgrimage to Ontario" with "his friend Robert Bruère of the *Survey*" magazine, which published a special issue on Giant Power and Ontario Hydro in 1924. As an associate editor of the magazine and a member of the advisory committee to the Giant Power board, Martha Bruère wrote an article for the same issue enumerating the positive social effects of electricity in urban and rural areas of Ontario. The article was illustrated by a cartoon showing a road sign advertising "Purple Pills for Tired Women," before electricity arrived, and "Tonight, Illustrated Lecture on Social Service," after electrification. She provided information on the domestic use of electricity and offered to estimate consumption in the farmhouse for the Giant Power board. Cooke recommended her to the editor of *Farm Journal* as an author who could "humanize" a series of articles on rural electrification that were being prepared by an engineer working for Giant Power. In 1932, Cooke wrote an article on the future of electrical power for *Survey,* which was illustrated by the before-and-after scissors picture (fig. 1) with which I began this book—a picture created by Martha Bruère.[10]

Public agencies tried several means to develop the rural market in the 1920s. When the Pennsylvania legislature voted down the Giant Power scheme in 1926, the state's Public Service Commission issued an order requiring utility companies to provide rural extensions in any territory where there were at least three signed-up farms to the mile. In less populous areas, the farmers shared the cost of the line, with the company paying three hundred dollars per customer. The order also permitted "rural distribution associations," similar to those of Ontario Hydro, to be formed. The order vaulted Pennsylvania into a leadership position in rural electrification in the United States, but the next year utility companies convinced the commission to rescind the order, on the grounds that it was too expensive, and to replace it with an order that required farmers to pay part of the cost of constructing lines. In Washington state, the Grange and public-power advocates persuaded the state legislature in 1930 to create public utility districts, similar to those in Ontario, which provided water and electricity to all classes of users, including rural areas. By 1940, thirty-three districts had been set up, aided by the Bonneville Power Administration, a "little TVA" established in 1937 to develop hydropower in the Northwest. Oregon passed a similar law in 1930, but only four of the twelve power districts went into operation in that state. Senator Norris's home state of Nebraska had much more success with a 1933 law establishing public utility districts, mainly because the REA later loaned money to these districts.[11]

Private utility companies took a much different approach by forming the Committee on the Relation of Electricity to Agriculture. Consisting of representatives from the federal government (Departments of Agriculture, Commerce, and Interior), farm organizations (American Farm Bureau Federation and National Grange), professional groups (American Society of Agricultural Engineers, American Home Economics Association, and General Federation of Women's Clubs), and industry (National Electrical Manufacturers Association, Individual Plant Manufacturers, and the NELA), the CREA utilized this vast network of the "associative state" to conduct research on rural electrification. Funded by private utility companies and headquartered in Farm Bureau offices in Chicago, the CREA's leaders insisted that the best way to advance rural electrification was to show skeptics that electricity could be used extensively on the farm. More usage would break the Catch-22 cycle of low revenues not paying the high cost of building lines, mentioned in chapter 3. Public-power groups advocated more usage, as well, but they criticized the CREA as a half-hearted attempt to electrify the farm, a smoke screen to make it look like the electrical industry was addressing the farm problem. An REA administrator later charged that the CREA had been "set up to fool the farmers and their leaders and the county agents and the extension people in the land grant colleges. . . . They made a grant of a few hundred or a few thousand dollars annually to a so-called research project in the colleges, a kind of subsidy" that allegedly prejudiced agricultural leaders against the REA.[12]

Although the charges are probably exaggerated, the CREA funded elaborate studies, aimed at high-income farms, that tended to keep agricultural engineers and home economists in the orbit of the utility companies. By 1931, twenty-five state committees had conducted 211 investigations that identified 227 uses for electricity on the farm and 190 uses in rural industries. Universities and manufacturers also conducted numerous short courses on rural electrification. In Oregon, agricultural engineer George Kable funneled utility funds into the local agricultural college's study of irrigation, dehydration (of prunes, walnuts, and hops), poultry lighting, silo filling, milking machines, and other farming applications. At Iowa State College, professor Eloise Davison, the first representative of the American Home Economics Association to the CREA, studied time spent operating household appliances and taught a course for home economists employed by electrical manufacturers and utility companies.[13] Both researchers left their local projects to become national spokespersons for the electrical industry. Kable became director of research for the CREA, where he headed the National Rural Electrification Project at College Park, Maryland. Davison moved to New York City to become chief home economist for the NELA in 1928, the same year that agency recommended that utility companies hire home economists to build up the domestic market. Yet both Kable and Davison switched to the public-power side of the fence in the mid-1930s after the Federal Trade Commission revealed the misdeeds of the "power trust." Kable joined the Tennessee Valley Authority (TVA) as its chief agricultural engineer. Davison started an educational program for the Electric Home and Farm Authority, the TVA's appliance-purchasing agency.[14]

The career of Emil Lehmann, a prominent professor of agricultural engineering at the University of Illinois, shows a similar ambivalence toward the electrical industry. As we saw in an earlier chapter, Lehmann conducted research on rural electrification while employed by a utility company during a sabbatical leave in 1928. Lehmann used a questionable methodology and analyzed data selectively in order to show that electricity reduced the time farm women worked in the house. Having previously chaired a CREA project leaders conference, Lehmann received six hundred dollars and travel expenses from the committee in 1930 to update its bulletin on electrical appliances. Lehmann drew on his Illinois study and gathered material from NELA home economists to contradict the growing body of evidence from Bureau of Home Economics studies, which showed that owners of electrical appliances spent more time on housework than did those without electricity.[15]

Yet Lehmann gained some independence from the electrical industry. During the depths of the Great Depression, he proposed writing two articles on inexpensive collapsible septic tanks and wind-power electric plants for the journal *Electricity on the Farm*, only to be told that advertisers of septic tanks would object to the first article. The editor could not accept the second article, either, because the magazine, in which large manufacturers advertised heavily, was "devoted solely to high-line service, and thus cannot endorse either the independent lighting plant or the so-called wind electric plant." Lehmann wrote an article on home plants in 1932 and came to respect the REA, even while still doing publicity work for the Public Service Company of Northern Illinois, a private holding company, as late as 1944.[16]

The debate between private and public power for rural electrification intensified during the early years of the Great Depression. In response to criticism that the CREA focused too much on research and that it was essentially a holding action against the rising tide of public power, the NELA proudly pointed to the progress in rural electrification in the 1920s. The Census Bureau independently reported that about 840,000 farms had electricity in 1930. Critics pointed out that the figure, though impressive in absolute terms, represented only 14 percent of the country's 6 million farms, a dismally small fraction when compared with the nearly universal coverage attained by government programs in northern Europe. The comparable U.S. figure was actually 9.5 percent, because the Census Bureau data included more than 200,000 farms that made their own electricity. One critic observed that "in parts of Switzerland, Germany, Denmark, and other countries, rural electrification is far ahead of railroads, and many farmers there who have never traveled on a train are able to read meters and turn switches." The NELA responded that, with the help of the CREA, the number of farms served by central stations had increased more than sixfold since 1921, that the national figure had climbed to 11 percent at the end of 1931, and that this percentage masked a wide variation by region: the proportion of farms that were electrified ranged from between 1 and 9 percent in the South to 20 percent in the more prosperous Midwest and to more than 50 percent in New England and in western states dependent on irrigation.[17]

Some utility companies took vigorous steps to develop the rural market. The Virginia Electric and Power Company nearly doubled the extent of its rural lines in this period, from 802 miles in 1929 to 1,580 miles in 1933. The Detroit Edison company introduced its Michigan Plan in 1928, which required farmers to deposit with the company a thousand dollars per mile of line. The utility would rebate $100 for each customer who signed up in advance and $60 for those who signed after the line was energized. At four farms to the mile, the initial cost could be as low as $150 per farm. Detroit Edison reduced the cost per mile to $750 in 1933, perhaps because the TVA was established that year, and then to $500 in 1935, when the REA was formed.[18] Although Alex Dow, the president of the utility company, had an interest in rural life, the program was not motivated by altruism. Detroit Edison and other private companies were in the business of making and selling electricity at a profit, not contributing to the social uplift of rural life.

The climate for government support of rural electrification improved dramatically during the early years of the depression. Franklin Roosevelt had impeccable public-power credentials. He had crusaded against the "power trust" as governor of New York, and in 1931 he established the New York Power Authority to develop the hydroelectric potential of the St. Lawrence River. Appointing Morris Cooke as an out-of-state member of the authority ensured that Roosevelt's long-standing interest in agriculture would be represented in the project. Roosevelt sounded these themes during the presidential campaign of 1932. In a speech at Portland, Oregon, seat of the public-power movement, Roosevelt railed against a predatory electrical industry, symbolized by utility magnate Samuel Insull's flight to Greece to avoid arrest under larceny and embezzlement indictments after his multitiered pyramid of holding companies fell and ruined thousands of stockholders. Public initiatives were needed to clean up the industry, lower rates, and electrify the nation's farms. During the New Deal's first hundred days—while Insull was still a fugitive from justice and after the NELA had been reconstituted as the Edison Electric Institute following disclosures of its unethical propaganda methods against public power—Roosevelt and Senator Norris pushed a bill through Congress in May 1933, creating the Tennessee Valley Authority.[19] Ontario Hydro had arrived at Muscle Shoals.

As shown by previous historians, Cooke spearheaded the drive to create the REA.[20] Although the TVA sold electricity to newly created co-ops that served urban and rural areas as part of its program of flood control, the improvement of river navigation, and regional development, it could—at best—serve only as another model for how to electrify the farm. Cooke wanted a national program. He broached the subject to Roosevelt, who appointed him head of a Public Works Administration (PWA) committee to study how to develop the water and land resources of the Mississippi Valley. Cooke used the opportunity to survey rural electrification in this large region, and he suggested a program combining public and private agencies to Harold Ickes, secretary of the interior and head of the PWA. Because Ickes thought private utilities were corrupt, Cooke revised his plans and proposed, in February 1934, a publicly controlled National Plan for the Advancement of Rural Electrification. Rather than beginning with a survey, Cooke recommended "an allotment of $100,000,000 actually to build independent self-liquidating rural projects" at an estimated cost of five hundred to eight hundred dollars per mile in areas not served by private utilities. The main methods of distributing electricity to farmers would be "electric service districts and consumers' mutual companies" (i.e., the same organizations proposed in the Giant Power report). They would be set up under "uniform state legislation" and receive expert engineering and management assistance from federal or state agencies. Power would come from either public or private sources. A Rural Electrification Section in the Department of the Interior, "manned by socially minded electrical engineers" (like Cooke), would run the program.[21]

Like the promoters of the telephone, automobile, and radio, Cooke drew on the traditional themes of Country Life rhetoric when he discussed the social aspects of rural electrification in the report.

> Agriculture is a major problem. It must evolve toward the status of a
> dignified and self-sustaining sector of our social life. . . . Both for the
> farmer and his wife the introduction of electricity goes a long way to-
> ward the elimination of drudgery. The electric refrigerator will effect a
> considerable change in diet—more fresh vegetables and less salt and
> cured meats. The inside bath-room, made possible by automatic electric
> pumping, brings to the farm one of the major comforts of urban life.
> Electricity will be a strong lever in keeping the boys and girls on the
> farm—in encouraging reading and other social and cultural activities.[22]

Cooke was not a stranger to the Country Life movement, as we have seen.
Although he did not use the rhetoric of rural uplift in the article containing
Bruère's scissors picture, he often spoke in those terms after the New Dealers
came into office. Cooke wrote Ickes in February 1934, "My own keen interest
in rural electrification is based on a growing conviction that the gravest of our
troubles can be traced to the increasing dominance of 'city people' in human
affairs and the lessening of the influence of those in material and spiritual
touch with the soil." He told a conference on rural electrification the next
year, "I began my interest in this thing because of the opportunity for taking
the load of drudgery off the backs of the farm women, and the farm man
too."[23]

Cooke's 1934 report—enclosed in a zebra-striped cover to gain attention
in the commotion of creating New Deal agencies—was received well by the
Departments of Interior, Agriculture, and Commerce. But FDR and other
New Dealers were, at the time, more interested in Cooke's claims that building rural lines would provide immediate relief to the vast numbers of the un-
employed. After conferring with utility executives and a colleague from the
Giant Power survey, Cooke drew up in October a "revised proposal for a
large-scale development of rural electrification intended to provide a *large volume of immediate employment* in the capital goods industries, and of local
skilled and unskilled labor." A central agency would run the nearly $2 billion
program to electrify 3.5 million farms by coordinating the work of one or
more of the following agencies: the Federal Emergency Relief Administration, the PWA, and private utilities. Cooke thought $50 million to $100 million could get the program off the ground quickly; lines could be built under
government supervision as early as the spring because the lines could follow
public highways. The government would sell or lease the lines to local public
agencies or private industry.[24]

The report seems to have had some effect. The Roosevelt administration, which had been lobbied by national farm groups and state authorities in the Carolinas to do something for rural electrification, included provisions for it in the work-relief bill sent to Congress in January 1935. Rural electrification would constitute $100 million of the $5 billion relief measure. Following a lengthy debate, a compromise bill was passed in April. On May 11, FDR issued an executive order that created the Rural Electrification Administration as a temporary relief agency. *Literary Digest* noted that the REA was the New Deal's "eighty-third alphabetical agency."[25]

Deciding to Use Cooperatives

Agricultural leaders, especially the editors of national farm magazines, were generally enthused about the agency. Clarence Poe, editor of the southern *Progressive Farmer* and a fan of Ontario Hydro, thought the REA marked the "beginning of a movement as far-reaching in its influence and benefits as the rural mail delivery service or Federal aid to road building." The Republican *Farm Journal*, which had published articles on Cooke's rural electrification plans in the 1920s, said, "We are not boosters for the Washington alphabeticals, as a general thing. Our opinion of the NRA [National Recovery Administration] is scarcely printable. But Washington occasionally does something right, and one of these accidents is the plan for helping to modernize farm homes, and particularly in putting at the head of it our friend Mr. Morris L. Cooke." *Wallaces' Farmer*, then owned by Henry A. Wallace, Roosevelt's secretary of agriculture, approved of the REA but devoted much more space to the debate over the constitutionality of the Agricultural Adjustment Act.[26] The National Grange and the American Farm Bureau Federation, both of which had resigned from the CREA in 1934, pledged their support of the REA at their annual conventions.

Appointed administrator of the new agency, Cooke recruited colleagues who had worked with him during two decades of promoting electric power from the public point of view, and he began to draw up a plan on how to best spend the $100 million allotment. There were strings attached. Congress stipulated that 25 percent of all relief funds had to be spent for labor and 90 percent of that portion for those on relief. The shortage of skilled electrical workers on the relief rolls handicapped the REA as a relief agency, as did the time required to design and build a rural system from scratch. For these and other reasons, Cooke—a moderate Progressive—thought utility companies would play a large role in the program. A confidential, tentative plan, drawn up in April 1935 in anticipation of FDR's executive order, said power would probably come from the lines of utility companies or municipal plants but advised farmers to apply to the government to get service. In a radio address delivered a week after the REA was formed, Cooke said, "Rural electrification can be extended under any of four auspices": private utilities, state power districts or municipal plants, farm cooperatives, and the federal government. He thought it was "very likely that some of this work will be carried on under each of these groups."[27] Cooke explored the first three options roughly in the order listed. He held a conference with utility executives in late May, which seemed to go well, then took up the other options while waiting for a detailed reply from the power companies. He met with managers of municipal plants a few days later and presided over a one-day conference between cooperative leaders and the REA in early June.

In leading off the conference, Cooke praised the cooperative movement and distributed a report written by Udo Rall, head of the Federal Emergency Relief Administration's Division of Self-Help Cooperatives. Rall noted an "alarmingly high" failure rate among electric cooperatives, because of poor capitalization, poor management, diseconomies of scale, unfavorable state laws, and uncooperative utility companies. He cited instances of several electric co-ops failing in Ohio, Iowa, and Nebraska, most of which sold their lines at a loss to local power companies. Although it was a rather bleak picture, Rall reasoned that federal sponsorship would solve many financial and organizational problems.[28] Cooke said that "in the past, cooperatives have been subject to considerable criticism, much of which may have been unwarranted. To merit a leading position in our program, the cooperative that applies to us for funds must toe the mark. Its management will be required to prove itself trustworthy, and the organization must have a vital character which will give promise of future growth and strength." J. P. Warbasse, president of the Cooperative League of the United States, outlined ways existing co-ops could handle the electrification task, but he also said that "constant supervision" by an expert agency like the REA "will be wholly acceptable and advantageous." An attorney for the TVA gave a glowing report on its Alcorn cooperative in Mississippi, which relied on cheap rates and an urban center radiating lines into the countryside.[29]

The minutes of the conference indicate enthusiasm from co-op leaders, but, according to later accounts, the Cooperative League advised Cooke not to form electric co-ops because it feared that federally funded groups would ruin the cooperative movement.[30] Other representatives were apparently more friendly, because the Farm Bureau and the Grange lobbied the REA to use cooperatives.

Cooke had already taken steps to push the co-op idea forward. A week before the conference, he drafted a letter to C. W. Warburton, head of the U.S. Department of Agriculture's Cooperative Extension Service, asking its county agents to help farmers get assistance from the REA. The co-op conference must have been encouraging, because Cooke's letter to Warburton, which went out soon after the meeting, described in some detail how county agents could assist farmers to establish organizations to apply for an REA loan to build electrical distribution systems. Here, Cooke revised the order of his four options to be the federal government, power districts and municipal plants, cooperatives, and private industry.[31] By late June, Cooke had hired Boyd Fisher, a former partner in his engineering-management consulting firm in Philadelphia, to investigate the cooperative and municipal-plant options. Fisher met with Warburton, who agreed that county agents in populous regions would be able to assist the REA in promoting co-ops, and explored the subject with state extension services and farm bureaus in the Midwest in July. A mimeographed pamphlet, sent to farm leaders at this time, still listed co-ops as third on the list of four options but noted that the REA had hired a staff person to advise farmers on how to form co-ops that would qualify for loans.[32]

The REA was thus moving in the direction of cooperatives when it heard from the committee representing privately owned utilities in late July. The proposal was a bombshell. The utilities proposed to borrow all of the REA's initial allotment of $100 million with long-term, low-rate loans, plus $13 million more, to build seventy-eight thousand miles of lines and service facilities for 351,000 rural residences, 247,000 of which would be farms. The REA would coordinate the lending of another $124 million to these rural customers to wire their houses, install service extensions, and buy appliances. The utilities did not address the issues of low rates and area coverage, claiming, instead, that most large users of electricity on the farm had already been served and that rural electrification was a social, not an economic, problem. Cooke wrote a courteous reply, thanking the industry for its proposal but announcing that the REA would deal with individual companies—public and private—instead of an entire industry. The letters were made public on August 1, along with an REA news release that put a positive spin on the utility company letter.[33]

Senator Norris hit the roof when he read newspaper reports that Cooke intended to lend $95 million to private power companies and the rest of the REA's allotment, that is, $5 million, to public agencies—in proportions roughly equal to the fraction each held of the country's generating capacity.

Cooke quickly soothed the feelings of his friend and strongest backer in Congress, as well as President Roosevelt, by saying he had been misinterpreted by the press and inserting a paragraph in a news release to explain the misunderstanding. The REA, which had just received permission to make loans, would give preference to "applications from public bodies and farm cooperatives." In an address broadcast on the *Farm and Home Hour* on August 9, Cooke moved cooperatives to first place on his list of options. Later that month, the REA added the preference for public borrowers to the published version of its pamphlet on how to participate in the program.[34] The Iowa-based *Wallaces' Farmer*, published in a state filled with telephone and farm cooperatives, quoted Cooke on his move toward co-ops. In September, the paper ran a full-page story, drawn from the REA pamphlet, saying that rural communities wanting to participate in the program should first approach their local utility companies for service. If that failed, they should write to Cooke. The REA might then grant a loan to the utility, or the farmers could work with the agency to establish an electric-power cooperative.[35]

The die was cast for cooperatives by the end of the year. In early November, representatives of municipal plants from all over the country met at Kansas City at the urging of the REA and apparently did not show much enthusiasm for building rural lines, because of legal and economic obstacles.[36] The REA made its first loans to cooperatives in November, when five of its first eleven loans went to co-ops in Indiana, Ohio, Mississippi, and Tennessee.[37]

Norris and Cooke then instituted a comprehensive program for rural electrification. Following an open letter by Cooke to the utility industry inviting it to reconsider its position, Norris and Representative Sam Rayburn from Texas introduced a bill into Congress in January 1936 to establish the REA on a permanent basis. Congress amended the bill and passed it in May, with little opposition from utility companies, who were still smarting from the passage of the Public Utility Holding Company Act of 1935. The REA began operations as an independent agency on July 1, 1936, with a ten-year mandate to electrify the nation's farms by loaning $410 million at low government rates (typically 3 percent for twenty-five years), principally to public bodies. The lack of interest in rural electrification on the part of municipal groups and the reluctance of most utility companies to provide area coverage meant the vast majority of borrowers would be nonprofit cooperatives, which would buy power from utility companies and sell it to their members over lines built with self-liquidating REA loans.

Cooke recalled the importance of this decision at a staff conference in early 1937. He described the unpromising meetings with utility companies and municipal plants. "Then, God leading us and Boyd Fisher helping, we called a meeting of the co-ops and there we began to see daylight. The co-ops have been our help in our time of great need, as you know, so today we are going ahead with the co-ops. . . . They seem to be a God-sent agency which makes it possible for us—the REA—to 'do its stuff.'"[38]

Organizing to Build Co-ops and Sell Electricity

Deciding to rely on cooperatives was difficult enough; implementing the decision was an enormous challenge. It became even more so when the REA realized it had to encourage farmers to use enough electricity to pay off the loans in time.

Although Cooke had included cooperatives in all of his rural electrification proposals—from Giant Power to the "zebra" report—he seems to have thought they were a last resort, not a primary vehicle for rural electrification.[39] Cooke may have vilified power companies for overcharging customers and not serving farms, but he admired their technical accomplishments and expertise and believed that through regulation and planning he could make them socially responsible. As the *Farm Journal* noted, "In spite of the opinion of many public utility men, [Cooke] is no enemy of the utilities, and he is a firm friend of agriculture." *Literary Digest* remarked, "Cooke is one of the few administrative heads in Washington liked and respected by both Democrats and Republicans."[40] But the prospect of relying on electrically illiterate farmers and their leaders to carry out his beloved rural electrification scheme seems to have been another matter for the patrician reformer and scientific-management expert. Murray Lincoln, executive secretary of the Ohio Farm Bureau, recalled that when he proposed, in the summer of 1935, to organize cooperatives throughout the state, Cooke asked him what he knew about the electrical industry. "Not a thing," replied the former county agent, "I was trained in dairying and animal husbandry." Cooke "suppressed a smile" and "indulgently" explained the intricacies of the electrical business to him.[41] Harvey Hull, general manager of the Indiana Farm Bureau Cooperative Association, who was at the REA co-op conference, recalled that Cooke first supported Hull's plans to develop co-ops but then gave him a frosty reception on his second visit in the summer of 1935.[42]

Cooke's attitude seems to have stimulated both men to prove him wrong by organizing REA-funded co-ops in their states. Hull had a head start. He was favorably impressed with the rural electrical cooperatives he had observed during a trip to Denmark and Sweden in the summer of 1934. He then visited the TVA, whose officials advised him to establish countywide associations along the lines of its Alcorn co-op, which was based on English cooperative principles and bought power from the TVA. Indiana co-ops could buy power from municipal plants. Hull convinced the Indiana Farm Bureau to push an enabling act through the state legislature two months before the REA was created and to incorporate the Indiana Statewide Rural Electric Membership Corporation in July 1935. Controlled by Hull's association, the Statewide organized co-ops on a county-by-county basis, loaned them seed money, provided legal, management, and engineering expertise, and helped them get REA loans.[43] According to Boyd Fisher, Cooke's liaison with farm organizations and a proponent of co-ops, the Indiana Farm Bureau made a "gentleman's agreement . . . with the utilities, by which in return for supervision of cooperatives by the Public Service Commission the power companies would stay out of cooperative territory."[44] In Ohio, Murray Lincoln, who thought the Indiana Farm Bureau had sold its birthright by placing itself under the state's Public Service Commission, bypassed the state legislature and quickly put his considerable experience organizing consumer co-ops to use. Founded in September 1935, the Ohio Farm Bureau Rural Electrification Cooperative Corporation sponsored five co-ops for REA loans by the next summer.[45]

Cooke was won over. Before the REA became a permanent agency, Fisher praised the organizations headed by Hull and Lincoln for their vigorous promotion of co-ops and welcomed the American Farm Bureau Federation's decision to give rural electrification the same level of support it planned to give the New Deal's Agricultural Adjustment Act. Cooke sang the praises of farm bureaus in Indiana, Ohio, and Minnesota to the head of the Farm Bureau Federation and to Sam Rayburn. He told Lincoln that he was doing "wonderful work in Ohio." The REA's in-house magazine supported both Hull's and Lincoln's groups and called the latter organization the "guiding genius of the movement" in Ohio. Fisher and Cooke (to a lesser extent) worried about the monopoly tendencies of the farm bureaus, the size of their promotional fees, Hull's proposal to charge competitive power rates, and the ties between the American Farm Bureau Federation, the Extension Service, and the private utilities. But they thought the farm bureau's statewide agencies were a godsend in the early days of the program.[46]

Establishing viable co-ops was not a trivial matter. As one staffer recalled, it required "effecting a legally adequate organization, securing an initial and expanded membership, designing a suitable system, negotiating with sources of wholesale power, securing certificates of necessity and rights of way, standing for their rights in conflicts with commercial companies before commissions and legislatures, and, when legislative authority was indefinite, introducing appropriate bills in legislatures."[47] Many power companies waged war on the co-ops by building "spite lines," in the terminology of the REA, across co-op territories to cut them up into unprofitable pieces. No wonder the REA initially welcomed the statewide organizations: they fought the spite-line wars in their territories and performed all of the above functions on site, with the advice and oversight of REA engineers and lawyers. Before approving a loan, REA economists and accountants calculated whether a projected co-op would be able to pay it off.

Cooke wanted to provide assistance to borrowers, not to create and manage them, as the statewides were doing. In early 1936, Cooke told the head of an advisory committee that the "REA has had a rule against actually organizing electric cooperatives in order to avoid the unethical position of creating an agency and then loaning it money." The REA explained this position publicly the following year by saying "that as a lending agency it could not properly call into existence proposed borrowers, if it wished to reserve discretion to refuse loans when necessary." Furthermore, advocates of the co-op approach advised that "healthy cooperatives cannot be called into existence by external promotion," they have to grow organically at the grass roots.[48] Yet very few farmers were organizing themselves into co-ops during the first years of the REA. Letters from farm people, many of them women, poured into Cooke's office, asking how to obtain electricity. But few acceptable loan applications from co-ops arrived, except from state organizations in the Midwest.[49]

The REA addressed this problem through the Development Division, which Boyd Fisher headed from the beginning. Cooke's main field agent during the early days of the REA, Fisher held views about electricity as a progressive social force that were shared by his boss. On a field trip in 1936, Fisher wrote Cooke that he had seen a "sign on a [railroad] station door, blunt, cruel, and stupid: 'Negro Waiting Room.' This is the Tennessee of the anti-evolution law, but light is coming to it on electric wires." Fisher also admired large-scale technocratic projects in the Soviet Union, on which several of his engineering-management friends had worked.[50] By the fall of 1936, the Development Division had a field staff of seven men and one woman, Oneta Liter, formerly a state home economist in Kentucky. Their job was to "initiate, stimulate, and develop [local] interest to the point where it results in a formal application for REA loans."[51] Although the agency distrusted some state bodies—like the Rural Electrification Authority in North Carolina, which seemed to be controlled by utility companies—Fisher's field representatives worked extensively with state organizations. The REA listed the North Carolina group alongside the Indiana Statewide as one of the eleven rural electrification authorities recognized as approved borrowers in late 1936. Field personnel contacted farmers, farm bureaus, and farm organizations directly. Fisher said in an article that they were not assigned to "any one State or region, but have been moved all over the country as occasion demanded. This tends to prevent their 'going native'" and getting mixed up in local politics.[52]

Fisher did not always practice what he preached. In October 1936, a month before he made this statement, the manager of a large organization of Minnesota cooperatives, which supported the REA, sent a telegram to President Roosevelt urging that Fisher "stay out of Minnesota" because his "actions are detrimental to the cooperative movement in Minnesota and the administration. Mr. Fisher has taken sides in a cooperative situation here which can only be settled by the cooperatives themselves." A Minnesota senator made a similar request to the president.[53] Furthermore, although the title of an article in the REA house organ, *Rural Electrification News*, claimed in January 1937 that "cooperatives use their own initiative in applying to REA for loans," Fisher admitted at a staff conference the next month, "We have done things that are dangerously close to organizing cooperatives."[54]

The practice reflected policy changes occurring at higher levels of the REA. Soon after the agency became permanent, Cooke brought in John Carmody, from the National Labor Relations Board, to be deputy administrator. A fellow scientific-management expert whom Cooke had known while president of the Taylor Society, Carmody had much experience in industry, professional organizations, and government. He had worked for two decades in engineering management in the coal, steel, and garment industries. As chief engineer of the New Deal's Civil Works Administration and the Federal Emergency Relief Administration in 1933 and 1934, he had learned about the problems of rural electrification. Carmody's letters reveal that he held a much stronger public-power position than Cooke. He was also a better manager than his boss, whose strengths lay in innovation and promotion. The REA, which had proceeded rather cautiously under Cooke, gained drive and momentum under the forceful Carmody.[55]

Whereas Cooke and Fisher had encouraged the (admittedly problematic) state organizations in order to get the program off the ground, Carmody disliked them intensely. After the sixty-five-year-old Cooke resigned as administrator in favor of Carmody in early 1937, the younger man took on the farm bureau statewides. Fisher paved the way in the spring when he said that the "statewide organizations . . . ought to take a back seat before long" because some of them were "demanding pure gold, and giving the cooperative back very small change," a reference to their fees for promotion, management, and engineering.[56] Finding out that the Iowa Farm Bureau had created a subsidiary to provide engineering services to the state's co-ops, Carmody wrote one co-op that he was "fearful of anything that resembles a holding company arrangement in view of the long struggle the American people have had to break the shackles of the utility companies." Because Carmody knew the REA had asked the Iowa Farm Bureau to "help get the program under way," he approved the engineering contracts but assigned them to the Farm Bureau instead of to the subsidiary.[57] By early 1938, Fisher thought the formation of an independent, REA-approved Iowa Statewide Association had signaled to "our Iowa projects the end of the State Farm Bureau influence upon rural electrification."[58] In Indiana, Carmody bailed the farm bureau out of a crisis by loaning it engineering staff in late 1937, then challenged the bureau's monopoly. He supported one co-op's right not to purchase management services from the statewide and approved an REA loan to an independent co-op.[59]

Carmody also balked at reimbursing state organizations for developing co-ops. Cooke had said that Murray Lincoln's organization could collect a membership fee of five to ten dollars and use the money to "make the showing which R.E.A. demands"; that the Farm Bureau could obtain quantity discounts on construction materials it resold to contractors as long as it passed the savings on to the co-ops; and that the bureau would receive 4 percent of construction costs as an "overhead" fee. Lincoln complained that at this rate his organization would not recover what it had already spent on promotional costs.[60] When Carmody took over the REA, he told Lincoln and state organizations in Indiana, Iowa, Illinois, and Minnesota that the REA's statute did not permit it to pay promotional fees. The debate over this issue in Ohio became so intense that Lincoln got out of the rural electrification business in the fall of 1937 because the engineering and legal fees paid by co-ops did not cover development costs. The REA tacitly agreed to allow the Ohio Farm Bureau to raise its engineering fee by 1 percent, which cut its losses almost in half.[61]

Carmody reorganized the REA to reflect his policies. In October 1937, Fisher became head of a new Division of Operations Supervision, which maintained tight control over the affairs of co-ops. More field staff were hired to monitor, advise, and—in some cases—temporarily run co-ops. By the end of fiscal year 1937, the REA staff had grown to nearly five hundred people.[62] In assuming the (euphemistically named) function of "local guidance to borrowers," Carmody made a policy decision that ranked in importance with

Cooke's decisions to make the REA a lending agency and to rely on co-ops.[63] Ironically, in taking over the functions of the statewides, the REA performed engineering, management, legal, and financial services similar to those of the utility holding companies Carmody despised.

Carmody borrowed another tactic of the utility companies in developing a strong mediating network to sell electricity to consumers. He came to understand the importance of using home economists to develop the domestic market because he realized that co-ops would not be able to pay off their loans in time if farmers purchased only "lights," irons, and radios, which they tended to do. Cooke had recognized this problem early in the program when he told the USDA that each farmer "in addition to his lighting and small appliance loads must use one of the three major household appliances, either a refrigerator, a range, or an electric water heater, or a milk cooler, feed grinder, utility motor, or some other piece of farm equipment of comparable current-consuming capacity." Cooke told Murray Lincoln that the REA was to "come in at this point [after line construction] and lure the customer into not being satisfied with a single drop light or any other superficial use of electricity. This thing will not succeed unless we can fairly promptly build up a high average use."[64]

The Utilization Section formed the basis of the REA's consumer network. Established in the fall of 1935, it was headed by Emily KneuBuhl, executive secretary of Business and Professional Women's Clubs of New York City and a political colleague of Eleanor Roosevelt. Paid the same salary (six thousand dollars per year) as Fisher, KneuBuhl was the highest-ranking woman in the REA. She and her small staff visited field sites and argued for more and better-trained field-workers to cooperate with the USDA's Extension Service. Within a year, her professional staff had grown to seven people. Two field men assisted cooperatives in merchandising appliances, Dora Haines did special field studies, and a husband-and-wife team operated a demonstration farm near Washington, D.C. Oscar Meier and Mary Taylor were hired to work with the Extension Service in its gendered categories of agricultural engineering and home economics. Taylor had the unusual distinction of being a woman with a bachelor's degree in electrical engineering, but her gender still landed her in the household sphere.[65] Despite the Utilization Section's small staff, Cooke began to worry about the increasing emphasis on "load building" (that is, encouraging consumers to use more electricity). In late 1936 he warned about the dangers of copying the "private power industry in unduly encouraging load-consuming devices to the exclusion of those which have social implications far beyond their power to consume current."[66]

Cooke's injunction fell by the wayside when Carmody took over in early 1937. The REA then emulated the utility companies by embarking on an aggressive load-building campaign to sell large home appliances and electrical farm equipment. The Utilization Section became the Utilization Division in April, when KneuBuhl was replaced by George Munger, commercial manager of the Electric Home and Farm Authority, a government agency first set up in the TVA area to loan money for household wiring and the purchase of appliances. Formerly sales manager for a gas and electric company in New York, Munger brought in Clara Nale, the TVA's chief home economist, to head the home economics section in the REA. Trained in extension work at Auburn University, Nale was well respected by Mary Taylor, who had visited the TVA. KneuBuhl took what turned out to be a permanent leave of absence in the summer of 1937.[67]

Munger and Nale hired eighteen field representatives and divided them into six regional groups, each having an agricultural engineer, a utilization representative, and a home electrification specialist, to cover the entire country. A large number of these "agents of modernity" had worked for the Extension Service. Oscar Meier had worked on rural electrification as a county agent, and Nale had come to the TVA after a stint with the Alabama Extension Service. Of the field personnel, Elbert Karns had been an Extension Service engineer with the University of Arkansas, and James Cobb had held a similar position in Louisiana. Oneta Liter had headed home economics departments at the University of Louisville and the Tennessee College for Women, and Elva Bohannan had been a home-demonstration agent with universities in Alabama and Maryland. A member of the USDA's Bureau of Home Economics commented that Nale had been able to hire nearly all of the nation's home economists who had extension experience and technical training in electrical household equipment.[68] Another indication that the REA had borrowed a page from the utility companies was that one of the utilization representatives, Lee Prickett, had been assistant director of the CREA.[69]

When the federal government was reorganized in mid-1939, the REA was transferred to the USDA. Norris and other friends of the agency fought for its independence—to no avail. Carmody resigned and accepted a position as head of the new Federal Works Agency. A public-power crusader who sought to break ties between the Extension Service and the utility companies, Carmody apparently thought the REA would lose too much autonomy under the USDA.[70] President Roosevelt offered the REA post to a young public-power congressman from Texas—Lyndon Baines Johnson. Adept at getting government projects for his district, Johnson was extremely proud of his efforts in bringing rural electrification to his constituents. He had developed a "little TVA" on the lower Colorado River to supply cheap power to a gigantic, low-density cooperative he had convinced the REA to fund by pleading his case directly with Roosevelt. But Johnson turned the REA job down to remain in Congress, a popular decision with the folks back home.[71] Harry Slattery, undersecretary of the Department of the Interior, a conservationist who had served under Pinchot, a public-power proponent, and friend of Cooke, took over the reins of the REA in the fall of 1939. Cooke was "dee-lighted."[72]

By that time, the agency had succeeded beyond Cooke's dreams. It had built 180,000 miles of lines that brought central-station electricity to more than four hundred thousand consumers. The vast majority of them were farmers served by more than five hundred REA cooperatives (see table A.9). The agency was a success, but creating federally funded cooperatives was a contentious matter—both for the REA and for the farm family.

CHAPTER 6

Struggling for Local Autonomy

When the Rural Electrification Administration made the decision to use cooperatives to electrify the farm, it embarked on an extensive experiment with a new form of cooperation in American agriculture. Although producer and consumer cooperatives had received advice and encouragement from the U.S. Department of Agriculture since World War I, none of these efforts involved federal financing—and the governmental oversight that went with it. Farm co-ops typically obtained their capital from membership fees and, increasingly, from seed money from farm organizations, not from the federal treasury. Instead of being run locally (like the telephone cooperatives) or operating under the guidance of county and state groups (like the Farm Bureau and Grange cooperatives), the REA co-ops were under the constant supervision of a federal agency.[1]

This supervision, which had increased during John Carmody's term as administrator, was a well-established policy by the time the REA was transferred to the USDA in 1939. The agency's chief legal counsel explained the situation.

> Owing to the character of 95 percent of our borrowers and their lack of past experience, very close supervision during construction is necessary. This has extended into the early days of operation and concerns retail rates, form of set-up under by-laws, management personnel, accounting systems, and utilization of electricity. As they get further into operation the problem is one of a proper balance between a desirable local autonomy and the help that we should give in protecting the government loan and guiding a program of national policy.[2]

Achieving this balance was difficult, especially when the agency felt besieged by the utility industry. The REA co-ops did not gain much autonomy until after World War II, when the program expanded dramatically, loans were paid down, and the passions of the public-power debate had subsided.

The early history of the REA co-ops is the story of two struggles: one between public and private power, with the agency fighting the spite-line wars against the utility companies, the other between local autonomy and federal supervision. The first struggle, which has been chronicled by commentators then and now, has tended to overshadow the second battle and to perpetuate the myth that the REA, its co-ops, and rural people worked in harmony against the "power trust" to modernize the farm. This chapter is concerned with local struggles for autonomy—how farm people interacted with Washington and its field agents to mutually construct the basis for rural versions of electrical modernity.

The cooperative was the main site of this (not so "cooperative") battle. During the long process of establishing a running concern, the REA supervised the organization, design, construction, and operation of co-ops. Field representatives explained the program to local leaders, encouraged them to form a board of directors, helped them select an attorney and project superintendent, and told them how to conduct membership drives, make project maps, submit loan applications, and write bylaws that satisfied state laws. The engineers at the REA helped select project engineers and contractors, approved the design of the system, and monitored its construction. Field representatives made sure the co-op used state-approved wiring inspectors. The REA taught co-op personnel standardized accounting and management techniques at regional conferences and monitored accounts, operational reports, and minutes of meetings of boards of directors. Agricultural engineers and home economists worked with co-ops to sign up more members and persuade those with electricity to use more of it by buying appliances.[3]

Rural people did not readily adapt their local cultures to the technocratic plans emanating from Washington. At the state level, farm bureaus tried to run away with the program, and many extension services resisted it. Locally, the REA complained that boards of directors often ran co-ops like private fiefdoms and refused to follow recommendations on how to operate the co-ops. The REA criticized staff and members of co-ops for not showing a cooperative spirit in matters ranging from granting rights-of-way for power lines to attending annual meetings.[4] Many of the REA's complaints against boards of directors indicate a struggle for local autonomy similar to what occurred with the telephone, automobile, and radio. Resisting REA recommendations, not granting easements, refusing to use more electricity—these are actions taken against the imposition of an alien culture. As in the case of our other technologies, rural people used age-old resistance tactics to "domesticate" the REA. They were able to balance local and federal control to create a framework for rural modernities based on electrification.[5]

Creating the REA Cooperative

Boyd Fisher's Development Division took the lead in working with state, county, and local leaders to organize cooperatives. Tensions between the Farm Bureau, the Extension Service, and the REA complicated this effort, especially before the REA became part of the USDA. Behind Carmody's fight against the statewides was his belief that the Farm Bureau and the Extension Service had not severed their ties with the utility industry. As we saw in the previous chapter, Carmody knew that the Committee on the Relation of Electricity to Agriculture (CREA) had been headquartered in the office of the American Farm Bureau Federation. He never forgot that it influenced state extension services by funding research in agricultural colleges throughout the country.

Evidence was not lacking of a continued influence after the REA was established. In late 1936, a member of Fisher's staff charged that "substantially all (the writer knows of no exception) of the State Agricultural Colleges are under obligations to the Private Utility Interests either for actual money contributions to the college or one or more members of its faculty, or for the customary favor of taking a large number of graduates into their employ." An REA field representative concurred: "In many States I found the Extension Service completely tied up with the power interests." At one college, utility companies subsidized a department of electrical engineering and told it not to cooperate with the REA. Another field representative reported that 1 percent of the Alabama Extension Service's funds came from power companies, which "was true in a number of States."[6]

These ties usually worked against the REA. In Illinois, Emil Lehmann, the agricultural engineering professor who worked for utility companies and the CREA, reminded county agents at a meeting in 1936 that they had no authority to organize REA co-ops unless they first cleared it with the university, a policy set by the director of the U.S. Extension Service. Fisher lamented that the "Illinois situation is the most disagreeable situation in the country relative to REA." Field reports showed that "our nominal collaborators, the Farm Bureau and the Extension Service, are openly and flagrantly carrying the ball for the power companies."[7] By the end of 1936, Fisher found that "utility-minded" professors at agricultural colleges had also hindered or refused to help organize REA co-ops in Iowa, Kansas, Michigan, Oklahoma, Texas, and Virginia.[8]

The most embarrassing situation, though, was within the REA itself. At the request of the secretary of agriculture, Morris Cooke agreed early in the program to hire someone to work with state extension services. Cooke settled on Professor David Weaver, who had made a state survey of rural electrification as the head of the agricultural engineering department at the University of North Carolina. Hired in early 1936, Weaver gave talks at state meetings and wrote at least one article on the REA. According to Fisher, Weaver was soon recalled by the state extension director. Weaver declined to head the state's CREA, but he discouraged the formation of REA co-ops. Later, the head of the state's farm bureau confronted Weaver, in front of Fisher and a Grange leader, with evidence that Weaver had been on the payroll of the Carolina Light and Power Company while working for the REA. Cooke wrote Weaver for an explanation in late 1936 but received no response.[9] Thereafter, the REA staff, which was already highly suspicious about the underhanded propaganda methods of power companies, double-checked the background of all engineers with whom they did business.[10]

Many agricultural engineers did help the REA. When the revelations about Weaver prompted Fisher to ask for reports from the field, his staff listed friendly extension engineers in Arkansas, California, Kentucky, Missouri, Nebraska, New Mexico, Oregon, and Washington.[11] Arkansas engineer Elbert Karns was hired by the Utilization Division, which began working with state extension services. Fisher condoned that policy as long as the division's Oscar Meier, a former county agent, and others friendly to the Extension Service confined their work to utilization. Changes were also in the wind. Meier reported in 1937 that a USDA committee had recognized the "unhealthiness" of the past arrangement whereby the CREA funded most rural electrification research in agricultural colleges. He thought the REA could use the committee's report "to good advantage in those states where too great an affinity seems to exist between the agricultural educational institutions and the utility."[12] The next year Fisher even made up with Professor Weaver. Because Weaver and the North Carolina Extension Service were now helping the REA with utilization work, Fisher accepted Weaver's offer to let "bygones-be-bygones."[13]

The REA's strained relations with the Extension Service can be seen in New York state, whose story is more complex than the mere conflict between public and private power. For more than a decade, from 1924 to 1936, the Empire State Gas and Electric Association, with the advice of the CREA, funded research on rural electrification at the New York State College of Agriculture at Cornell University. Moreover, the Federal Trade Commission's investigation of the utility industry revealed that the association had paid travel expenses for agricultural engineering professors in the college to give presentations at one of its conventions.[14] After the REA was created, the association, Cornell's colleges of agriculture and home economics, and farm organizations joined forces with the state's Public Service Commission to run their own rural electrification program from the college. The leaders of these groups opposed the long-term central-planning aspects of New Deal agencies, which they considered to be economically unsound and insidious forms of government control. Another factor was the continued research funding provided by the Empire State Gas and Electric Association to the agricultural college. When asked why New York farmers did not form REA co-ops, Lincoln Kelsey, the coordinator of the Cornell-based program, told his colleagues that an alumnus on the Public Service Commission agreed with him that REA co-ops were "unnecessary" and "illegal" in New York.[15]

Yet in the summer of 1936, when Kelsey was worried about the financial ability of the New York State Electric and Gas (NYSEG) company to build rural lines, he turned to the REA—in a telling fashion. Kelsey wrote the dean of the college of agriculture that "we may have to set up some corporations which are not technically cooperatives under our State Laws but which will pass with the Rural Electrification Administration as cooperatives, then go after some of their money, of which about a half million dollars is allotted to us under the 1936 Act."[16] The phony corporations were not set up, but the REA did loan a substantial amount of money to NYSEG for rural-line extensions. Fisher seems to have agreed with this policy. When an electrical contractor attempted to form a co-op in Sullivan and Ulster Counties in mid-1936, Fisher asked Cooke, "Do our good will arrangements in New York make it advisable to attempt to throw this business to one of the private companies in that area?" Only in the early 1940s, when the REA became dissatisfied with NYSEG's rural programs, did it establish a small number of co-ops (six) in the state.[17]

At the local level, many county agents opposed the REA, whether by following the policies of their employers—state extension services and farm bureaus—or for more personal reasons. The state farm bureau in Vermont supported a private utility plan. In Illinois, many county agents agreed with the state extension service that the REA was just another "New Deal Dream" and

worked with the power companies. Kansas agents disrupted the development
of a co-op in Jewell County by spreading false information; they opposed another co-op near Topeka by employing a "passive or 'do-nothing' form" of
resistance.[18] A few county agents in Iowa, Kentucky, and Texas spoke against
the REA at organizational meetings or remained aloof. Carmody recalled that
some agents helped survey co-op projects, then turned the maps and rate
schedules over to private utility companies to build spite lines, a practice that
was apparently prevalent in North Carolina.[19]

But hundreds of county agents supported the REA. Agents organized
co-ops not only in Ohio and Indiana, where the farm bureau statewides ruled,
but in Arkansas and Iowa, where the state farm bureaus adopted the REA as
a regular project. Agents were the mainstays of the program in Georgia and
Texas.[20] Although several Kansas agents opposed the REA, at least seventeen
others helped organize twelve co-ops in that state from 1938 to 1947. Part of
the reason may have been Carmody's efforts to negotiate a truce with the
state's extension service in early 1939.[21] At least three Illinois agents defied the
university and contacted the REA to organize co-ops. Once co-ops were established, the state farm bureau got into the act by providing engineering
services.[22] At one Iowa meeting early in the program, the "County Agents
were very enthusiastic and all of them expected to begin making the necessary
surveys and appointing the committees of farmers" to organize the co-op. In
mid-1937, Carmody told the director of the U.S. Extension Service that the
earlier "hostility" to the REA by agricultural colleges "has faded in a good
many areas. . . . In some places, County Agents, for instance, are the most aggressive promoters of cooperative electrification." Later in life, Carmody said
of an agent who had organized fourteen co-ops in Iowa that he "richly de-
serves a monument." In 1938, Lyndon Johnson was so impressed with the success of one county agent in signing up his constituents for the proposed Pedernales co-op that he told him to go to one of the best stores in Austin, buy
himself a new Stetson hat, and "send me the bill."[23]

The REA grew closer to the U.S. Extension Service after the agency became part of the USDA in 1939. Administrator Harry Slattery reached an understanding with C. W. Warburton, the new director of the service and an old
friend, that the two subagencies would cooperate. Oscar Meier was appointed
the official liaison in mid-1940. Within a year he had worked out joint agreements between REA and state extension directors in South Carolina, Georgia, and Oklahoma for sharing expenses to hire an extension rural-electrification specialist in each state. Selected by the state extension director and
approved by Meier, the specialists worked with county agents and farm

groups to develop REA projects and increase the utilization of electricity. In keeping with the new spirit of cooperation, a survey of extension service directors in late 1940 deemphasized their criticisms of the REA to report that farm and home county agents had helped develop co-ops in thirty-four states since the beginning of the program. Slattery tried unsuccessfully to get Boyd Fisher, who had been replaced as division head in 1938, on board with the new program. Although Fisher had made up with Weaver in North Carolina, it is not surprising that the declared enemy of extension services and critic of Oscar Meier left the REA in 1940.[24]

Grassroots Resistance

The REA spent a good deal of time negotiating with federal, state, and county agencies to develop co-ops, but community leaders and local workers were crucial to the success of the program. The historian of the REA in Louisiana concludes that county agents "were not much more than administers of information. The real initiative came from prominent local citizens." Farmers, wealthy landowners, and businesspeople worked with county agents to organize Louisiana's thirteen co-ops. The chief promoter in Otter Tail County, Minnesota, was Albert R. Knutson, who had graduated from an agriculture college and had worked as a county agent. Returning home in 1929 to run the family's dairy farm, he tried to organize an electric co-op. Once the REA was established, Knutson worked with a local attorney and an electrical engineer to convince farmers to establish the Lake Region Cooperative Electrical Association. After many delays, Knutson finally got electricity for himself and his neighbors in November 1938. Outside of Lyndon Johnson's district, the state Grange was the catalyst for forming co-ops in much of Texas. Local Grangers, however, did the footwork and founded such groups as the Guadalupe Valley Electric Cooperative in December 1938.[25]

Their job was made easier by the high demand for rural electrification. The REA reported in October 1935 that President Roosevelt and Eleanor Roosevelt had received hundreds of letters from farm people, especially women, demanding electricity. Mrs. John B. Merritt, a "rural citizen of the state of Alabama, Chambers County," wrote Morris Cooke early in the program, saying her house had been wired when it was built in 1929 in anticipation of getting electricity from a utility company's line located a half-mile from the house. Her family did not hook up because of the high cost ($17.50 a month). The REA seemed her best hope. "Can you help us," she asked Cooke. "This is an S.O.S. call. We want power." An Indiana woman was heard to exclaim at an REA meeting, "Daddy had better sign up or things are going to get hot around home. If there is any chance of getting electricity I am going to get it. This is one time when he is going to make up his mind in a hurry." An Iowa man told the REA in 1936, "We farmers in Cedar County want and need electricity. Electricity is something I have been dreaming of for a long time."[26]

But when REA canvassers solicited members for the local co-op, many farm people resisted the new technology, just as the previous generation had initially opposed the automobile. The first point of resistance, which was widespread and not limited to one region or economic class, was the farm-to-farm membership campaign. An organizer in Arkansas recalled that "about half of them would sign up and the other half wouldn't. They would say, well, they might not get their five dollars [membership fee] back" if the co-op failed. Many Illinois farmers refused to sign when a co-op required that they install two outlets in the kitchen and one in each room. "Who ever heard of putting electric lights in the bedrooms?" one woman asked. "We can carry a kerosene lamp to our bedrooms and cut down on the cost of electricity." A project attorney in Washington state advised the REA that "these people all want electricity, but some of them are luke warm on the idea, and a membership fee would scare them off." In these last years of the Great Depression, a minimum monthly bill of two to four dollars was too high for a large number of cash-poor farmers. During the organization of the Texas co-op founded by Lyndon Johnson, only three of the twenty-one families in one community signed up in 1939 because of the monthly minimum of $2.45. An organizer wrote Johnson that the "other eighteen stated they wished the lights and would sign for a minimum of 1.50 to 1.75 per month for twelve [kilowatts]. There are about three hundred signed in this [county] where there should be a thousand or twelve hundred." Economic considerations were surely a contributing factor: eggs, for example, were selling for only thirteen cents a dozen. Many people feared losing their farms if the co-op folded (a fear fostered by the private utility companies), some had lost money in disreputable co-op schemes, and most farm owners were extremely reluctant to sign up for their tenants. A canvasser in Iowa recalled that owners "took a rather negative attitude for they had milked by lantern light and read their newspapers by kerosene lamps so why should their tenants get the luxury of electric light and power." In 1941, some prosperous farmers in western Kansas were waiting until their home plants wore out before joining the co-op.[27]

Although most of the sign-up resistance seems to have been motivated by economic concerns, there were other factors. An REA staff member reported in late 1937 that "many projects do not serve 30 to 50 percent of the people along their lines merely because the people do not feel that electricity is worthwhile to them even though they can often well afford it." Obtaining other technologies seemed more important. An Iowa organizer heard lots of farmers say, "Listen, I would rather give you ten dollars to help bring us graveled roads than to become a member in an electric cooperative that may never get started. We need the graveled roads worse than we do the electricity."[28] Other farm people (especially McKinley Republicans in the Midwest) did not trust a New Deal agency, many thought the simple scheme of paying a co-op membership fee to get electricity was too easy, and the older generation often resisted giving up familiar technologies. Some would not sign a membership application for religious reasons. More than a few farm men and women were afraid of an invisible force that was still considered mysterious in the preelectric age. Some thought power lines would attract lightning. Illinois and Texas farmers worried that lines would fall across barbed-wire fences and electrocute their cattle.[29]

The fears were not groundless. The *Rural Electrification News* devoted a whole issue in April 1939 to electrical safety, with articles on caring for household appliances and "accident prevention on REA projects." Unfortunately, the articles did not help four co-op linemen in Kentucky, Michigan, Illinois, and Nebraska who were electrocuted while working on energized lines that August. Three of them were killed because they did not wear rubber gloves; the other lineman wore rubber gloves, but his arm touched the "hot" case of a faulty transformer. The REA advised farmers and co-ops of the dangers, but problems persisted. A Texas co-op warned its members in 1941 "of the danger in flying kites near electric lines, as a damp metallic string will conduct enough current instantly to cause death. Never let any kind of metallic string or any kind of wire touch the high lines!! *All children should be warned of this danger.*" The REA reported that the number of disabling accidents per mile of line had been cut nearly in half from 1939 to 1941. The *Rural Electrification News* ran a short article in late 1943, entitled "Let the Co-op Fix the Lines," in a further attempt to prevent accidents when farmers repaired the lines, a tradition that probably carried over from the self-maintenance of cooperative telephone lines. The article described the case of a southwestern farmer who tried to replace a blown-out fuse in a 7,200-volt transformer on a pole. When his hand came into contact with the high voltage, the "current raced through his body from his left leg into the transformer case. It seared a path from his right leg and foot to all three wires on the secondary side." The farmer fell to the ground, which revived him, but he died before the doctor arrived because his neighbors did not know how to apply artificial respiration. The lesson REA wanted members to learn from this example was to stay away from the lines and to learn first aid. In 1949, despite all the REA's precautions, a high-voltage line fell across a fence in Montana, electrocuting a hired hand who unsuspectingly opened a metal gate connected to the (now electrified) fence. News of the accident caused widespread alarm among co-op members.[30]

In addition to not signing up for the co-op, large numbers of farm people engaged in a more active form of resistance by refusing to grant rights-of-way (easements) for constructing power lines across their land. The nonprofit co-ops enjoyed many advantages over private companies—including exemption from taxes and the prescriptions of regulatory commissions in most states—but they lacked the ability to condemn property to obtain rights-of-way. John Carmody told a newly incorporated co-op in 1938 that after organizing about four hundred projects the REA had "learned that one of your biggest difficulties will be getting rights-of-way for your lines. Everybody says he wants electricity, but when it comes to locating the lines and locating the poles, many people either refuse to hand out essentials, thereby denying their neighbors electricity, or make it so difficult and so costly that certain lines cannot be built at all."[31] Farmers expected the government to pay them large sums for easements; lines crossed farms in ways that renewed family quarrels; and poles interfered with the plowing. One Iowa woman got upset with her husband for granting an easement after holding out for six months. An organizer recalled that the man "stamped his foot and shouted 'Now, mama, it's not you who is going to farm around those REA poles, it's me.'" In a Texas household with a different division of labor, a co-op director did not want poles and guy wires placed in his fields because it would interfere with his wife's plowing.[32]

Violence was not uncommon. An Illinois woman guarded a freshly dug hole with a shotgun to prevent the REA from completing the pole-setting job; Iowans tried forcibly to prevent REA crews from erecting poles along a highway; and a Minnesota man said he would chop down poles if they were put on his land. A farm couple carried out the threat in Monroe County, Wisconsin. When the co-op put up poles on their land without obtaining an easement, the husband chopped them down, parked the family car on top of the (unelectrified) fallen wires, then brought food to his armed wife stationed inside the car to keep the REA at bay with a shotgun.[33]

Farm people more enthusiastic about the REA—those who signed up for the co-op, paid their membership fee, and granted easements—still faced a number of problems in getting electricity. Delays of well over a year were common because of the long process of getting the government loan, hiring co-op personnel, and designing and constructing the system. A field representative reported that delays on an Oklahoma project were "creating dissatisfaction among the farmers, and in many cases making it impossible to get service contracts from customers, because they have no faith that the lines are going to be built." Evelyn Smith, daughter of a family in Lyndon Johnson's congressional district who had waited more than a year for power to be turned on, recalled that neighbors criticized her family for their "city ways" in helping to organize a co-op that had not delivered on its promises. Returning to the farm from Johnson City one night in late 1939, her mother thought the house was ablaze. "'No, Mama,' Evelyn said. 'The lights are on.'"[34] In more wooded sections of the country, some contractors engaged in what the REA criticized as "tree butchery" in order to run lines along the highways. Carmody sympathized wholly with the indignation of the governor of Illinois against contractors who "appeared to be ruthless in cutting away trees to set poles." Although Carmody said he would cooperate with the governor to save trees, he told a highway official in Ohio that "this is not possible where only tree-lined highways are available for line construction" because farmers would not grant easements across their fields.[35]

Wiring rural houses was a simpler technical and political task, but it had its share of difficulties. For those who did not have ready cash to pay the relatively high cost of wiring farmhouses, which were usually larger that urban dwellings, the REA arranged for loans, either from the local co-op or from the Electric Home and Farm Authority. But that did not guarantee smooth sailing. The building of hundreds of co-ops created an unprecedented demand for wiring and a shortage of qualified electricians who could pass the state inspection codes. J. Warner Pyles, an REA field representative, observed (rather condescendingly) in 1939 that early in the program, "everybody thought they knew how to do wiring when the REA project came into the territory. That meant turkey-raisers, farm hands, and manure throwers, so from that we have all kinds of prices for wiring, ranging from fifty cents an outlet to five dollars an outlet. The farmers became confused, and did not know who to give the wiring to, with the result that nobody did any wiring." Another field representative asked in 1937, "What can be done to overcome inspectors' tendencies to condemn knob and tube wiring [the system of running parallel wires across walls and ceilings] which has been serving farmers with battery lighting systems for from one to twenty years?" Pyles thought his wiring program would solve these difficulties, but it also required that three-fourths of signed-up members wire their houses before line construction could begin.[36]

These requirements point to yet another form of resistance to the REA: co-op members reneging on their commitment to connect to the lines after they were built. The practice was widespread early in the program. Thelma Wilson, one of the first field home economists hired by the REA, reported in late 1937 that "on every Project I have visited [in Indiana, Ohio, and Pennsylvania] it has been found that the number of customers taking service is below the figure upon which the Project was approved."[37] Although projects experienced this problem to varying degrees, data collected between 1937 and 1939 indicate that many co-ops, in all regions of the country, were particularly hard hit by farmers who joined the co-op but did not take electricity. In the Midwest, barely one-third of the members of a two-year-old Indiana co-op had connected to its lines (422 out of 1,215), and two years after energizing its system an Iowa co-op had connected only about one-half of those who signed up (1,970 of 3,928). In the South, a co-op in Alabama was serving about the same percentage of members (538 of the 1,000 required by the loan contract), while a Georgia co-op was in slightly less trouble at about a 60 percent connection rate a year after power was turned on (1,496 of 2,503 members). Further west, a Texas project fared a bit better by hooking up two-thirds of the farms that joined the co-op (about 800 of 1,200), but an Idaho co-op had connected only one-half of its members a year after completing its lines (125 of 248).[38] The problem was so worrisome in the Steele-Waseca Electric Cooperative in Minnesota that the co-op's newsletter asked members in October 1939 to "try and get that neighbor of yours 'WHO IS STILL OUT IN THE DARK' in this winter. When the REA was started, you were told over and over again that the low rates you now enjoy are ONLY POSSIBLE with three users to a mile and an average of one hundred kilowatts being used each month per member." This co-op had fewer than two farms to the mile, and its members averaged only sixty-four kilowatt-hours per month.[39]

Mary Taylor, the electrical engineer working as a field home economist for the REA, asked, "Why do they not put in electricity? What can be done to get them to connect to the lines?"[40] The questions were becoming familiar in the REA. As in the case of farm families not signing up for the co-op, economics seemed to be a major factor when members decided not to wire their houses and not to hook up to the lines. Two co-ops in Texas and New Mexico said their minimum monthly bills ($2.50 and $3.00, respectively) were too high for many farmers. Landlords would not wire tenants' houses. An REA field representative reported that some farmers in Indiana "have become members of the cooperative to get service by their farms and thus increase the value of the property with no intention of taking service."[41] There were fac-

tors other than money. Members of the Harrison County Electric Cooperative in Iowa postponed wiring their farms because of delays in getting the
REA loan. Several California farmers would not hook up as long as a project
superintendent, known for his wheeler-dealer activities in the county, was involved with the co-op. A project superintendent in Georgia explained his
poor hook-up rate by telling the REA that "our rural people are skeptical and
need to be coaxed along."[42]

Even some board members needed to be coaxed to hook up to their own
co-ops. A field representative observed in 1937, "When board members are
encountered so frequently who don't have their own premises wired or who
think electric ranges are foolish luxuries, it seems that something should be
done." When an Iowa board member did not connect in 1940, "in spite of the
fact that the line goes by his premises and has been energized for several
months," the REA reminded the board that its bylaws mandated that every
member must buy electricity from the co-op. The REA suggested that the
board member should resign if poor finances prevented him from taking electricity. The practice was common enough for Udo Rall to include the following admonition, under the entry, "Board," in a dictionary of REA cooperative
terms, which he wrote in late 1938: "No one should be on the board who does
not or cannot take electric service from the cooperative when it was available."[43]

Another widespread and serious problem occurred when power lines
were turned on. Much to the chagrin of the REA and the disgust of farm
people, the same electric current that coursed through high-lines along the
highway or across fields to bring the promised light and power to the farm often induced a loud "hum" in nearby telephones. A federally sponsored network thus played havoc with older local networks and caused a modern "range
war" of sorts between the two technological systems. In Alabama, nineteen
telephones along twenty miles of the Coffeeville Telephone Company's lines
were rendered useless when the Clarke-Washington Electric Membership
Corporation energized its lines in 1937. The owner of the telephone company,
C. J. Clanton, whose sister operated the switchboard out of their house, could
not afford to pay two hundred dollars to metallicize the line (adding a "return"
wire in place of the common practice of using the earth as the "ground" wire
on rural systems). When Clanton and his father refused to grant rights-of-
way to the electric co-op for its line extensions, the REA agreed to metallicize
the telephone lines in exchange for the easements.[44]

But the REA usually held fast to a policy drawn up by its legal department, which dictated that as long as the agency was not technically negligent,

it would not pay to modernize or relocate telephone lines. Although individual cases were not expensive, the cost of fixing the lines of several hundred telephone companies was too much. Carmody was adamant about the matter, saying he did not have a "moral or legal right to saddle farm people with debt or expense to modernize obsolete telephone lines." The REA's position led it to fight numerous battles in local, district, and state courts for nearly a decade. In Clovis, New Mexico, a District Court judge ("a delightful old gentleman who took snuff from a silver box, commented volubly, and frequently quoted Shakespeare—a scholar, but not a legal one," according to an REA lawyer) ordered in 1938 that a co-op pay two small telephone companies $783 to metallicize their lines. The judge thought the matter should have been settled locally rather than by "higher-ups" coming out from Washington. He then "renewed his allegiance to the New Deal and its principles, but added that the New Deal would have to comply with the law in New Mexico." The REA appealed, as it did in several other states. By 1941, it had won cases in most states and had seen the decisions upheld by state supreme courts, except in Iowa, Nebraska, and North Dakota, where telephone cooperatives were a strong political presence. In those states the REA usually managed to get waivers from telephone companies. Yet the Glidden Rural Electric Cooperative in Iowa did not resolve the matter until a "stubborn Englishman" was replaced as secretary of the telephone cooperative.[45] The REA won the telephone-interference war co-op by co-op and state by state, but the victory probably contributed to the demise of rural telephony in the 1930s.[46]

Electric lines interfered with another network technology—the radio. But in this case it was the REA that was usually at fault, both technically and legally. Unlike the steady hum induced in telephone receivers, radio interference hissed, whistled, and crackled much like the other "noise" rural (and urban) people often heard on the radio, making it difficult to locate the source of trouble. A common procedure was for linemen to patrol the system in an automobile, using a portable radio to locate problems (faulty transformers and insulators or loose tie wires that could cause the high-frequency interference). Specially designed locators with directional antennas were available in 1940. A co-op operations memo instructed linemen, "When the suspected pole is located, strike it with a heavy hammer and note any momentary interruption in the interference." An interruption marked the source of the interference. The REA adapted another familiar rural technology, the phonograph, to the task by using special recordings to help troubleshooters distinguish between noises caused by power lines and those caused by faulty home appliances.[47]

The REA program was thus shaped by the agency's response to the passive (and sometimes active) resistance of farm people and its efforts to solve the manifold technical and legal problems of electrifying the countryside. In conducting sign-up campaigns, securing easements, preventing accidents, wiring houses, and reducing telephone and radio interference, the REA created a culture of close government supervision of co-ops, which began to serve thousands of well-to-do farm people in the late 1930s. Of course, the power relations were not evenly distributed. Before energizing the lines of a co-op, the recurrent conflict between federal supervision and local autonomy seems to have been usually resolved on the side of the REA. Besides holding the upper hand in regard to technical, financial, and legal expertise, the agency could (and did) play its trump card by withholding allotments if its requirements were not met. But the operation of several hundred supposedly democratic cooperatives presented a set of personnel and management problems that threw into sharp relief the issue of local autonomy and helped increase the influence of local people.

The main point of contact between the REA and its co-ops in this regard was Fisher's Division of Operations Supervision. Established in late 1937, the division was created to provide the type of oversight Carmody, Fisher, and others thought was necessary to make co-ops succeed. In defending this function, Fisher said in an internal memorandum, "the REA Act clearly contemplates, and the REA program demands, not merely lending money, but the *supervision* of the expenditure of REA funds and the *supervision* of the property in which the funds are invested, and the *supervision* of the agencies operating those properties until the assurance of the repayment of the Federal funds invested in these properties is made reasonably if not doubly sure." The division's functions included assisting in the approval of all allotments, suggesting the consolidation of projects, initiating the selection of boards of directors and all co-op personnel, prescribing the duties of those employees and training them, monitoring the board's activities, auditing project accounts, and authorizing the turning on of power to projects. The division, in other words, closely monitored all operations of the co-ops and could even "send a Supervisor to take over a project temporarily."[48]

Many cooperatives did not like the division's management style. In early 1938, Fisher recommended that the REA create a pool of qualified personnel from which it could recommend project superintendents to co-ops before allotment. Thus, "we could start breaking our bronco while we have him in the corral, instead of carrying the saddle to him out in the open." A year later, Fisher told a staff conference that the building of spite lines through the richest territories by private utility companies meant the REA had to build lines

quickly "by the cooperative method before the farmers are prepared to operate the cooperative method. That is to say, we are building from the top
down."[49] This approach led to local opposition. Fisher acknowledged that co-
ops had complained about the REA's "dictatorship," had charged that it was
"setting up a bureaucracy permanently to supervise the operation of these
projects as if they were a Government enterprise," and had "got 'sore' at us" for
arguing whether a bookkeeper should be paid eighty dollars or ninety dollars
a month. The REA "will get off their backs," Fisher told a staff conference,
when the co-ops "can meet their amortization payments as an efficient business." Until then, "the farmers needed to be protected against themselves."[50]
Carmody approved of these statements and even ordered his staff to eliminate
the common phrase—"It is not an REA project. It is your project"—from official correspondence because it had come back to haunt him. He found that
"when some of our people ask the borrowers to do something that they normally would be glad to do they refer to this sentence which they got from me
and which proves conclusively to them that our field people apparently are
asking them to do things they do not need to do."[51]

A prominent site for these struggles over autonomy was the co-op's board
of directors. Often chosen by Fisher's representatives, state farm bureaus, and
other promoters, the initial board consisted of the incorporators of the cooperative. Ideally, once the sign-up campaign was completed, the membership,
exercising its rights under traditional cooperative principles, would elect a
new board. Often, however, the original incorporators remained in place, and
the board became self-perpetuating. Fisher complained in 1937 that after two
years of operation, no co-op in Indiana, Iowa, and Ohio—seat of the Farm
Bureau statewides—had called an annual meeting to elect a board of directors. The first complaint on a laundry list of co-op problems prepared by Dora
Haines and Udo Rall in 1939 was "self-perpetuation of virtually self-appointed
boards of directors." Although Rall agreed that the "situation indicates an unhealthy tendency," he could "see no possibility of attacking that situation directly without unfavorable reaction on REA itself." Only members could
change the picture; the REA could not "police every annual meeting to make
sure that the nominating and election procedure is carried on democratically
without any shenanigan[s]."[52]

What were shenanigans to the REA were, to some boards, weapons with
which to fight federal supervision and maintain local control. Field representative Elva Bohannan gave a detailed report of this phenomenon at the annual meeting of the Northwestern Rural Electric Cooperative, held in Cambridge Springs, Pennsylvania, in January 1941. Bohannan gave a short talk in

which she recommended that the co-op adopt utilization and cooperative education programs and elect some fresh faces on the board, including women. Then,

> Mr. Proctor, the chairman and President of the Board, leaped to his feet —screaming, "These accusations made by Mrs. Bohannan call for an answer." He grabbed the book "Rural America Lights Up" [and] screamed—"The only reason Washington wants to dictate to us about who shall stay on the Board of Directors is—This book was sent to us to *sell* (waved book frantically in the air) every member a copy of this thing. *We refused!* A man by the name of Harry Slattery wrote it." Then he looked at the book and stammered, "He's the administration." He accenticated [sic] every word with a pound on the table with "Rural America Lights Up." He further screamed. "Washington *Cannot Dictate* to us because we are men of our own opinions and none can change us." He claimed that Washington was not the friend of the members of the Cooperative because they now have on hand . . . [enough money to make an advance payment, but Washington] . . . "wouldn't take it. They want your interest payments." . . . He then screamed—"Do you know what Washington Demanded we do—? Why, stick a one-can milk cooler down your throats. But did we refuse—? I'll say we did." Then he rambled from one issue to another until 4:30 P.M.

By that time, most of the members had left the meeting and, after the project attorney ruled that the board had to be elected together, the old board was returned for another year. When the meeting adjourned, the attorney told Bohannan that she "had ruined the meeting—[that] I had told people things they weren't supposed to know. I reminded him that 'this is a Cooperative and to succeed they needed active support from the membership.' He stated 'This is not a Cooperative, it is a Utility Corporation.'" The REA's programs for low-income farmers, like the one-can milk cooler, were "just Washington's way of *hoodwinking* the *customers* into using [more] electricity."[53]

This case reveals the complexities of the issue of federal autonomy versus local control. Bohannan employed the rhetoric of rural uplift to complain that the co-op's board was under the domination of a president and project attorney who tried to thwart the REA's "progressive" measures of co-op education (e.g., by selling Slattery's book), democratic elections, and agricultural modernization (e.g., by selling the one-can milk cooler). President Proctor called on patriarchal populist rhetoric to resist the inroads of Washington and uphold an independent rural way of life; the project attorney cited "no-non-sense" corporate business principles. It is difficult to link this rhetoric to actions, but we should note that the criticisms of both sides had some basis. Proctor and the attorney seem to have controlled the board. Field representatives from the REA pushed Slattery's book to other co-ops,[54] and, as we shall see in the next chapter, they promoted appliances like the one-can milk cooler primarily to increase the usage of electricity.

The problem of one person controlling a co-op existed in several states. In 1940, Fisher reported that a Georgia co-op had been "virtually the private property of one man, C. V. Ellington, who was president, manager, and owner of the project all rolled into one."[55] But the REA seems to have looked the other way when influential Washington politicians were involved. The agency knew that Representative Lyndon Johnson, protégé of Senator Sam Rayburn who cosponsored the act creating the REA, was the guiding spirit behind the Pedernales Electric Cooperative. Johnson took a hand in every phase of organizing and operating the gigantic six-county co-op. He persuaded the REA (through President Roosevelt) to finance a system that had fewer signed-up farms to the mile than required, obtained cheap power from his own Lower Colorado River Authority, which also took care of the sign-up and easement problems, oversaw the building of a large headquarters building with funds from the Works Progress Administration, kept an eye on the co-op's activities (through regular reports from the co-op superintendent), and camped out in Carmody's office to lobby for line extensions.[56]

The REA tried to exercise its authority. Upon reading the Pedernales co-op's initial bylaws, Udo Rall noted in mid-1938, "it is obvious to me that this Board is looking forward to running things to suit themselves for at least a year. . . . Now is the time to show them that they can't get away with any monkey business." A year later a state legislator informed Johnson that farmers throughout the district were complaining that they had not been allowed to elect the board of directors, as promised. After Johnson turned down the appointment as REA administrator a few months later, in the fall of 1939, his involvement with the co-op raised no alarms at the REA. In fact, it became so commonplace that when the minutes of one board of directors' meeting in 1943 reported that Johnson entered the room and declared that new customers should not be required to make contributions, Rall simply asked an REA field representative what Johnson meant; he did not question Johnson's presence at the meeting.[57]

Rall and others at the REA tried to keep a tighter reign on other co-op boards. Nepotism in hiring co-op employees was discouraged in West Virginia and elsewhere (although relatives could sit on the same board). In Virginia, the agency did not look kindly on the board of the Shenandoah Valley Electric Cooperative's discussing the possibility of paying itself old-age pensions. Fisher obtained the resignation of the president and six other board members in Nebraska for running for public office. Clarence Winder, the former chief engineer of the Michigan Public Utilities Commission who replaced Fisher in the REA, told a Georgia co-op that the problem in these cases was that a board member usually "attempted to utilize his position as a director of an REA cooperative for the purpose of advancing his own personal ambitions." Field representative Evelyne Bloome reported that a project superintendent in Iowa warned her to stay away from a board meeting because a "couple of the Board Members were usually pretty well drunk by the time the meeting started and it wasn't a fit place for a woman. I agreed in every way, for I knew my patience wouldn't last long in a place like that."[58]

Some co-op boards had financial problems, some of which were the result of malfeasance, pure and simple. Fisher obtained the resignation of President Ellington in Georgia—whom he thought had "cheated, impeded, defrauded, and defied" the REA—by threatening to "wind the case up by turning Ellington in to the U.S. Marshall in Atlanta." Fisher suggested that the REA could avoid a scandal by advancing the co-op the money it owed Ellington, who could then get a bank loan to rescue him from bankruptcy and pay the co-op the larger amount he owed it.[59] The extent of Ellington's misdeeds seems to have been atypical, but the general difficulty was prevalent enough for Haines and Rall to place it third on their laundry list of co-op problems, as "directors exploiting their authority to gain personal financial benefits." The REA held that a board member on the Tallapoosa River Electric Membership Corporation in Alabama should not accept a large manufacturer's discount on an electric range from the Crosley company. George Munger, head of the Utilization Division, told Tallapoosa's superintendent that the decision was, legally, up to the board, but the proposal "seemed like modified bribery. . . . What are your project members going to think?" Fisher complained about "systematic rakeoffs on all purchases" on a co-op in Nebraska and a "truck body manufacturer who gave 10 percent kickbacks to [the] director" of a Missouri co-op. One of his field men reported that "charges of collusion between the board of directors and the engineer, as well as racketeering, were being made" on a co-op in Minnesota.[60]

Two related concerns were that many board members benefited financially from their position by selling electric appliances and by being employed by the co-op. The REA was concerned enough about both practices to include questions about them on a Special Membership Report questionnaire sent to co-ops in 1940. The first practice often offended local dealers. A businessman in Welasco, Texas, complained that the Magic Valley Electric Cooperative and the Magic Valley Electric Supply Company were run by the same directors, resulting in the "most unfair competition that I have ever heard of in my thirty years of Merchandising." Carmody informed a Michigan co-op in 1938 that the REA had found "several cases" of board members' setting up an appliance or wiring business after allotments had been made. "Wherever we have discovered these chiseling operators we have asked them to get off the boards or get out of private business, or, when it appears to be in the best interest of the community and the farmers, to get out of both." Carmody said it was not "ethical or honorable" to make a profit out of a government loan to a nonprofit group. But he did think it was permissible for a board member to be a director of an appliance consumer cooperative because of the high prices charged by manufacturers and dealers.[61] Fisher seems to have been lenient regarding directors' being paid for obtaining easements and other development work, but not after the co-op energized its lines. Rall, Winder, and their staff, however, adamantly opposed the practice of co-ops employing board members. They reported that a director was being paid as a project attorney in Tennessee, that directors were receiving five dollars a day plus mileage to obtain easements in Alabama and Oklahoma, and that two directors were receiving a consulting fee of five dollars a week in Tennessee.[62]

The REA had similar problems with full-time personnel. A California manager intermingled the financial affairs of the co-op and his own business, a superintendent in Pennsylvania was rumored to have a shortage of thirty-five thousand dollars, and a co-op bookkeeper in Illinois disappeared with about a thousand dollars.[63] Under the provisions of its loan contracts, the REA removed project superintendents and managers in several states, including Iowa (for too close a financial arrangement between the electric co-op and a merchandising co-op), Arkansas (for paying insufficient attention to managing the co-op and for not hooking up enough consumers), Georgia (for misuse of funds), and Nebraska (for getting involved in state political campaigns). When the president of the Farmers Electric Cooperative in Arkansas wanted to retain the superintendent because he felt all the co-op's problems were not his fault, the REA "played hardball." Fisher told Carmody, "I notified the project president that requisitions would be withheld until they acceded to our request for the superintendent's removal."[64]

The manifold problems with boards of directors and co-op leaders led some REA staff to complain about the lack of a cooperative spirit on the projects. Dora Haines and Udo Rall observed that farmers "joined not because it was a cooperative, but merely because they wanted 'lights' and couldn't get them in any other way. They feel no personal loyalty to the enterprise." Instead of blaming farm people, though, Haines and Rall faulted the REA for organizing co-ops from the top down, a criticism supported by reports from the field.[65] Home economist Victoria Harris "inquired a number of times as to why there was that lack of feeling [toward co-ops in Minnesota, which was known for a strong cooperative tradition planted by Scandinavian settlers], and the only answer I have got is that the Government is lending the money, and these lines are mortgaged, they feel it isn't theirs." The manager of a co-op development company in Wisconsin reported that "directors and members [in Minnesota, Michigan, and parts of Wisconsin] are beginning to view the projects as 'government' owned instead of cooperatively owned" and thus were losing the "cooperative enthusiasm" they originally had for the projects. The president of the Denton County Electric Cooperative in Texas complained that "through the removal of the management from the hands of parties selected by the membership and the placing of management selected by Rural Electrification Administration this fine spirit of cooperation was killed, and the members began to look on the Project not as their own, but merely as a Government Agency, with all authority and ownership vested in Washington."[66]

One indication of this attitude was the extremely poor attendance at annual meetings, at which the board of directors was elected and other important business of the co-op was conducted. Three co-ops in Mississippi, for example, drew very small crowds in 1940 (16 attendees out of 947 members, 129 out of 2,024 members, and 31 out of 3,202 members).[67] There were many reasons why farmers did not attend annual meetings, of course; often they assumed that the co-op was doing its job. But the REA worried about the poor showing. Fisher explained it in 1938 by saying that apathy was inherent in service cooperatives. Even the members of electric co-ops formed before the REA was established did not take as active a part in them as they did in consumer co-ops, mainly because once they obtained electricity, people tended to take it for granted (much as they did the telephone). Fisher concluded that the cooperative spirit of REA projects "will grow as their equity grows," an acknowledgment of one source of the difficulties the agency experienced in supervising co-ops.[68]

The REA's response to the co-op personnel and management difficulties was different from that toward the problems of safety, tree butchery, and telephone and radio interference. In addition to selecting co-op personnel and replacing recalcitrant board members and co-op employees, Fisher's division established legal and educational mechanisms to prevent these problems from arising. In 1938, Rall and Fisher created model cooperative bylaws for each state—by modifying those drawn up by the REA's legal department—which outlawed such common practices as board members' selling appliances, being employed by the co-op, holding public office, and employing their relatives.[69] The REA established national and regional training schools for project superintendents, bookkeepers, and linemen. When Winder replaced Fisher in the fall of 1938 (perhaps because of co-op complaints about Fisher's dictatorial style), Winder began sending co-ops a series of REA memorandums, written by Rall, on such matters as establishing an office library of cooperative publications, using a bulletin board to communicate with the membership, planning the annual meeting, paid employment by board members, and joint membership of husband and wife. The REA also distributed Rall's dictionary of cooperative terms to co-ops in a further attempt to educate them in the democratic principles and style of operations sanctioned by the agency.[70]

Carmody thought the joint membership policy would be an effective means to "civilize" the co-ops. In mid-1938 he wrote an article in *Rural Electrification News* calling for the inclusion of more women on co-op boards. He told a women's club the next year, "Electricity in rural areas means more to the women than it does to the men." Rural women "know what electricity is about; they have social vision; they have community interest at heart; they are not likely to get bogged down in petty disputes as so frequently happens to men." Winder claimed in a 1940 statement on joint membership of husband and wife that "frequently, it is the women who are more active and enthusiastic than the men in promoting a rural electrification project. We know of several instances where the men despaired of getting an electric cooperative developed until the women went to work and sold the idea to their neighbors."[71] Carmody, Winder, and other REA leaders thought that joint memberships, which still carried only one vote per family, would nevertheless help bring more women into the co-op as voting members. They apparently hoped that official recognition of the behind-the-scenes influence of farm women would help obviate some of the "petty disputes" among board members (and between them and the REA) that had plagued the program. By 1940, the REA had put some force behind this policy by including in its preallotment procedures a statement that at least three women should be on the incorporating board of new projects.[72]

Although the REA seems to have struck a blow for women's rights in this matter (at least a limited, nuclear-family version of women's rights), the agency's practices reveal a more complex story. In-house, as we have seen, the REA hired Emily KneuBuhl as a section head at the same salary as Boyd Fisher, and it hired several women in the Development and Utilization Divisions. Fisher favored gender equity, telling Cooke in 1936 that a prospective employee "is the kind of man who can adaptably work under a woman executive without strain upon either her or himself. This may seem trivial, but I regard it as initially important as a means of backing up Miss KneuBuhl's work." But the REA followed common gendered practices of the time by segregating most of its professional women (except Dora Haines) into the "feminine" job of home economist. Like other government agencies, the REA followed a law that permitted only one spouse to be on the federal payroll at a time, a law that tended to discriminate against women. Cooke ruled that "women candidates whose husbands are employed, particularly by the Government, shall receive no consideration for employment" in the REA, "women in the lower salary grades whose husbands secure Government work in the higher salary grades shall be released," and "women who marry men receiving high salaries in the Government service shall be released."[73]

Despite its promotion of joint memberships and women's election to co-op boards of directors, the REA often extended its in-house discriminatory policy to women's participation in co-ops. In Texas, the Limestone County Electric Cooperative, acting on the suggestion of field representative Frank Robinson, adopted a resolution in 1940 stating that, thereafter, all memberships would be joint memberships between husband and wife. Udo Rall—the REA's indefatigable monitor of the minutes of board meetings—agreed with the motive behind the resolution. But he told Robinson "that the board has exceeded its authority in making such a rigid ruling. After all, it is up to the applicant to decide the kind of membership he wants. Where farmers are not in the habit of drawing their wives into organization activities, such a ruling may meet with a good deal of opposition and may create considerable dissatisfaction." Robinson passed this gendered theory of cultural relativism almost verbatim to the Texas board and asked it to ensure that the "local conditions on your cooperative" corresponded with the "spirit of your resolution."[74] In Kansas, the Arkansas Valley Electric Cooperative (Ark Valley) proposed to hire Margaret Anderson as a utilization specialist. George Munger, head of the division that approved these hirings, agreed that Anderson was well qualified because she had worked as a home economist for a gas company. But because the job involved adding new members to the co-op and promoting farm as well as home appliances, Munger replied that a "man is usually preferred for this position." When the Ark Valley co-op argued that its "big crop" farmers had a higher demand for home appliances than for electric farm equipment, Munger relented and approved the "employment of Miss Anderson (against our better judgment) for a ninety-day probationary period." Alva Davis, the male co-op superintendent, replied (rather sarcastically?), "We fully appreciate your generosity in allowing us to place a lady in this position even though it may be against your better judgment." Munger closely monitored Anderson's performance, which—from all accounts, including his own—was exemplary.[75]

What we know about the women who served on co-op boards before 1945 corroborates this impression that some people in the hinterlands were ahead of the REA in promoting gender equality. The Thumb Electric Cooperative in Michigan had elected Ruth Brandmair secretary of its board of directors by May 1938, shortly before the *Rural Electrification News* published Carmody's article encouraging co-ops to elect women trustees. Two co-ops in Kansas elected one woman each to their boards in June, the month Carmody's article appeared.[76] From 1938 until 1945 at least thirty-six more women were elected to co-op boards, making a total of thirty-nine whom I have located (see table A.10). Only four of these women were elected after the (male) labor shortage following Pearl Harbor. Several women were officers: two presidents (in Illinois and Louisiana); four vice presidents (in Arkansas, Illinois, and Kansas); five secretary-treasurers (in Kansas, Montana, Ohio, and Texas), and five secretaries (in Arkansas, Kansas, Minnesota, and Montana). Reflecting the ingrained character of rural gender identities, almost all of the women were listed at the time as "Mrs." (the rest were probably not married).[77]

Biographies of three female board members show the flexible system of rural gender relations. Ruth Stevenson, the first women to be president of an REA co-op, served on the board of the Western Illinois Electrical Cooperative for eleven years, from its incorporation in August 1938 until February 1950. A member of the Home Bureau (the women's auxiliary to the Farm Bureau) who helped solicit co-op memberships when the local utility company quoted too high a price for hooking up, she became the first president of the co-op and turned on the switch to energize the first line. Stevenson headed the board for three years (another woman, Ruby Hurst, was vice president for one of those years), then served as vice president for six more years. She later recalled, using an apologetic tone common in women writing about themselves in this period, "I was just a woman and didn't know anything about the operation of an electric cooperative . . . [but] . . . none of the other officers and directors did, for that matter."[78] Anna Boe Dahl of Montana was a school teacher before she got married, then taught night classes in English and farm economics for the Works Progress Administration in the 1930s. A field-worker with her husband for the Farmers' Union, which organized REA co-ops in the upper Midwest, she canvassed farms to organize the Sheridan County Electric Cooperative in 1941 and served as the first secretary of its board of directors.[79]

Mrs. O. E. Bolon also had an activist career. A country schoolteacher and home demonstration agent in Montana, Bolon organized the Farm Women's Market in her town, conducted a local housing survey for the New Deal, gave a paper on Swedish cooperatives to her women's club, and was the woman member of the County Rehabilitation Committee. Carmody was so impressed with her credentials that he sent Bolon's letter (telling her story) to Eleanor Roosevelt to use in one of her "My Day" newspaper columns to show the type of woman involved with the REA. Bolon was elected, with two other women, to a nine-member board. Carmody confided to Roosevelt that Bolon's board replaced a five-member, all-male board because the earlier board "had given us some difficulties." But Carmody's hope that women would be a calming influence on co-op boards was not borne out in this case. Within a month, Bolon wrote Carmody that the process of building her co-op would improve "once we get rid of all the engineers"![80]

CHAPTER 7

Lights in the Country

Farm women and men responded to the efforts to modernize rural life electrically in an active rather than a passive manner, just as they had with the telephone, the automobile, and radio. Farm people not only helped (re)invent and run the REA cooperative, they also acted as innovative consumers in creating versions of rural modernity based on the way in which they chose to use (or not to use) electricity. Most preferred to have a few lights rather than a lot of energy-consuming appliances. Creating co-ops and modernizing the farmstead were thus processes of mutual construction in which farm people interacted with outside agents of modernity to create new forms of rural life. Many of their actions involved resistance, particularly to the middle-class, urban domestic ideal of separate spheres. Like the U.S. Department of Agriculture, the REA transferred this gendered ideal to the countryside by hiring male agricultural engineers to demonstrate electric farm equipment to men and female home economists to demonstrate home appliances to women. By the start of World War II, there were many more lights in the country, but the pastoral image evoked by that phrase masks a highly contested process of (re)shaping electrical modernity in the countryside.

Pushing Electricity

The REA's ideas of electrical modernity often clashed with those of farm people. At the start of the program, Morris Cooke remarked that "our big job is to build up the psychology of the generous use of electricity—a few lights in a home is not rural electrification. . . . Really to electrify rural America we must adopt every possible means for building up its use."[1] Most rural folk did not want to use that much electricity, and low usage (from the viewpoint of the REA) became a big headache in the early years of the program. In order for cooperatives to pay off their loans on time, the REA estimated, consumers on most projects would have to use an average of 80 to 100 kilowatt-hours per month.[2] This figure was well above the national average for all residences, which had been increasing steadily in this period, from 22 kilowatt-hours per month in 1912 to 31 per month in 1924 to 61 in 1936. A 1937 study showed that usage in nonfarm residences averaged 65 kilowatt-hours per month. This meant that farms would have to make up the difference with agricultural uses of electricity to meet the REA goals. The same study indicated that this was feasible because U.S. farms averaged 90 kilowatt-hours per month, with regional averages ranging from 65 to 125 hours. The figures, however, seem to be inflated, by irrigation farming in the West and by other prosperous farms, especially dairy farms, that had had central-station power for some time. Because the average usage for other areas in the 1920s and 1930s rarely exceeded 60 kilowatt-hours per month, the REA goal was out of reach for most co-ops at this time.[3]

In fact, in the early years of the REA, most cooperatives fell far below the 90-kilowatt-hour average. Statistics gathered for the spring of 1937 show that the average monthly usage for farm customers on twenty-two co-ops ranged from 23 kilowatt-hours for a project in central Texas and another one in north-central Kentucky to 79 hours for a co-op in the Tennessee Valley Authority's cheap-electricity region of western Tennessee. Two co-ops in the more prosperous farm states of Indiana and Iowa averaged less than 30 kilowatt-hours per month, and the only other co-op that used more than 50 hours was also in the TVA area.[4] These data support the observation, made by REA staff members and other commentators, that most farm families joined a co-op mainly to get lights, not to use electrical appliances to their fullest extent.[5] Most farmers bought only a radio, an electric iron, and perhaps one large appliance, like a washing machine or refrigerator, and many—if not most— stayed under their monthly minimum bills early in the program. The 1937 figures show that more than one-half of the members on thirteen of the twenty-two co-ops studied were minimum-bill customers. The REA gave these customers less than 100 kilowatt-hours per month (from 13 to 68 kilowatts) in order to keep minimum bills low enough to attract members to the co-ops.[6]

The REA responded to the problem of low usage primarily through the efforts of the Utilization Division. As we saw in a previous chapter, the agency brought in George Munger, an experienced load builder with the Electric Home and Farm Authority, and Clara Nale, the TVA's chief home

Figure 8. The REA's First Home Economists, 1938. From left to right: Victoria Harris, Louisan Mamer, Mary Alice Willis, Thelma Wilson, Elva Bohannan, and Enola Retherford
Source: Rural Electrification News, Feb. 1938, cover.

economist, to run the division in 1937. They hired a staff of agricultural engineers and home economists—many of whom were former USDA county agents—to go out into the field and work with extension services, co-ops, and farm people to connect more farms to co-op lines and increase the utilization of electricity (see fig. 8). The former agents used the traditional methods of extension work—bulletins, pamphlets, filmstrips, skits, and demonstrations (including "kitchen parties" and trial installations of such equipment as chicken brooders). But they usually worked with large groups because of their limited numbers: only seventeen (female) home-electrification specialists and an equal number of (male) agricultural engineers covered the entire country in late 1939, at the peak of the prewar program.[7]

In addition to demonstrating electrical equipment, the field representatives worked with their counterparts in the extension service and with appliance dealers on special promotions. They also monitored the status of co-ops in their regions—including the performance of boards of directors, project superintendents, and other co-op personnel—and conveyed REA policy to recalcitrant co-ops.[8] The prospect of women giving instructions to project officers upset at least one co-op superintendent in Texas. In 1938, Oneta Liter, who had worked two years with co-ops as a member of the Development Division, asked a co-op superintendent, whom she felt had not sufficiently publicized a meeting, to run electricity into a building for demonstrations of farming and household equipment. According to Liter, the superintendent replied, "I am accustomed to women asking me to do things as I am a married man," but said that she should ask one of the community leaders to do it, because he was busy. Liter told the home office that she did not think her request had been unreasonable.[9]

The Utilization Division's major promotional effort was the Farm Equipment Tour. Informally called the "REA Circus," especially by local promoters,[10] the Farm Tour was a popular traveling show of equipment demonstrated (in a gendered fashion) by representatives from manufacturers, the Cooperative Extension Service, and the REA inside trailers and under huge circus tents. Usually set up at farm fields near recently energized co-ops, the Farm Tour drew crowds ranging from fewer than a thousand (in Iowa and Texas) to eight and ten times that many (in Illinois and Georgia) during its brief existence from 1938 to 1941. Nearly 175,000 people attended the tour in 1939; by the summer of 1940, total attendance had reached 350,000 at 120 two-day stops in twenty-two states. In the carnival setting, male agricultural engineers demonstrated water pumps, milking machines, milk coolers, feed grinders, ensilage cutters, corn shellers, chicken brooders, soil-heating cables, safe wiring, household plumbing, and such novel devices as a rotating drum with rubber "fingers" that plucked chickens—a popular attraction. Female home economists demonstrated household lighting, "efficient" kitchen arrangements, and electric refrigerators, hot plates, roasters, ranges, coffeemakers, washing machines, irons, vacuum cleaners, and radios. They also supervised the lucrative all-electric lunch tent. During a tour in Iowa, the *Rural Electrification News* reported that an "ambidextrous farm girl comically scoots a sad iron [a heavy flat iron heated on a stove] up and down a shirt very fast but she is no match for the right-handed girl with the electric iron who only irons one way." Skits, written by the REA and put on by county 4-H and Home Demonstration Clubs, and cooking duels between local male celebrities took advantage of the carnival spirit to draw on traditional themes from village and rural humor to educate farm people in the value (and uses) of electrical appliances. A short film, made by General Electric or Westinghouse, or a New Deal social documentary like director Pare Lorentz's award-winning movies *The Plow That Broke the Plains* (1936) and *The River* (1937), was often shown on the first night of the show.[11]

Ceremonies held to celebrate turning on the power to co-op lines had a similar function as the Farm Tour and used many of the same techniques, because the Utilization Division often helped organize them. Dealers, agricultural engineers, and home economists demonstrated electrical equipment, 4-H Clubs put on skits, and local men in aprons fought the popular cooking duels at these usually one-day affairs. But because the events celebrated the modernization of a rural community, instead of being a stop on a national traveling tour, the organizers tended to draw more heavily on local customs to get their message across. Speakers included Grange leaders, local politicians, members of Congress, and governors (President Roosevelt even spoke at a ceremony near the summer White House in Warm Springs, Georgia), as well as representatives from state extension services and county farm bureaus. High-school bands and other musical groups entertained their neighbors, 4-H Club members read their prize essays on the benefits of rural electrification, skits had more of a community flavor, local residents paraded in "pageants of light," Texans put on barbecues, and rural folk competed in traditional picnic sports (a fat-man's race and a woman's baseball throw in Michigan) and regional outdoor contests (wood-chopping and sawing contests by hand and electricity in Arkansas).[12]

Two celebrations stand out for their attention to local concerns and customs. The highlight of the Woodruff Electric Cooperative's ceremony in Arkansas was the electrification of the house of a white sharecropper, Ed "Tater" Smith, and his wife. Newspapers published photos of crowds ringing the farmhouse, watching Ed (his smiling, gaunt face showing the marks of a hard life) pulling the circuit breaker on the outside of the house and then, inside, a nervous Mrs. Smith reaching up to turn on a bare bulb hanging on a cord from the ceiling. One paper quoted Ed as saying that "it will be just like living in town, only better." Watching were the governor, power company officials, and REA dignitaries, who praised the landlord members of the co-op for wiring all of their tenants' houses, an act worthy of news in 1938 because of its rarity. The common irony of one New Deal agency trying to improve the lives of sharecroppers while the policies of another New Deal agency, the Agricultural Adjustment Administration, was forcing them off the land is evident here.[13]

That same year, 1938, the Bluegrass Rural Electric Cooperative in Kentucky symbolically eradicated a part of rural culture by burying a kerosene lamp. Gloria Allender, a Jessamine County 4-H Club girl, read the "solemn funeral rites" for an "old-fashioned coal oil lamp" buried in a white casket as a choir sang hymns during an energizing celebration held at Daniel Boone's Cave. A Texas co-op held a similar ceremony in 1941.[14] The funerals symbolized a rite of passage into the modern world, achieved by burying a symbol of the technological distance separating urban from rural life. As with most ceremonies, it is difficult to gauge the response of onlookers. Co-op members undoubtedly viewed the burial of the lamp as a modernity ritual, but it went against the grain of a self-sufficient, make-do rural culture to discard anything of value. Farmers probably had a similar reaction to co-op officials' smashing a kerosene lamp, filled with water, against an electrical pole during pole-setting ceremonies in Indiana.[15] Kerosene lamps may have seemed old-fashioned to the agents of modernity, but they had been part of rural life for several generations and often saved the day, even in the electrical era, when storms and equipment failures caused power outages.

Pageants and skits also served many of the functions of modernity rites. During a Kentucky pageant a "minister from one of the Roman Catholic Churches reviewed the development of lighting as each era was portrayed by an individual appropriately costumed, and carrying a particular kind of lamp," walking through the darkness of an auditorium to the stage. "At the last moment all lights in the building were turned on, giving a very brilliant effect and something for the audience to really cheer about (AND THEY DID!)." The audience may not have realized that these events relied on material provided by the REA and other outsiders. In 1938 a field representative wrote that an eight-act "Pageant of Light," directed by the agency's Mary Taylor and put on by eight women's farm clubs in Arkansas, "reflected the training she had received at Nela Park," General Electric's center for lighting research and home economics training near Cleveland, Ohio.[16] Thelma Wilson, an REA home economist, reported in 1939 that the skit, "Country Lights," was "unusually effective and went over big" during a celebration in Georgia. "These different skits which are being sent out from Washington to us in quantity are filling a long-felt need."[17]

As with most popular media, the plots of these skits reveal as much (if not more) about their authors as about the intended audience. In "Tombstone or Washing Machine," which was performed at several celebrations in the South, a farm woman wants to follow her family's tradition of buying a tombstone for herself while she is alive to see what it looks like. On the way to the mortuary, she and her husband stop at the shop of the electrical dealer, an old friend. While the farm woman looks at pink lamps for the daughter's bedroom, her husband decides that the money set aside for the tombstone would be better spent on a washing machine, which will keep her alive longer. The washer does not appeal to the woman until she overhears one of her neighbors (who will get a free "marcel wave," which is being given away with every washer purchase) comment on how tired the farm woman looks. According to an REA account, "that bit of gossip sealed the matter. The dealer made a sale of a washing machine and everyone was happy." Several of REA's ideologies are evident here: the Country Life tenet that technology will save the overworked farm woman; the understanding of modernity as the replacement of traditional values (buying a tombstone) with those of consumer culture (getting a marcel wave and buying a washing machine); and the perception of rural gender relations as a flexible system that can change with the times but yet retain the public authority of the man as the head of the household. Later skits communicated more matter-of-fact messages like the ease of operation and affordability of an electric range, that flickering lights can be caused by trees rubbing against the co-op's lines, and how easy it is for (even) a woman to read an electric meter.[18]

The skits show that much of the REA's promotional material employed gender-related schemes aimed at farm men, who usually controlled the purse strings. Cooking duels were popular because of the role reversal of having prominent men in the community, including the head of the co-op, don aprons to do the cooking. The message was that cooking with electricity is so easy that even a man can do it.[19] In another event, supported by the REA, the journal *Prairie Farmer* sponsored a Men Folks' Washday in Wisconsin in 1937. The philosophy behind the stunt—which Maytag had used in a 1930 advertisement—was that once a farm man experienced the back-breaking work of hauling water and scrubbing clothes by hand, he would buy his spouse an electric washing machine.[20] Several cartoons that appeared in the *Rural Electrification News* in 1938 and 1939 and were made available to publishers reflected the perceived system of male-dominated gender relations and gave credit to the husband for having the bright idea to buy the "tired farm wife" such laborsaving devices as a water pump, an electric range, a washing machine, and a vacuum cleaner. In only one of the twelve cartoons did a woman have the idea to electrify her house, and that was to provide running water for her husband's bath, so that he would no longer slip and fall out of a tub on a soapy kitchen floor![21]

In fact, women may have had more power over these types of purchases than these cartoons implied, which did not go unnoticed by the REA. A 1936 report of an extension meeting in the TVA area noted that "women are a

power in the family. . . . Men would sign up for a range without consulting
[their wives] in order to get electricity onto the farm. Then, when the [wives]
found out about this range and should [they] object to it, the men would retreat." An REA annual report noted that a Pennsylvania farm man gave orders to have his electricity shut off. When workers came out to the farm, his
wife replied, "You take away that meter and see what happens." She went
home to her mother's, leaving her husband with a note and a cold supper. After a few days, he gave in and had the electricity turned back on. "By this
time, she had her back up. As the price of peace, he had to buy her a new electric washing machine." When shown photographs of the Farm Equipment
Tour in 1939, REA administrator Harry Slattery "was particularly impressed
with how many women were in attendance at the demonstrations and viewing the exhibits. It is like the old Barry play 'What Every Woman Knows.'
Translated this means that she runs the pocketbook and makes the purchase."[22]

The REA received help from home demonstration agents in the Extension Service, many of whom had cooperated with the agency before it merged
with the USDA in 1939. Farm and home agents gave demonstrations at
REA-sponsored kitchen parties, co-op energization ceremonies, and the
Farm Tour. Because most home agents did not have expertise with electrical
equipment, REA home economists organized training schools for them or
participated in extension schools for agents and local leaders in most states
from 1936 to 1946. C. W. Warburton, the director of the Extension Service,
encouraged more cooperation between the REA and his agency while the
REA was a separate organization. Although the Extension Service hired
many agents in this period to handle the New Deal agricultural programs,
more help was needed because of the big jump in rural electrification. The
number of families installing electricity, for example, rose from five per home
agent in 1930 to seventy in 1940.[23]

The REA cooperated with USDA agencies in a more formal manner in
regard to utilization when it became part of the Department of Agriculture.
(It probably helped that John Carmody and Boyd Fisher, the main foes of the
Extension Service, left the REA at this time, Carmody right away and Fisher
in 1940.) Clara Nale worked with the USDA's Bureau of Home Economics to
develop educational filmstrips, even though she thought the bureau focused
too much on long-term research and was "under the direction of people who
are too far removed from the field of rural electrification and its attendant
problems to be of much help in a program that is as new and widespread as
REA is." Milburn L. Wilson, the new director of the Extension Service,

praised the REA's annual reports in 1940 and sent them to all state extension directors. Wilson also supported the appointment of former county agent Oscar Meier as the liaison between the REA and the Extension Service in that year.[24]

Although the REA and the Extension Service cooperated on many fronts, before the war it was an uneasy partnership. Extension directors complained that REA field representatives did not always work with county agents, and the REA noted that many agents neglected the co-ops after they helped organize them. One factor was the continued "inexperience of home demonstration agents with electric household equipment," an endemic problem in the profession. Some agents were also criticized by their superiors for helping to organize co-ops.[25] But county agents in Alabama, Texas, and elsewhere helped the REA demonstrate farm and home appliances.[26]

The utilization programs were carried out on a day-to-day basis mainly by utilization specialists hired by co-ops. Typically, George Munger would monitor the amount of electricity used by co-ops and suggest (strongly) that those with low usage hire a specialist. Munger's Utilization Division and the Division of Engineering and Operations approved each hire, and field representatives often recommended someone in the area for the job. Munger reported in January 1940 that "fifty-five projects now have such a man on their payrolls," and by March there were seventy-one.[27] Despite Munger's preference for men for the position, the former home economist Margaret Anderson in Kansas performed her functions well.

The REA expected quantifiable results. A weekly report required the specialists to list the number of "calls on prospective members, memberships sold and fee[s] collected, wiring contracts secured, new members connected to lines, calls on prospects for home appliances, home appliance prospects reported to dealers, calls on prospects for farm equipment, farm equipment prospects reported to dealers, general promotional calls on contractors and dealers, contractor-dealer meetings held [and number in] attendance, member and prospect meetings held [and number in] attendance." The form also asked for the number of appliances dealers sold each week and the average number of kilowatt-hours thereby added to the co-op's load. Among the other duties of Cecil Beyler, a utilization specialist in Indiana, were to "maintain a constant and complete display of electrical equipment by local dealers" at the co-op, "conduct tests on all types of electrical equipment," "act as agent and inspector for EHFA [Electric Home and Farm Authority]," and "make surveys from time to time showing the amount of electrical equipment on the project lines."[28]

The specialists employed many of the sales techniques used by utility companies, manufacturers, and home economists. Beyler carried out sales campaigns for house and barn wiring, the electric range, refrigerator, poultry lighting, plumbing, small appliances, feed grinders, and Christmas lighting—all in the last half of 1940.[29] In the same year, Margaret Anderson created a co-op display window with the theme, "Do It Electrically and Save." She gave every member who attended the annual meeting in 1941 an "attendance gift" of a 100-watt lightbulb (with the twin goals of increasing usage and improving lighting), placed "free publicity" stories in local newspapers, and surveyed appliance ownership while reading 151 meters in one week. She then prepared a meter-reading card with blanks on the back for farmers to indicate which appliances they owned. Like her counterparts in urban utility companies earlier in the century, she did a lot of door-to-door solicitation, but out in the country. In the summer of 1940, she reported carrying a "roaster, waffle iron, grill, etc., in trunk of car, in addition to catalogs from name manufacturers. Believe it a definite asset in creating desire and arousing interest in appliances they may or may not have considered. Most of these appliances are good load builders, particularly the roaster. Naturally this is shown where the possibility of range installation is small. These prospects, of course, are turned over to dealers to get 'action' plus 'interest.'" That December, she was "out among the members talking up the idea of buying electrical gifts for Christmas in line with our group purchase plan."[30]

Utilization specialists often wrote co-op newsletters to send their modernizing messages to members on a regular basis. By mid-1941, more than five hundred co-ops (about two-thirds of REA co-ops) published monthly newsletters.[31] Although the REA ran training schools on how to write newsletters and provided boilerplate for them, newsletters reflected many aspects of rural culture. The titles—*Radiant Eye Announcer* in central Alabama, *The Nodak Neighbor* in Grand Forks, North Dakota, and *REA News: The Monthly Messenger of Craighead Electric Co-op* in Jonesboro, Arkansas—recall a rural rather than an urban way of life.[32] Other names, based on puns, were probably chosen by utilization specialists to show an acquaintance with electrical modernity: the *Line Up* in Georgia, *Kilowatt Ours* in Indiana, *Current News* in Sleepy Eye, Minnesota, *The Lighter Way* in Texas, and *The Live Wire*, also in Texas.[33] Other local features included news of project construction and operation, columns by co-op managers, news of annual meetings, essay contests, user endorsements, problems and warnings, cartoons and jokes, names of new members, and honor rolls proudly listing those who used more than one hundred kilowatt-hours per month. Many of the modernizing messages came from translocal sources, such as the "ten necessities for a modern kitchen, as listed recently by an authority on home modernization." Advertisements such as a co-op ad for "electrical wife savers for the farm home" reflect the gender and technology themes prevalent for all supposedly urbanizing technologies in this period.[34]

Appliance dealers did not respond enthusiastically to these initiatives by co-ops in the early days of the program. Mary Alice Willis, an REA field home economist, reported in 1937 that the "apathetic attitude of small-town appliance dealer[s] has impressed me on occasions too numerous to mention. Many are not interested in displaying their stock at meetings. Many have no finance plan to offer their customers. Some even have said they would sell appliances if people came into the store and asked for them, but they would not solicit business in any way, and would not send merchandise out on trial." An attorney for a Georgia co-op complained that the "dealers merely sit around in their stores and wait for customers to come in instead of going out on the lines and trying to sell them what they need. In fact there are not any progressive lighting or fixture dealers in this town." Other resistance was more active. Dealers in one area of Minnesota would not carry major appliances like ranges, refrigerators, and water heaters; Kansas dealers in the Ark Valley co-op area were "not interested in small appliance prospects as the profit is not great enough to warrant a drive in the country."[35]

Many dealers did work with co-ops in these early years.[36] But a measure of their resistance is that one purpose of creating the Farm Tour was to convince reluctant dealers to get on the REA bandwagon. Oscar Meier told a staff conference in 1939 that the "principal purpose of the Farm Equipment Tour was to get more dealers interested in handling farm equipment," as well as domestic appliances. Daniel Teare, the director of the tour, said that in the beginning, "it was necessary that something be done to break down the passive resistance of both consumers and merchandisers. The Farm Show was the first large-scale effort on my part and was designed to awaken farmers and merchandisers to the advantages of rural electricity."[37] The REA thereby adopted another technique of the private utility companies, forging close links with electrical manufacturers and dealers.

One reason dealers did not cater to the rural market was low farm income during the Great Depression (see table A.1), but dealers also disliked competition from co-ops selling appliances in their area. George Munger told the president of the Cherokee County Electric Cooperative in Alabama that the "apathy of your local dealers is partly due to discouragement because of the fact that you contemplate selling appliances, and further, that if you do engage in this business it will probably be on a basis with which they cannot compete." Because of the rumor that the Ark Valley co-op in Kansas was going into merchandising in 1941, "dealers were hesitant in stocking a great quantity of merchandise and were showing a lack of interest in farm sales." An REA field representative in Indiana reported, "I do not suppose the dealer relations will be any too pleasant since the Co-Op is located in the Farm Bureau Store."[38]

The REA was partly to blame for this state of affairs. Early in the program, both Emily KneuBuhl and Boyd Fisher encouraged co-ops to go into the merchandising business. Although the REA began to advise co-ops against merchandising in the late 1930s, John Carmody still promoted appliance co-ops because he thought dealers overcharged consumers. This policy did not endear the REA to many dealers, as we saw in the case of the Magic Valley co-op in Texas. The REA established a policy along lines laid out by George Munger, who had worked as a load builder for a private utility company. *Electrical Merchandising,* the trade journal for appliance dealers, published a long article in late 1937 praising Munger's efforts and encouraging dealers to get more involved in the REA program. In June 1939, Munger told a co-op that most attempts by co-ops to go into merchandising "were disastrous." Merchandising, he suggested, should be handled by a professional sales force, not by farmer co-ops.[39]

Power and the Land

Many of these debates about selling techniques were evident in the making and reception of the REA film, *Power and the Land.* Impressed by the persuasive power and visual beauty of Pare Lorentz's prize-winning government-sponsored films, the REA contacted Lorentz in 1939 in the hope of making an equally successful "educational" film (or "propaganda" film, according to critics of the New Deal). The REA thought such a movie would supplement its other promotional activities aimed at inducing often-reluctant farm men and women to participate in its program. Hiring an avant-garde director was not out of character for the REA, which had commissioned the graphic designer Lester Beall to produce a set of modern-art social-realism posters for the agency in 1937 and 1939.[40]

In April 1939, the REA asked Lorentz and his newly formed U.S. Film Service to make a documentary film, along the lines of his *The Plow That Broke the Plains* and *The River,* to be shown mainly on the Farm Tour.[41] Because the Farm Tour created goodwill among members of Congress when it toured their districts, and because it was also designed to convince appliance manufacturers and dealers that farmers were potential customers, a film shown at the Farm Tour would promote the REA program to several of its target audiences: farm people, politicians, and the electrical appliance industry. Both of Lorentz's previous films had been shown commercially, and the REA undoubtedly hoped that a rural-electrification film would also reach a wide audience and thus gain more support for its program.

One obstacle was uncertainty as to whether the REA's charter allowed it to make a promotional film. The issue threatened to scuttle the project until the agency's legal department split enough hairs to conclude that the REA was permitted to show the "many advantages which *have been and are being enjoyed* through the use of electrical appliances" but not what advantages "*could be enjoyed*" through this technology.[42] In other words, the REA could educate its audience using established "facts" (showing what electricity had done and was doing for rural life), but it could not engage in promotion (advertising what electricity could do). Armed with the authority of this legal opinion, the REA began negotiations with Lorentz to produce a "documentary film portraying the conditions and progress" of rural electrification. In the process, the construction of facts and propaganda merged across a line of tension between education and sales, which home economists had experienced since the beginning of the REA program.[43]

These tensions are apparent in a script prepared by the Film Service in consultation with members of the REA's Information Section and in comments on the script made by other agency officials in August 1939. Intended to guide the director in shooting the film, the script laid out some general principles and suggested three sequences. An average, "middle-class" farm and an "intelligent and pleasant-looking" farm family from Ohio were preferred over more picturesque uplanders or southerners. To get the right combination of gender and age, the Film Service was even prepared to create an ideal farm family by mixing people from different biological families. Three sequences would show a technologically backward, isolated, and tiring farm life before electricity; a middle period of community decision making to organize an REA cooperative; and a bright, consumerist, fast-paced utopia that left time for leisure, after electricity. The before-and-after contrasts would be sharpened by comparisons with the bright lights of the city and factories at the end of the first sequence; music emanating from a radio in the well-lit farm home and yard would signal the end of rural isolation in the closing scene. The script strayed slightly from its technologically determinist theme by addressing the resistance of farm people and private utility companies to the REA during debates in the middle sequence.[44]

Although REA leaders liked the script, they questioned its accuracy and wanted it to be more promotional in certain areas. The main problem was that Ohio was an above-average farm state and the suggested farm family, having ten to twenty cows, was too well-off. Harlow Person, the REA's chief economist, pointed out that Ohio farmers would not take kindly to the primitive conditions pictured in the first sequence. As we saw in chapter 3, even without electricity, many prosperous farms had several "conveniences": gasoline engines for grinding grain and pumping water; windmills to pump water for livestock; running water in the house and barn; kerosene stoves in place of hot wood stoves in the summer; and battery-powered radios charged by small, roof-mounted windmills.[45] Udo Rall, the REA's expert on cooperatives and rural life, also suggested balancing the picture by showing the long workday of farm women in the house, dairy, and garden. Both Person and Rall argued that invidious comparisons between urban and rural life would do more harm than good by increasing rural resentment toward the city.

Rall, more than his colleagues, suggested that the filmmakers always keep the farm audience in mind. "This does not mean playing down to inferior intelligence," Rall said, "but a recognition of farmers' peculiarities due to less formal education, to relative geographical isolation, to dependence on and acceptance of tradition, to a generally conservative attitude that seems to go with nearness to the soil." Along this line, Rall advised that "care should be exercised to avoid anything that might offend or humiliate farmers or expose them to ridicule. If any scene must depict carelessness, stupidity, ignorance, or pettiness, it must be made clear that this is not typical of farmers." In a proposed fire scene, suggested by novelist John Steinbeck, who had not committed himself to writing the film's narration, Rall advised that the fire must not be the fault of the farm family.

Although Rall was deeply concerned about accuracy and a sympathetic portrayal of rural life, he wanted to promote his favorite part of the program—the cooperatives. Rall thus suggested idealized scenes depicting the potential democratic and community-building aspects of co-ops, which had not yet been achieved. As we saw in the previous chapter, Rall co-wrote a scathing report about the undemocratic nature of REA cooperatives, showing that because they were not grassroots cooperatives, they had been plagued by domination by a small board of directors, nepotism, and financial malfeasance, among other problems. Rall's ideology infused his suggestion that the blatant consumerism of the last scene, in which the farm family would proudly display its new appliances to a neighborhood couple, should be replaced by scenes of the improved community life fostered by working together to build a cooperative from the bottom up rather than from the top down.[46]

What type of film was produced? Marion Ramsay, head of the Information Section, recognized that the "type of film that we will get depends vastly more upon Messrs. Lorentz, Ivens, and Steinbeck than upon this rather sketchy outline."[47] Ramsay was largely correct: the end result depended heavily on Joris Ivens, whom Lorentz—who was busy producing *Fight for Life*—had hired as the film's director. An internationally acclaimed Dutch filmmaker, a pioneer in documentary films, and the first foreign director to be invited to work in the Soviet Union, Ivens started work on *Power and the Land* in September 1939. The project interested him because it illustrated the social benefits of a government program. More than likely, he had faith in the progressive ideology of electricity, which V. I. Lenin had expressed in his well-known aphorism equating communism, socialist power, and electrification in 1920, just a few years before Ivens worked in the Soviet Union. The REA film also gave Ivens an opportunity to explore more fully the use of reenactments in a documentary, which he had first used in *Song of Heroes* in 1932. Unlike some filmmakers, Ivens did not apologize for using reenactments. He argued instead that because documentaries were not strictly objective—expressing as they did the viewpoint of the filmmaker—reenactments allowed him to film the truth as he saw it, to shape "reality into the dramatic structure of the growing work."[48]

Working with the guidelines negotiated by the REA and the Film Service and sharing large parts of their ideologies as well, Ivens started shooting the film in October 1939, under the supervision of Lorentz, and completed it, a little behind schedule, in late November. Ivens and his partner, Helen van Donge, began preliminary editing back in New York City; Lorentz and his crew then began cutting and polishing the film and started on the narration in the early part of 1940. There was some urgency, because the Film Service was under attack for its "propaganda" activities from senators and members of Congress unfriendly to the New Deal. Shortly after Congress eliminated funding for the Film Service in the spring of 1940—partly because of the vivid scenes of slum life and maternal mortality portrayed in *Fight for Life*—Lorentz told the REA that he could not finish *Power and the Land*. On Lorentz's suggestion, the REA hired Ivens to complete the cutting and editing, which he did in late May. The REA then hired the Pulitzer Prize–winning author and poet Stephen Vincent Benét, rather than Steinbeck, to write the narration and Douglas Moore, head of the music department at Columbia University, to compose the musical score. The veteran radio actor William Adams was the narrator. The film was first shown to President Roosevelt in August 1940—as was the custom with new films by Lorentz. *Power and the Land* premiered in Ohio in late August, a year behind schedule, began to receive bookings from RKO theaters in mid-November, and showed in Washington, D.C., during the first week in December and then in a Broadway theater in New York City.[49]

The result of these collaborations was a visually appealing and quietly dignified "before-and-after" film of an Ohio farm family, the Parkinsons, getting electricity. The thirty-six-minute "short" opens before dawn with the older Parkinson men (father and two sons) beginning their morning chores by the light of kerosene lamps in the barn of their dairy farm. Unremitting toil fills the daylong first segment of the film. The men milk the cows, pump cold water into a tank to cool the full milkcans, load them on the dairy truck, saw and chop wood by hand, mow hay using horses, and cut corn by hand for silage. Even the youngest child, Bip, the eight-year-old, carries wood to the cookstove, but he is then free to play. The women (mother and thirteen-year old daughter) cook, scrub clothes by hand, hang the washing, iron clothes with a heavy flat iron, and fill kerosene lamps. The day ends with a family almost too tired to eat the laboriously prepared evening meal, complete school work, and perform still more chores—all by the dim light of kerosene lamps.

The pace picks up dramatically in the next segment when men from neighboring farms help cut the Parkinson's corn. This age-old cooperation—accompanied by uplifting narration and music—leads "naturally" to farmers getting together at a schoolhouse to discuss forming a cooperative to electrify their farms. A county agent answers their questions and explains the REA program. In no time at all—following a brief shot of a power plant in the city—poles are raised ("liberty trees for farmers"), lines are strung into the country, transformers hoisted, and the Parkinson's house wired and plumbing installed. When the "juice" is turned on, Hazel rings the dinner bell for Bill, who's riding a tractor in a field near the house. (If one looks closely, the tractor is visible in one of the preelectric scenes showing the men taking the horses out of the barn.) The narrator proudly says that Bill—now a modern farmer—has joined his local REA co-op.

The final segment depicts the utopia of an electrified farm. The men come smiling into a well-lit kitchen for breakfast, work happily in a bright barn (where electricity milks the cows and pumps water to cool the milk), saw wood, and do other farmwork with electric motors. The women have plenty of hot running water and cook, wash, and iron with the aid of electricity. A modern bathroom brings city comforts to all. The day ends with a grateful family enjoying a sumptuous meal cooked on an electric range.

As we can see from this brief description, the filmmakers took into account many of the suggestions made by the REA staff. Care was taken to portray farmers as intelligent, hardworking folk—a view that Ivens shared in any event. Scenes of city life were played down, much attention was paid to the work of women, and the benefits of cooperation were stressed in a subtle way. Yet other concerns were not addressed. In the desire to emphasize the technological extremes in the before-and-after format, Ivens probably removed such common technologies as a telephone, battery-operated radio, automobile, portable gasoline engine, and windmill from the first segment. As Person pointed out, such a prosperous family could afford these technologies and use them without having electricity. Other concerns were muted, such as the fierce opposition to the REA by private utility companies and the resistance of many farm men and women to electrification—which was one reason to make the film. The Parkinsons did not show off their new appliances to the neighbors in the final scene, but happiness was still defined in terms of consumerism, of buying an electric range rather than building a cooperative.

In making a documentary that promoted the sponsor's point of view (Ivens thought this was a valid function of all documentaries as long as a socially desirable truth was portrayed), Ivens took a few liberties in shooting what was essentially a reenactment with nonactors. He tried to construct a "synthetic farm family," in his words, by mixing and matching farm people as the Film Service had suggested, but this fell through. He then returned to the Parkinsons, whom he had passed over because the husband seemed cool to the project, and saw a quiet dignity in Hazel Parkinson, a college-educated woman who had lived in town. Yet he did not portray the Parkinsons, who were already members of a local REA cooperative, exactly as he found them. The film crew rented horses to replace the family's mules, which the REA considered to be old-fashioned; the crew had to remove electric lines into the farm, which had been electrified for two years; and the REA gave the family many of the appliances shown in the movie, including the water pump and bathroom that the Parkinsons had not been able to afford. After making the film, the REA took back the milking machines and electric range.

The film leaned toward a promotional brochure in other ways, as well. *Power and the Land* does not indicate the lengthy time (often two years) it took to organize a cooperative, get the government loan, and install electric wiring and lines. From the viewpoint of recent scholars, it also reinforces the erroneous belief that household technologies not only eliminated drudgery but also saved time—an impression belied by the time-use studies of full-time home workers (see chaps. 3 and 8). Although the film has many inaccuracies, Ivens later recalled that he spent time getting to know the Parkinsons so that he could get the small details right in what he did shoot, knowing this would be effective with rural viewers.[50]

Power and the Land was a success. After opening on August 31, 1940, in St. Clairsville, Ohio, in the movie house closest to the Parkinson farm and its REA cooperative, the film received a fairly wide showing. By paying for prints of the film, the REA was able to sign a contract with RKO Radio Pictures to distribute it without requiring a rental fee to its system of theaters across the country. More than one thousand theaters booked the film in the first month.[51] Reviews were generally positive. *Variety* and *Time* praised Benét's "earthy" commentary; *Commonweal* thought the film was "particularly brilliant in its documentary aspects in its first half which has a nostalgic, back-to-the-land quality." Reviews praised the musical score, the fine photography, and Ivens for making a quietly effective social documentary. Some reviewers complained that the contrast between the "before" and the "after" segments was too sharp because of the unnecessarily "primitive" portrayal of the Parkinson's standard of living and that the technological fix of electrical appliances seemed too pat and too easily affordable.[52]

Popular support for rural electrification helps explain why *Power and the Land* did not face the political problems of Lorentz's other films. Both critics and supporters of the New Deal agreed with the strong statement of technological determinism that formed the basis of the progressive ideology of electricity, an ideology that crossed class and political lines. With so many powerful groups unquestionably accepting the efficacy of rural electrification, it would have been difficult to call a middle-class pastoral film about it "New Deal propaganda," even when the director was a leftist who had made films for the Soviet Union. It was doubly hard to do so when a New Deal agency was working to bring electricity, an urban technology symbolic of modern life, to a social group that represented the equally strong agrarian ideology of the family farm.

(Re)shaping Modernity at the Grass Roots

Despite the best efforts of filmmakers, magazine writers, and home economists, farm men and women did not buy the middle-class, urban form of electrical modernity pushed by these agents. Lights and radios were the main objects purchased for the house. A brightly lit farmhouse (along with a telephone, an automobile, and a radio) showed that its owners were not the hicks city people joked about when they ridiculed life in the country as primitive. This resistance to urbanization—not purchasing a full complement of electrical appliances for the house and not installing running water and a modern bathroom—worked against the efforts of REA and Extension Service agents to create their version of electrical modernity. Like other forms of rural resistance to urban-style technologies, the grassroots opposition to electrical appliances was widespread.[53] But in this case it was endemic not only among farm people at large but among co-op superintendents and board members, two mediating groups who were supposed to transfer the new technology from the city to the country.

Many project managers and superintendents resisted the full-blooded commercial approach of the REA's Utilization Division. The field home economist Thelma Wilson reported in 1937, "On each of the Projects I visit I am impressed with the fact that these Project Officials need the Utilization program which we are presenting more than the customers do themselves." Unlike co-ops that sold appliances, the eleven projects she had contacted in Ohio, Indiana, and Pennsylvania were mostly interested in constructing lines. The REA promoted wiring and plumbing loans, group bidding, and the financial plan from the Electric Home and Farm Authority, but field representative Elbert Karns noted in 1937 that "project officials and leading farmers [in Missouri, Oklahoma, and Texas] don't accept them as workable plans," a phenomenon also observed by a colleague working in Minnesota and Wisconsin that year. Field-worker James Cobb was so disgusted with the apathy of one project superintendent in Texas that he laid out an aggressive Christmas promotional campaign for him in 1938; Cobb and Oneta Liter prodded him into doing more load building the next year. Some superintendents objected to these high-pressure sales techniques. In 1942, Liter reported that one in western Indiana had written the REA to request "suggestions on activities other than 'peddling appliances.'"[54]

Board members who did not sell appliances themselves were often reluctant to steer the co-op onto the path of electrical modernity marked out by the Utilization Division. In a revealing report about a Wisconsin co-op, field representative Walter Zervas said that the

> project sponsors [were] loathe to push any degree; they believe [the] project will be successful in due time, but development must take its natural course; they feel that demand must come entirely from farmers

themselves. In no way do they want to force anyone to wire or to buy equip[ment]. This does not mean that they are adverse to load-building activities, through education, but [they] want to avoid strong persuasion on grounds that the farmer will develop an attitude of contrariness and [then] becomes a poor and unenthusiastic user with no resultant benefit.

But, Zervas continued, "they plan to have GE and West[inghouse] and other outfits cover their territory with their trailers, and then follow this up with some appliance dem[onstration]s by REA."[55] Other boards were opposed to load building in general or to specific applications. A field representative noted that "there seems to be a general feeling among the directors [of a Texas co-op] that they should advise customers not to exceed the minimum bill, instead of educating the customer to use the electricity more fluently [sic]" to their economic advantage. The REA's plumbing expert, Willard Luft, told a regional conference, "Sometimes it has taken a couple of years to get the idea over to them that their organization is concerned with farm modernization, and that plumbing is a basic part of the program."[56] In 1939 and 1940 at least six boards in Alabama, Indiana, Georgia, and Texas balked at hiring a utilization specialist, or fired theirs after a trial period, even though the REA strongly recommended this load-building technique to pull them out of economic difficulties.[57]

The agents of modernity (including reluctant appliance dealers and co-op officials) employed a multitude of methods to transfer their (urban) version of electrical modernity to the countryside, but farm men and women had the final say in what that modernity would look like. First of all, they used more electricity as the REA program progressed. The average usage for seventy newly energized projects (with an average service time of six months) improved to 45 kilowatt-hours per month in 1938. By 1939, the average for all REA projects had risen to about 60 hours per month.[58] Many successful co-ops exceeded that figure: the Ark Valley co-op in Kansas (62 kilowatt-hours), the Steele-Waseca co-op in Minnesota (65 hours), the Claverack Electric Cooperative in Pennsylvania (72 hours), and the Gibson County co-op in Tennessee (118 hours).[59] But before 1940 many were below par. Scattered evidence reveals an average of 24 kilowatt-hours per month for the Upshur County co-op in Texas, 32 hours for a co-op in Wisconsin, 42 hours for one in Indiana, and a range from 39 to 46 hours per month for the Denton County Electric Cooperative in Texas. Two co-ops in Missouri and Minnesota averaged less than 42 kilowatt-hours a month. The variation in Denton's figures was a result of increased appliance purchases and seasonal usage.[60]

An indication of what appliances farm men and women bought is given by surveys of farms on REA lines (see table A.11). Although the figures undoubtedly understate the eventual number of appliances purchased by these co-op members before the United States entered World War II, this is counterbalanced by the probability that people who owned fewer appliances chose not to participate in the surveys (response rates were about 65 percent) and the fact that the average length of service for members surveyed rose from about nine months in 1938 to nearly twenty months in 1941. The table shows a remarkable consistency in the choice of which appliances were first purchased in these years. Because the order is not based simply on purchase price or operating cost (compare with table A.12) and because the order is similar to that of pre-REA surveys, the data give an idea of which appliances were valued on the farm.[61] Despite increased personal farm income in this period (see table A.1) and the best efforts of REA agents, new co-op members bought mainly radios and irons. Only about one-half bought washing machines and refrigerators. All four items were familiar technologies that prosperous farm people had run on either batteries (for radios) or petroleum products (for irons, washing machines, and refrigerators) before they had electricity. In contrast with the earlier mechanization of agriculture, the house was electrified long before most farming operations. Lighting in the barn was universal, but electrical farm equipment went begging—except in regions dotted with dairy farms, large-scale poultry operations, or irrigation pumps. Other items that had a more middle-class urban aura, such as vacuum cleaners and coffeemakers, were often seen as "foolish luxuries."[62]

Table A.13 illustrates the contrast between urban and rural ownership of some of these technologies. The percentage of new REA customers buying radios, irons, washing machines, and hot plates in 1940 was comparable to the national figures for these goods (i.e., for all households, farm and nonfarm). Yet farm people purchased fewer refrigerators, toasters, vacuum cleaners, and coffeemakers. Running water and a modern bathroom (with a water closet, sink, and bathtub) required not only a water pump but a dependable and adequate water supply and a sewage system—common utilities in the networked city but rare in the country, as we have seen. The expensive electric range (expensive to buy and, especially, to operate) was unpopular in the city and the country.[63]

There were some regional differences in rural purchases. A persistent one, owing to differences in climate, was that new co-op members in the South bought considerably more refrigerators than those in the Northeast and the North-central states (e.g., 40 percent to 17 and 30 percent, respectively, in 1938), whereas the figures were reversed for washing machines (in 1938, Northeast, 68 percent; North-central, 75 percent; and South, 21 percent). The difference narrowed to about 10 percent for refrigerators by 1941 but stayed at 40 percent for washers. The contrast was particularly vivid in some states. Newly energized Minnesota cooperatives reported the highest percentage of washing machines in 1941 (93 percent), but they were next to the last in refrigerators (20 percent). South Carolina co-ops were the fifth highest in the percentage of households owning refrigerators that year (49 percent) but next to the last in the ownership of washing machines (11 percent).[64]

The substantially lower farm income in the South does not seem to explain these figures, because washing machines cost a third less than refrigerators and refrigerators were more expensive to run (see table A.12). The much hotter weather in the South was undoubtedly a factor for the higher number of refrigerators in that region, as were lower operating costs because of cheap electricity in (and near) the TVA area. Fewer washing machines also correlates with the fact that well-to-do southerners often sent laundry out, usually to African American women. The sales manager for the Delco home-lighting business alluded to this reason when he told Morris Cooke in 1935 that southerners bought refrigerators instead of washing machines for their Delco plants "because of the low labor cost in the South."[65]

Why did farm people make these consumer choices? Why did most families mix up-to-date electrical appliances such as radios, irons, and washing machines with old-fashioned coal stoves, spring houses (for refrigeration), and outhouses? V. L. Heid, superintendent of the Hancock County Rural Electric Cooperative in Iowa, repeated a common explanation given by REA field representatives and other co-ops when he said that his customers had used electric power "almost entirely for lights, because of the expense." Hancock suggested calling the monthly statement an "electric bill," instead of a "light bill," to change their way of thinking about electricity. A project superintendent in Indiana indicated a common difference between urban and rural consumer practices in this period when he reported that "more than 90 percent of our customers refuse to purchase equipment on a payment plan but will buy readily as soon as cash is available from farm income." Although resistance to debt was a common economic strategy of small family farms, a co-op manager transplanted to Minnesota attributed this tradition to the "psychology, temperament, and backwardness of the members and prospects along the lines. . . . The Board felt, as I have felt since being here, that the members are all very conservative and of stubborn stock and will not make any moves until they are financially able to do so. The experience has been that when they are ready to make the move of wiring, they will go the entire way." Another Minnesota manager complained to the REA that the Finnish farmers in his area were very "conservative" and would not listen to outsiders, whether they were from the REA or the Extension Service.[66]

Yet none of these reasons adequately answers the question of why some appliances were purchased much more frequently than others. The survey data show that farm people valued a common set of electrical appliances above others. As we have seen, radios were popular in the country because they provided entertainment and reports about the weather and farm markets. High-line electricity enabled farm people to save the expense and trouble of batteries, but this was offset by the need to purchase an AC set. The popularity of electric irons undoubtedly stemmed from their well-known differences from sad (flat) irons: they did not have to be heated on a hot stove in the summer, they could be heated up quickly and used in any room of the house that had an outlet, and they were generally lighter in weight. An Indiana woman recalled recently, "Oh my lands, I couldn't wait to iron, as bad as I hated to iron, when electricity came in."[67] Electric washing machines must have seemed more desirable than the popular, but smelly and hard-to-start, gasoline-engine washers, especially because many a woman had to find her husband or another male and have him start the engine if it quit in the middle of a wash.[68]

Yet farm men and women were often reluctant to part with a working washer (or any other large appliance) for a new one just because it was "modern." Another Indiana woman recalled, "We got electricity in the summer of '42. It had taken that long to find enough money. But if you started out with a gasoline motor on your washer, at that time, you kept on using it until it wore out. You just didn't trade." The Maytag company addressed this situation by bringing out a washer in the late 1920s that would work with either the company's standard kick-start gasoline engine (described in chapter 3) or an electric motor. Electrical dealers added gas-engine washers to their line by the 1930s and would take in the engine in exchange for an electric motor when the family got electricity. The sales of gas-engine washers were not insignificant; they made up 8 to 10 percent of all washer sales from 1936 to 1939.[69] In 1939, the project superintendent of an Iowa co-op told the REA, "we have added a number of hot plates, as many of our members have bottled-gas ranges and will not give them away."[70]

The lower percentage of refrigerators purchased by farm families with electricity, especially in the North, in 1940 seems to be a function of both economic and cultural considerations. Even in urban areas, the refrigerator was initially viewed as a luxury item. Although the percentage of electrified households in the United States having electric refrigerators grew from 34 percent in 1935 to 56 percent in 1940, only 21 percent of all homes in Muncie, Indiana (the "Middletown" made famous by the Lynds), had gas or electric refrigerators in 1935.[71] The cost was high for large farm families. A study of seventy-four rural households in Rhode Island in 1932 found that the average purchase price of an ice-box was $35, compared with $275 for an electric refrigerator. (Refrigerators designed for smaller families in urban areas sold for about one-half of this price in the mid-1930s.) From laboratory tests of the refrigerators, the Rhode Island researchers calculated operating costs of thirty dollars per year for ice and an electric refrigeration bill of forty-seven dollars per year. A kerosene refrigerator used only six dollars' worth of fuel in a year. The annual cost for ice was calculated for only six and one-half months, because that was the average length of time iceboxes were used by these rural families.[72]

As the purchase price of refrigerators came down in the 1930s and manufacturers and the Bureau of Home Economics began to push the new technology, urbanites still bought more mechanical refrigerators than did farm people. One reason was the difference in food preservation and storage patterns on the farm. Farm people had limited access to ice delivery and electricity, but more of them had springs, cellars, and wells. Most farms also had a daily supply of fresh milk and an abundance of fresh vegetables in the summer to be canned for later use. Thus, the transformation in shopping patterns—from daily buying of groceries and home delivery to weekly purchases at larger stores—occurred much later on the farm, well into the 1960s in most areas.[73] Outside the South, year-round refrigeration was seen as something farm people could do without when it came time to decide which electrical appliances to buy in the 1930s. The cost was so high and the usage so marginally beneficial that many who bought electric refrigerators turned them off in the winter months, just as they had used the icebox on a seasonal basis (see below).

The unpopularity of the electric range seems to have been a matter of economics. Although the electric range cost about one-fifth less, on average, than the more popular refrigerator, it had a much higher operating cost, using about four times as much electricity per month (see table A.12). Usually there was also the added cost of bringing a 220-volt service into the house. Furthermore, the range did not fare well against its nonelectric competitors. According to a 1930 study by Purdue University of fourteen hundred Indiana farmhouses, the average purchase price of an electric range was two to five times higher than any other type of stove (coal, wood, gasoline under pressure, or kerosene) and it was two to three times more expensive to operate, as well.[74]

By the time the REA got rolling in the late 1930s, a new competitor had come to the fore: bottled gas. Because the high energy usage of the electric range promised to bail out a lot of financially troubled co-ops, the REA spent a good deal of time promoting it against the new competitor. Bottled gas was developed in the 1920s, when Phillips Petroleum and Union Carbide developed ways to bottle propane and butane. They and other companies then developed the liquefied-petroleum-gas business by lowering the prices of customer storage tanks, stabilizing gas prices, selling bottled-gas appliances at discounted prices, and starting a trade association to promote the new fuel.[75] To meet the challenge, the REA often corrected (from their point of view) erroneous cost comparisons between electricity and bottled gas. In its 1939 range campaign in Iowa, for example, the REA stood by the earlier Purdue study because it showed a gas usage that, at the bottled-gas rate, resulted in an operating cost only seventy cents less per month than electricity, at the REA rate. The bottled-gas dealers claimed that the Purdue study showed too high a usage for gas.[76] The "range" wars intensified after World War II and ended with a victory for bottled gas for cooking and heating, but not for lighting, in the countryside.

There were other factors influencing the decision of whether or not to buy an electric range—ones the REA seems to have ignored. The 1936 report about an extension meeting in the Tennessee Valley recounted that some farm "women were afraid to operate their ranges which perhaps the husbands had bought in order to get the line." "Some farm women shy from the electric range because they have never cooked on one and others are really afraid of them."[77] Furthermore, the coal or wood range (whether used in town or on the farm) served a variety of purposes. In addition to cooking and baking, women used the range to heat their (usually large) kitchens and to heat water for washing dishes, doing the laundry, and bathing. Using the stove in this manner continued on the farm longer than in town because furnaces, bathrooms, and gas water heaters were rare in the country. Many farm families also needed to clean dairy utensils in hot water. For these reasons, most farm women were extremely reluctant to give up their coal or wood stoves. Even when northern women bought a kerosene or gasoline stove, they did so primarily for summer use (for canning and increased cooking during harvest time) and kept their coal or wood stove for the colder months. Two proponents of rural electrification at the Nebraska agricultural experiment station explained in 1934 that "many of the farm homes [studied over a fourteen-year period] have little provision for heat for the kitchen except that given by the range, and most Nebraska farms have [corn] cobs and wood about the farmstead. These two facts will continue to prevent the general use of the electric range."[78] Another disadvantage of electric cookery was the unreliability of electricity owing to frequent outages. Two farm women interviewed recently in New York listed this as a reason for not buying an electric range.[79]

An innovative response by appliance manufacturers was the combination coal or wood stove and electric range. The 1930 Purdue study reported that five of the forty-one farms with electricity had purchased a combination coal stove–electric range. The women told the researchers that they were satisfied with the hybrid device, with which they cooked mostly by electricity and heated the kitchen and hot water by coal. Three women interviewed in Indiana had combination stoves (wood and gas, coal and gas, and wood and electric), as did two men and one woman interviewed in New York (two wood and electric, one wood, coal, and electric). The New York woman, Thena Whitehead, recalled that she and her husband bought a combination wood-coal-electric stove when they got electricity in the mid-1940s for heating purposes but also because she thought gas stoves were too dangerous.[80] Although the REA usually recommended an electric roaster as an alternative technology to the electric range, at least one REA home economist, Elva Bohannan, suggested a hybrid device. In 1941 she asked two Pennsylvania farm women who did not have central heating to consider buying a "wood or coal occasional companion heater finished in acid-resistant enamel to match the electric range." The range could be purchased with or without water-heating coils.[81]

These combination technologies help us parse what modernity meant to farm people. As we saw in chapter 3, there is nothing essentially "modern" about using wood, coal, gasoline, kerosene, or electricity as a fuel: all of them were, at one time, considered to be the latest, up-to-date fuel, that is, they were socially inscribed as "modern." As the coal or wood stove began to seem old-fashioned in the 1920s and 1930s, kerosene, gas, and electric stoves looked more "modern" to both urban and rural eyes because they were more compact, had knobs, seemed cleaner, and usually were made of enamel and steel. When looking at photographs of these devices today, it is difficult to distinguish the combination coal-wood-electric stove from these newer stoves, but not from the coal or wood stove (see fig. 9). The combination stoves looked similar to the newer ones because of their smooth-line cabinets and the fact that cook-

Figure 9. Combination Coal Stove and Electric Range. Combination range developed by manufacturers in response to consumer resistance to buying electric ranges and discarding their
coal stoves, which also heated farm kitchens
Source: J. P. Schaenzer, *Rural Electrification,* 4th ed. (Milwaukee: Bruce,
1948), p. 313.

ing and baking duties could be relegated to the smaller electrical part of the
stove.

The hybrid stove is another good example of an urban-style technology
being woven into rural life. In this case, the material culture (and economic
conditions) on the farm at the time encouraged rural folk to heat the kitchen
and hot water with the same source used for cooking and baking. One way to

bring electricity into this part of rural culture—without upsetting the entire farmstead—was to combine the new with the old technology (an almost literal "weaving" together of so-called urban and rural technologies).

The invention of the combination stove shows the interactive aspect of the contested transformation of the family farm. In response to widespread resistance to the electric range, entrepreneurs developed and sold combination stoves. The result was a change in technology (new forms of stoves) and changes in rural life (new summer cooking patterns and the addition of a symbolically "modern" device in the kitchen). Yet the changes occurred within a familiar structure. Winter heating habits and the size and design of the farmhouse remained the same. Thus farm people's consumer resistance contributed to the shaping of a technological system and the creation of a new form of rural modernity—as occurred with the telephone, the automobile, and the radio.

A related issue surrounding the replacement of the coal or wood range by newer types has to do with cooking skills and a preference for the aesthetic and heating qualities of the older stoves. The evidence on this score, however, is mixed. The recollections of women who cooked with the older stoves vary widely. On the one extreme are statements like those from Elizabeth McAdams of Alabama, who recalled, "I really enjoyed cooking on a wood stove. There's something cozy about a wood stove, and I never minded it at all. I had a modern range that baked beautifully. It was cozy in the kitchen. Even after we left [our rented farm] and bought this place, I moved my wood stove with me. I finally bought my first electric stove in 1951. I hated to part with my wood stove, but I didn't have room in the kitchen. If I had a big kitchen now, I would still want a wood stove." At the other extreme, Anna Sorenson of North Dakota recalled that the new appliance she enjoyed the most was "the electric stove. Push that button and there you had heat. Didn't have to chop wood or carry coal to get some heat to cook on." More toward the middle of the spectrum are the sentiments of Marjorie Whitney from Indiana: her two-tone wood-burning range "was very modern. . . . I thought it was beautiful, but I never thought the stove and myself were compatible. If I wanted it hot, it just sort of simmered along and didn't get hot enough. If I wanted it just medium, it would just get red hot. I never really mastered the big old range. Many people loved them and used them a long time."[82]

The fact that two of these women called their wood stoves "modern" illustrates the extreme relativity of this term. But there was one area in rural life where "modern" had a definite meaning: sanitation. An inside bathroom with a toilet and running water was modern, an outhouse was not. As we have seen, the outdoor privy (and the well-worn path between it and the farm-house) was probably the main symbol of the "backwardness" of rural society in the twentieth century, the object of countless jokes that helped keep rural-urban tensions alive in this period. Morris Cooke so prized the reform of this rural institution that he "told the Committee of Congress [in the mid-1930s] that I wanted inscribed on my tombstone simply the words, 'He believed in the rural bathroom.'"[83]

Believing in the rural bathroom was one thing, creating it was quite an-other. K. E. Miller, senior surgeon in the U.S. Public Health Service, admit-ted as much in the *Rural Electrification News* in 1937. "Unfortunately the rural homes do not readily lend themselves to urbanization from a health stand-point," mainly because of the lack of adequate water sources and methods of sewage disposal.[84] Problems just as serious were high prices of plumbing fix-tures, the necessity, in most cases, to remodel the farmhouse to add a bath-room, the reluctance of REA staffers to push plumbing because of its poor load-building potential, and the opposition of some co-op boards of directors to plumbing loans for this urban luxury. Although the REA made some prog-ress on these matters by negotiating group discounts from manufacturers, making plumbing loans from its own funds, and educating its own staff and the co-ops to this type of farm modernization, it did not meet Cooke's goal until well after World War II.[85]

Trying to fathom farm people's attitudes toward the rural bathroom in this period is difficult. Many prosperous farmers, especially those who owned Delco plants, had installed plumbing before the arrival of the REA. They un-doubtedly made up most of the steady 6 percent of newly electrified farms that purchased water closets up to 1940.[86] Cost was enough of a factor for other families for the superintendent of an Indiana co-op to tell the REA in 1938 that at the "present prices of farm products many farmers do not feel jus-tified in an expenditure of the amount needed to install a bath room." The cost was substantial. Until the end of 1939, the REA loaned an average of $212 per farm just for the installation of plumbing fixtures. A water pump, water heater, and bathroom facilities could easily add another $300, which did not include the cost of renovation to create a bathroom inside the farmhouse.[87]

Another factor was the not inconsiderable cultural change associated with using a toilet indoors. Farm people took baths in the house, often near the kitchen stove, but discharging human waste was traditionally viewed as an activity that "naturally" belonged to the great outdoors, or at least to a special facility built outside the main house. A report by the REA's Willard Luft in-dicates how difficult this cultural shift was to make (and also provides an example of the ploy of using the young to initiate change). Having set up a plumbing demonstration and water-carrying contest in a North Carolina high school in 1938, Luft proudly told an REA conference, "Those boys and girls went home at the end of the week, [and] we heard the farmers in town on Saturday saying, 'Well, we are not going to have any peace at home until we get these things.'"[88] But most farm people did install plumbing, usually after World War II, and they later remembered the event fondly. An Indiana women recalled, "We had to wait several years, and finally we got a bathroom, and that was the highest part of my life, was when I got a bathroom in the country."[89]

Yet when farm people bought urban-style electrical appliances, they often used them in rural ways. Refrigerators were run on a seasonal basis, reflecting agricultural rhythms rather than those of the city. Two of the four test farms that purchased their refrigerators during Emil Lehmann's study of electrification in Illinois in the late 1920s, for example, used them only in the summer months. When the Belfalls Electric Cooperative in Texas reported that its average usage dropped from thirty-five kilowatt-hours per month in November 1938 to thirty-one hours in December, A. R. Tucker, the project superintendent, explained, "The only reason I can account for this is that so many of our customers disconnect their refrigerators at this time of year."[90] Rural neighborliness also came into play. George Kable, the TVA's chief agricultural engineer, told an REA staff conference in 1939 that he had

> heard the story of a lady who was doing her ironing on a minimum bill and [it was] along toward the end of the month, and she had learned to read the meter—that is an educational program, don't teach them to read the meter—she was watching this, and when she had about finished she noticed that she had used up her monthly minimum and she went to the phone and called up her neighbors and found one woman who hadn't used the minimum so she took the iron over there to finish it. (Laughter.) That isn't paying back for the line.

Even lights were used selectively. Tucker told the REA, "We have a good many members [who] when they use their minimum kWh they go back to the old coal oil lamp."[91] As late as 1941, the *Rural Electrification News* reported the anecdote of a Kansas farmer who only turned on his electric lights to find his kerosene lamp.[92]

Alternative uses of electrical equipment were not uncommon. Farmers discovered that a radio playing in the barn increased milk production and decreased milking time. A North Dakota farmer put a radio in a potato-cutting room, which increased the production of his men by more than 10 percent. A South Dakota farmer even used a chick brooder to ease his rheumatism. A Missouri farmer rigged up a piece of angle iron on the shaft of an electric motor to kill rats in a barn. The REA reported in 1939 that Mrs. I. E. Murray of Ash Grove, Missouri, "has found that the wringer of her electric washing machine made an efficient pea sheller. Her cooperative superintendent visited the Murray home one day and she shelled and canned thirty-two quarts of peas between eleven o'clock in the morning and five o'clock in the afternoon," without crushing any peas.[93] Although the selective and alternative uses of electricity do not involve the radical altering of artifacts that occurred with the telephone and the automobile earlier in the century, they show that before World War II, some farm people did not use electrical appliances in exactly the manner prescribed by manufacturers.

All data indicate that electrical farm equipment found little favor with the majority of farmers. Dairy farms bought electrical milking machines, milk coolers, and water pumps, chicken ranchers purchased electric brooders and lights for the factory-like henhouses, and large western farms dependent on irrigation installed electric water pumps. But farmers who did not specialize in these areas saw little reason to adopt the new technology. Even dairy and poultry farmers were reluctant to "go electric" because of their large investment in oil brooders and gas-engine-driven pumps, milking machines, and milk coolers. A co-op superintendent in Iowa told George Munger that most farmers in his area had oil brooders and "they feel discarding the brooder on hand would be a waste. They have expressed a desire to have an electric one when the old one is to be replaced." There were technical concerns as well. The frequent loss of power on the Denton County Electric Cooperative lines in Texas hindered sales of brooders, as did the disastrous experience of a neighbor who shifted from an oil to an electric brooder and lost most of his chicks in this delicate art.[94] Designed to be affordable to the average farmer, the one-can milk cooler was not popular in Pennsylvania (as we saw in the last chapter) nor in Kansas in 1941, when farm income was rising.[95]

The Response from Agents of Modernity

Promoters responded to the choices made by farm people in a variety of ways. Private companies tended to focus on economic matters in dealing with consumer resistance in the countryside. A 1938 Sears, Roebuck booklet advised the company's agents that "sales resistance to electrical equipment can be removed if it is shown that the prospect is paying plenty for the service he gets on the non-electric farm and that electricity will save him money as well as do the job better. A simple chart in which his present expenditures are totaled and compared with what electricity would do the same job for, will convince anyone. Such charts can easily be made up from the data given in this booklet."[96]

The REA's chief home economist, Clara Nale, had difficulty understanding the appliance choices made by farm people. A survey of ten Ohio co-ops in 1938 showed that 40 percent of families had purchased vacuum cleaners but only 17 percent had bought water systems, even though salespeople at equipment dealers and the REA offices seem to have pushed both appliances with equal vigor. Nale thought farm women should plan better. They should save their money for a laborsaving water pump rather than buy a vacuum cleaner when they had "comparatively few rugs."[97] The paternalism in Nale's plea was also evident when she wrote that "it is the privilege of home economists to study this [REA] program in all its broadest implications and to *interpret* its possibilities to the rural homemakers of America."[98]

By interpreting the artifacts of rural people for them in their best interests, home economists and agricultural engineers did not always endear themselves to a social group known for its pride and independence. Although rural people appreciated information on wiring the farmstead and selecting and repairing appliances, they probably resented the occasional humorous article in *Rural Electrification News* on the ignorance of farmers toward electricity. Minnesota women politely watched the REA's Victoria Harris cook a dinner for fifteen hungry harvest hands on an electric range in 1938, then kept their wood stoves, which warmed the kitchen on cold winter days.[99] Indiana women emphatically disliked the attitude of the REA's Enola Retherford, who gave them a "balling out" at the Farm Tour in 1941 if "they were noisy or had gotten up to leave" during her demonstration.[100]

Located at public nodes between the producer and the consumer in the rural electrification network, the REA's home economists faced a major problem in responding to rural resistance. Although their profession stressed adult education, the REA had hired them to convince rural folk to use more electricity—resulting in a divided loyalty between professional ideals (education) and business demands (sales) similar to that experienced by salaried engineers and by home economists employed by electrical manufacturers and utility companies in this period.[101]

Although increased rural electrification was also a goal of the Extension Service,[102] many of its home economists, following a time-honored tradition in their field, discussed the relative merits of competing household appliances. An Alabama agent told farm women how to make technical and economic comparisons between gas, kerosene, and electric refrigerators. Home economists also took into account the fact that many of their clients had less cash income than city folk. A Tennessee agent declared in 1936 that "much poor lighting is due to lack of forethought and care rather than a lack of electricity." Louise Peet, a specialist in household appliances and professor of home economics at Iowa State University, wrote an article as late as 1944 suggesting ways to improve kerosene and gas-mantle lamps because a "large percentage of farm homes are still dependent on liquid fuels for light."[103]

Home economists at Cornell University faced a particularly difficult situation because of the close association between utility companies and the university, described in the last chapter. Seeing rural electrification as a means to carry out the goals of rural reform and to strengthen its expertise in household technology, the Cornell home economics department encouraged Professor Florence Wright and her colleagues to develop the new field as they would any other, by teaching classes, writing pamphlets, and training extension agents. From 1936 to 1939, they also conducted statewide electrification meetings in conjunction with home service directors from utility companies. Wright, who served on the university committee that coordinated the rural electrification program in the state, was fully aware of potential conflicts of interest and sought to avoid them. She told a conference in 1936 that she always made it clear to utility company home economists "that when working on a program with us they must keep entirely on an educational basis."[104]

Clara Nale held this philosophy when she worked for the Tennessee Valley Authority. According to the REA report of a 1936 visit to the TVA area, "Miss Nale stated that in educating the rural people, she talks from the standpoint of use not dollars. In her first contacts with men and women she talks about the selection of equipment in general. She is not sure that farm families should have large refrigerators. She thinks that the size and habits of the family should determine the size of refrigerator, and one smaller than that which urban people use may be large enough." They should buy what they could afford, she suggested. Nale said that training for field-workers "must be bigger and broader than just selling household equipment. . . . The Home Economist's work is more the Extension type of work than the utility type. High-powered salesmanship does not work with rural people in the long run," as was shown in the TVA's Alcorn co-op. Field-workers "must be educationally minded rather than commercially minded."[105]

Nale may have maintained this philosophy after joining the REA's Utilization Division in 1937, but complaints about the division's high-pressure sales techniques were often heard within and outside the agency. Dora Haines and Udo Rall complained in 1939 that "the effectiveness of load-building activities has suffered from wrong approaches and misplacement of emphasis. . . . High-pressure salesmanship is more costly and less effective in the long run" than educating "a rural community to understand the importance of consumer-controlled electrification as a factor in solving its social and economic problems." Rall noted two years later that "county agents, being educators, have been alienated by REA co-ops' disregard of the educational approach in favor of a commercialized selling and load-building program with a close dealer tie-up."[106] In 1940, five state extension service directors reported that "their leaders objected to the present pressure methods employed in building up load on REA lines. Certain activities, such as electric cookery programs, did not seem to fit in with the needs of local people. On the other hand electric brooder and electric feed-grinding programs were commended."[107]

The criticisms made their way to Congress in the spring of 1941 and marked the end of an era. According to an unfriendly electrical trade journal, the House Appropriations Committee rebuked the "ambitious load-building program" and the "excessive paternalism" of the REA. This led to a cut in the agency's administrative funds. The REA had already taken a precautionary move in announcing in early 1941 that it had abolished the Utilization Division and moved its personnel into the new Cooperatives' Operation Division. Oscar Meier became head of that division's Cooperatives' Education Section, which took over the functions of the Utilization Division.[108] Meier had attended to the problems of poor farmers by promoting self-help wiring programs and criticizing REA personnel for concentrating on the upper two-thirds of farmers traditionally served by the Farm Bureau and the Extension Service. He had worked with 4-H Clubs, encouraged REA field representatives to cooperate with county agents, and persuaded the Extension Service to hire state rural-electrification specialists. In the spring of 1941, he moved six field home economists to his Washington office to develop programs in the areas of lighting, kitchen design, laundry equipment, school electrification, 4-H Clubs, and the Farm Tour.[109] The move was propitious, because the entry of the United States into World War II that December placed a hold on rural electrification, which after the war proceeded apace into a new era of consumerism.

Part Three

POSTWAR CONSUMERISM

CHAPTER 8

Completing the Job

By the start of World War II, large numbers of American farmers had adopted new communication, transportation, and household technologies. Although many farm people gave up the telephone during the Great Depression (only one-fourth of farms had telephones in 1940), about 60 percent of farms owned radios and automobiles at the start of the war. The Rural Electrification Administration increased the number of electrified farms dramatically, from 13 percent in 1930 to about 30 percent at the start of the war (see fig. 2). The war brought a renewed prosperity to American agriculture. The net income from farming tripled, from $706 per farm in 1940 to $2,063 in 1945 (see table A.1). Good times continued in the immediate postwar years. Historian Gilbert Fite observed in the early 1980s that the "period from 1940 to 1952 was the longest period of sustained prosperity in American agricultural history."[1] Farm prices fell and income declined slightly in the 1950s, but not to the depths reached in the depressions of the 1920s and 1930s.

The postwar years are generally seen as the climax of the great transformation of rural America. The sustained efforts to industrialize agriculture and to urbanize the culture of rural life, which began in the mid-nineteenth century, came to fruition in the 1960s. Farming became a capital-intensive enterprise pursued by fewer and fewer people on larger and larger farms. The productivity of agriculture increased dramatically—owing to a fever pitch of mechanization, the planting of higher-yielding crops, extensive use of chemical fertilizers and pesticides, increased specialization, and the adoption of new management techniques. Facing government policies that favored larger farmers and a "cost-price squeeze" that started in the early 1950s (the high cost of agricultural equipment and supplies overwhelming low farm prices, owing largely to overproduction), marginal and near-marginal farmers left in droves.

The farm population, which had remained at a fairly constant level of 30 million people living on 6 million farms throughout the first half of the twentieth century, shrank to one-half its size by 1964, to 13 million people on 3 million farms (see table A.1).[2]

Those who farmed in the postwar era demanded the "city conveniences" of electrification and telephony. Having survived the low priority given its project during the war, the REA geared up for the task of finishing the electrification of rural America. At the same time it began to rebuild another networked service—rural telephony—which, ironically, the REA had harmed when its growing number of power lines disrupted telephone calls on small systems. Its programs completed these two infrastructures essential for postwar efforts to modernize the farmhouse. Although the general farm prosperity following the war made the REA's task easier, the agency relied on the organizational and promotional methods it had developed before the conflict, mainly because farmers and local leaders still resisted its aggressive tactics. The REA's practices thus had a good deal of continuity with the past, during a period usually known as a watershed in American life.

Rural Electrification Achieved

Although World War II accelerated many changes in science and technology, the entry of the United States into the war in December 1941 braked the expansion of rural electrification. "Because electric-power systems and home plumbing installations use the same materials as battleships and aircraft and bullets," explained the REA, "a substantial part of this money [allotted by Congress] will not be translated into construction until after the war." The "manpower" demands of the armed forces and wartime agencies also cut sharply into the REA program, which was not a high priority during the emergency. With office space at a premium in Washington, in March 1942 the REA moved its headquarters to St. Louis for the duration, thereby releasing engineers and other professionals for the war. By midsummer, more than two hundred staff members had transferred to war agencies and about one hundred had joined the armed services.[3]

The utilization side of the house was also hit hard during the war. Shortly after the bombing of Pearl Harbor, the REA replaced the popular Farm Tour with traveling "electro-economy" demonstrations at co-op annual meetings; the exhibits were canceled after a few months because of the lack of rubber tires for the caravan's trucks. Oscar Meier, head of the Cooperatives' Education Section, joined the War Production Board (WPB). Oneta Liter and two colleagues were loaned to the Bureau of Home Economics to perform research on food conservation. Louisan Mamer, one of the first group of field home economists hired by the REA, recalled that she and the other regional home economists "were brought into headquarters in St. Louis, Missouri, to substitute for men being taken out of the agency by the war program." She also did some postwar planning in home electrification under the supervision of chief home economist Clara Nale. In the fall of 1942, Mamer and three other home economists transferred to the Applications and Loans Division to do work that had generally been assigned to men, including trying to save six co-ops in New York "from dismemberment by power companies." The REA then loaned its version of Rosie the Riveter to the War Production Board in 1943 "for liaison work between REA and WPB's Scrap Metal and Rubber program."[4]

The war opened new doors for women in the REA, but paternalism prevailed. In 1942, the *Rural Electrification News* called seven women working in its Technical Standards Division "petticoat engineers." The journal also publicized cases of women temporarily taking men's duties in REA cooperatives. Another 1942 headline read, "Her Husband Went to War, So Mrs. Hinton Took His Super's Job." Hinton was president of her local garden club and had worked ten years as a bookkeeper with a large dairy farm before becoming acting manager of the Fayette Electric Cooperative in La Grange, Texas. A 1943 headline told readers, "Tennessee Co-op Finds Girls Can be Trained to Do Technical Jobs," like meter testing. The journal did not reveal that almost forty women had sat on co-op boards of directors since the beginning of the program. A more typical story was entitled "Electricity Frees Farm Women for Defense Work."[5]

The productivity argument enabled the REA to resume some of its operations during the war. Although the agency built lines and added consumers at a pace far behind prewar levels, its wartime expansion of 3 to 10 percent per year was not insignificant (see table A.14). The REA acquired and refurbished power company lines, served war plants and military facilities located in rural areas, organized co-ops, and even built lines to some farms. The War Production Board, which controlled the use of wartime materials, was reluctant to cooperate at first. The board issued an order in March 1942 granting "permission to complete construction on thirty-four REA projects which had been 40 percent or more completed" before Pearl Harbor, then froze construction in August. The REA acted quickly. Having overcome charges in late 1941 that it had hoarded copper wire in a vacant cotton field in Texas, the vindicated agency teamed up with the newly established National Rural Electric Cooperative Association (NRECA) to lobby the War Production Board to resume construction on a limited basis. The board accepted the NRECA's arguments that rural electrification would increase agricultural production and make up for "manpower" lost to the war, and it ruled in January 1943 that the REA could build extensions to eligible farms.

In April, the board agreed that another thirty-two projects could finish building their systems. Under the January order, farmers who had enough "animal units of production" (one unit was allowed for each milk cow, six hundred broilers, seventy-five laying hens, three brood sows, etc.) and the necessary electrical equipment qualified for an extension of up to one mile to be built from a high-line to their farms. County agricultural boards certified the applications, which were approved by the board and the REA. Farm families needed ten units (soon reduced to five) to qualify, plus one unit for every hundred feet of line. Farmers responded enthusiastically. The REA approved twenty-six thousand applications by mid-1943 and another sixty-seven thousand the next year.[6]

The new lines were important to the stalled REA program in many areas, especially in New York state. The six co-ops that Mamer had worked to save from "dismemberment" by private utility companies had their loans approved shortly before Pearl Harbor. The co-ops got a break the following spring when Governor Herbert Lehman signed the Rural Electric Cooperative Law, exempting co-ops from regulation by the state's Public Service Commission. The law shielded co-ops from utility companies, who used the commission to attack them. But because the co-ops did not energize their lines before the War Production Board froze construction in August, the REA had to shut them down. There things stood for a year; the board then used its new guidelines to approve the construction of more than seven hundred miles of line for three co-ops: the Otsego Electric Association in Oneonta, the Delaware County Electric Association in Delhi, and the Oneida County Electric Association in Clinton. The board permitted two more co-ops to build lines in 1944. These actions enabled the REA to survive in New York. The Chautauqua-Cattaraugus Electric Association in Cherry Creek energized its lines in February 1944, the first co-op in New York state to do so; the Otsego cooperative turned on power to farmers in August of the same year.[7]

Rural electrification boomed after the war, aided by several decisions made during the conflict. After investigating charges of influence peddling over copper-wire contracts and listening to complaints against Administrator Harry Slattery from many quarters, Congress exonerated Slattery in 1944 and helped lay the groundwork for the postwar program by passing the Pace Act that September. The act, drafted by Representative Stephen Pace of Georgia, removed the ten-year limit on the REA program, set the agency's loan rate at the low figure of 2 percent, and increased the amortization period from twenty-five to thirty-five years. Slattery's weak leadership, indecisiveness, and absenteeism because of illness had made him many enemies in the U.S. Department of Agriculture, the REA, the NRECA, and the White House. He finally resigned after President Roosevelt was reelected in November. Roosevelt proposed an experienced New Dealer, Aubrey Williams, to replace Slattery, but the Senate turned Williams down in March 1945. On May 12, four days after V-E Day, the War Production Board lifted its major restrictions on line construction, and by midsummer virtually all such restrictions had been removed. On May 23, Harry Truman, who had been president for only a month following the sudden death of FDR, appointed Claude Wickard as the fourth administrator of the REA. A prominent Indiana farmer, an embattled secretary of agriculture who had had to share the management of the USDA with food administrators during the war, a fierce critic of Slattery, but a strong supporter of the REA, Wickard looked forward to running an agency that directly helped farm people. The REA moved its headquarters back to Washington in early 1946. Its staff, which had reached a low point of 650 in mid-1943, nearly doubled to supervise an unprecedented construction program.[8]

The REA and private utilities finished the job of rural electrification in a remarkably short time. In less than a decade, from 1945 to 1954, the percentage of farms that were electrified grew from 48 to 93 percent. The figure stood at 96 percent in 1956, the last year data were compiled for what was essentially a completed task. Postwar prosperity and the "pent-up" demand for consumer goods, which was apparently strong among GIs returning to the farm, help explain this phenomenal growth.[9] The rural exodus from family farms to the town and city, the enclosure movement in the South, which forced sharecroppers off the land, and the general consolidation of farms into larger units were significant factors, as well, because most large farms were already electrified. The traditional argument of rural reformers, that supposedly urbanizing technologies like the telephone, the automobile, and electricity would keep people on the family farm, finally came undone in 1954. That year, the REA reached sight of its long-sought goal of universal rural electrification, but the next few years saw a sharp drop in the number of farmsteads: a decline of 1.1 million farms (22 percent) in five years. Rural electrification had gained full force during the late 1940s, but it could not stem the economic and social forces pulling men, women, and youth to the brighter lights of cities and towns.

The REA played the major role in rural electrification in these years by embarking on a massive construction program that stimulated private utility companies to increase their efforts in the countryside. In nine years, from the end of 1945 to the end of 1954, more than nine hundred REA co-ops constructed almost nine hundred thousand miles of line, connecting more than 1.6 million new consumers (see table A.9). The co-ops served other rural customers, such as rural industries, churches, schools, and townspeople, but they concentrated on bringing electricity to farms. At the end of 1954, about two-thirds of the REA's 4.1 million consumers were farms, accounting for 61 percent of the electrified farms in the United States. (The percentage of the REA's consumers who were farms declined sharply in 1950, partly because the Census Bureau changed its definition of a farm that year in a way that removed almost half a million suburban homes from this category.)[10] The growth in the number of lines and consumers rose sharply after the war, and the rate held at about 20 percent from 1946 to 1949, the peak year of the REA building campaign (see table A.14). The rates tapered off in the economically leaner 1950s as the agency strove to complete its goal of electrifying all farms in a co-op's area, what it called "area coverage." The REA established about two hundred co-ops during the war, in preparation for the postwar building boom, and about the same number during the middle of the boom in the late 1940s. The REA essentially stopped organizing co-ops in the early 1950s and eliminated many of them through consolidations.

Data published to celebrate the twenty-fifth anniversary of the REA in 1960 provide a statistical snapshot of the role of co-ops in the mature program. At the end of 1959, REA's 927 cooperatives served a little more than 2.5 million farms, representing about 70 percent of the nation's electrified farms. The agency operated in Alaska and in all of the continental United States except Massachusetts, Connecticut, and Rhode Island, where private companies prevailed. Cooperatives powered more than three-fourths of the electrified farms in ten states: North Dakota, South Dakota, Minnesota, Wyoming, Nebraska, Kentucky, Missouri, Mississippi, Oklahoma, and Georgia. The REA was particularly strong in prosperous states with strong cooperative traditions (serving 64 percent of electrified farms in the Midwest) and in poor regions that private companies had not developed (serving 57 percent of electrified farms in the South). The REA was weakest in areas that had achieved a good deal of rural electrification before the agency was established (serving 35 percent of electrified farms in the West) and near urban centers (serving only 8 percent of electrified farms in the Northeast). The number and size of co-ops varied greatly by state. Texas led the way with ninety-six co-ops; Iowa was next with fifty-five, but thirteen states had no more than ten co-ops. The average cooperative served twenty-five hundred farms, but the size of co-ops varied widely, from those with fewer than three hundred farms in Maine and Arizona to those with more than seven thousand farms in Arkansas and Minnesota.[11]

All was not smooth sailing for the postwar program. Shortages in materials with which to build lines plagued the REA every year until 1949 and reappeared with the onset of the Korean War in 1950. By mid-1951, the agency had worked out an agreement with the National Production Authority and the Defense Electric Power Administration whereby the REA administered a program for control of pooled materials for its borrowers. Although it was a marked improvement over the agency's relationship with the War Production Board in World War II, Wickard reported "shortages of varying degrees in twenty-one states" as late as February 1952. Rising prices and shortages of labor and electric power also hampered the agency. In 1946–47, power shortages in some areas of the country "became so acute that co-ops were forced to close down entire sections of line at peak-load hours of the day, causing their members to relight their discarded kerosene lamps and perform farm chores by hand." Power shortages and the cost of power eased in 1950, as the REA and utility companies built more power stations (urban centers also lacked enough electricity in these years). But the utilities did not stop fighting the program in many areas and continued their prewar tactics of serving prosperous farmers before anyone else and building spite lines.[12]

Utility companies tried a new tactic after the war: buying out cooperatives. The first instances occurred in 1946–47, when the Jordan Valley Electric Cooperative in Oregon and the Long Valley Power Cooperative in Idaho sold out to utility companies. It appeared that the Craig-Botetourt Electric Cooperative in Virginia would meet a similar fate. Founded in 1936, the co-op started with a little more than 100 miles of line and nearly 150 consumers in 1938 and grew to 520 miles of line and 1,830 consumers a decade later. The co-op, though, had many problems. In 1940, the REA cited its bylaws for violating eleven principles of a good cooperative. The attendance at its annual meetings averaged fewer than thirty members (an objectionable bylaw permitted unlimited proxy voting), it had poor relations with appliance dealers, and material shortages prevented it from expanding after the war. When a nearby power company offered to buy the co-op for a half-million dollars in 1948, REA field representatives got into the act and mobilized a publicity campaign against taking the offer. Other Virginia co-ops loaned materials to the struggling co-op so it could connect two hundred customers before the annual meeting. A record crowd of seventeen hundred voted down the buy-out. The REA also stopped the takeover of the Southeastern Michigan Rural Electric Cooperative the following year by conducting another extensive ed-ucational campaign.[13]

Such campaigns were becoming increasingly necessary because utility companies raised the specter of that venerable bogeyman, socialism, during the anticommunist fervor of the early cold war to combat the growth of the REA and other forms of public power. Utility companies claimed in 1947 that "cooperatives are the mildest form of socialism or commies." As the REA be-gan to invade the turf of private companies by building more generating plants, the utilities waged a stronger publicity campaign against the New Deal agency. One piece of propaganda for 1949 was a fake dollar bill, labeled "One Tax-Free Buck," attributed to the "Co-operative Commonwealth," whose motto was "In Tax Exemption We Trust."[14]

A public opinion poll in Texas at this time found that respondents were not overly concerned about the taxation issue. But more education was neces-sary, because two-thirds of those surveyed did not know REA co-ops were owned by their members. Other areas of the country were more alarmed. Carl Wild, manager of a North Dakota project, told a meeting of co-op leaders in early 1950, "The utility theme, 'business-managed, tax-paying electric compa-nies,' with its implication that co-ops are not business-managed, tax-paying systems, is continuously being spread through utility advertising in the press and over the radio. Utilities label rural electric co-ops as socialistic and un-American in an effort to destroy public confidence and respect for the REA program." The leaders listed the following items under "Misinformation Fre-quently Heard": "A co-op is just a slick way to put the Government in busi-ness. That's the first step toward Communism. Co-ops are a foreign idea, brought over here a few years ago by a crowd of left-wingers. Co-ops operate at the expense of tax-payers. That's a direct Government subsidy. . . . Co-ops violate the American tradition of business-managed enterprise." The REA advised the leaders to respond that yes, co-ops received government loans, but they were privately owned by farm families, run in a business-like manner by paid managers, and were exempt from federal income taxes only because they were nonprofit organizations.[15] Another common defense was that co-ops upheld America values because they fostered democracy at the grass roots, a variation on the theme of TVA chairman David Lilienthal's popular book, *TVA: Democracy on the March* (1944).[16]

The utility company attacks resonated with the results of the Commission on Organization of the Executive Branch of the federal government. Established in 1946 by President Truman and chaired by former president Herbert Hoover, the Hoover Commission, as it became known, recommended in its final reports, published in 1949, that the REA become self-supporting by securing its funds from private sources, like other purportedly "subsidized" lending agencies of the USDA. Critics of the commission said the proposal would virtually liquidate the REA.[17]

Although Congress allotted the REA record amounts of money to build electrical distribution systems in the postwar era, it kept a keen eye on the management of the agency, especially because Republicans managed to gain control of both houses in the Eightieth Congress (1947–49), a goal that had eluded them during the Roosevelt years. The House cut 419 positions from an REA staff of more than a thousand in May 1947; the agency managed to restore 218 positions, but only 130 employees returned.[18] In 1949, Representative Porter Hardy Jr., chair of a government operations subcommittee, initiated a thorough investigation of the REA. Hardy looked into the financial health of troubled co-ops, alleged kickbacks, and almost every aspect of the agency's operation. Hardy wanted to know why Wickard continued the REA's extension activities, which the Hoover Commission, in one of its preliminary reports, had recommended abolishing. Wickard replied that the "REA does not carry on extension work in the usual sense of the term." He thought the Hoover task force did not understand that the co-op's power-use and member-education programs were conducted not by REA field personnel but by electrification advisers hired by the co-ops. Wickard admitted that the agency oversaw these programs, but only to increase electrical consumption so the co-ops could pay off their government loans in time, the traditional argument used to justify its oversight role. Mindful of Hoover's criticism, Wickard said his field personnel did not work directly with farm people but only with co-op employees, the same policy followed by private utility companies. Hardy sent his own detailed questionnaire on this and other (mostly economic) matters to all co-ops, then issued a report in 1950 that was generally favorable to the REA.[19] But the REA reduced the size of its Information Services Division in 1951 to comply with the 1952 Agricultural Appropriation Act, which cut the federal government activities in press, radio, and popular publications.[20]

The changes at the home office were accompanied by a multitude of challenges in the field when the REA took up the task of meeting a farm demand for electricity that was at an all-time high. The agency put a human face on its backlog of a quarter of a million dollars in applications in 1945 and 1946 by reporting such anecdotes as that of an Arkansas farmer who, when told his family lived too far from the high-line, put skids under the house and moved it a half mile to get electricity. When farmers requested larger service entrances than a New York cooperative provided, the co-op decided in 1947 that it could not meet the expense of satisfying this demand and stayed with its standard 60-amp service. An Alabama co-op decided in 1946 not to connect new customers to its overloaded lines until a "new substation is built or [the] trouble is corrected." In response to a USDA survey in 1950, Mrs. Larry Hamilton, who helped run her dairy farm in New York with her nineteen-year-old son and his wife because her husband was sick, said they had paid six hundred dollars for a Delco plant, which they could afford to use only "in the barn to earn our living." They had signed up for the REA, but the co-op had run out of line to serve them.[21]

Although the REA undoubtedly preferred such problems to the prewar difficulties of persuading skeptical farm people with little cash income to sign up for its program, the agency faced many of the same obstacles it had before the war. Postwar consumerism and the desire to modernize the farmstead seem to have erased many forms of resistance prevalent in the 1930s—such as cutting down poles, not signing up for the program, and refusing to grant easements—but the REA faced familiar difficulties in running co-ops and in convincing farm people to increase their electrical consumption to levels the agency desired.

Referring to the near takeovers of cooperatives in Virginia and Michigan, Kermit Overby, director of the Information Services Division, told Wickard in 1949 that the long-standing (and now dangerous) problem of apathy among co-op members was mostly the result of the rapid growth of the REA. The doubling of co-op membership, the addition of one thousand to two thousand board members, and the high turnover rate among managers brought into the system hundreds of thousands of people who had no, or very little, knowledge of cooperative principles. A study of the cooperative education program in 1950 made similar complaints. Deputy Administrator William Wise admitted in 1952 that "in far too many of our co-op borrowers, member apathy appears to be still rather widespread."[22]

Field representatives reported the old problems of autocratic and self-perpetuating boards of directors in Florida and North Carolina. The president of a North Carolina co-op was not taking electricity as late as 1952.[23] The REA thought financial malfeasance was enough of a problem in all areas in 1951 for it to consider revising an administrative bulletin, "Borrowers Records and Accounts: Irregularities," to require the co-op to take more responsibility for detecting fraud rather than rely on the REA to audit almost a thousand projects. An Alabama co-op addressed the problem of collecting tenant's bills in 1946 by making landlords responsible for them.[24] Many other problems grew out of the program's rapid expansion. Sheridan Maitland, author of the 1950 study of cooperative education, observed that "examples of indifference or even hostility [by co-op officials] toward the ideas of area coverage, patronage capital, and democratic processes in general are all too common. Field reports indicate that REA field men obtain similar impressions in face-to-face dealings with many co-op officials."[25]

There were good reasons for resisting. The REA had pushed for area coverage since the days of Morris Cooke, but it did not have to face the issue squarely before the war because of the large number of relatively prosperous areas that could be readily served. Once co-ops had accomplished this task, many were understandably reluctant to extend their lines into less affluent areas, even after the Pace Act lowered the loan rate and extended the amortization period. The REA insisted and in 1946 began to require co-ops to take four steps to achieve area coverage: (1) make a boundary survey of the eventual extent of the co-op area, including available power sources and possible integration with existing systems; (2) survey all potential customers in this area and sign them up for the co-op if possible; (3) prepare an engineering design to achieve full area coverage; and (4) implement the design section by section. By 1948 practically all borrowers had completed the first stage, most had embarked on the second stage, and more than five hundred co-ops had finished the engineering design. Under political pressure from Congress to complete the program, the REA reported in 1949 that "nearly all" co-op boards had committed themselves to area coverage and that it "is doing all it can to stimulate the rest of its borrowers to go on record as adhering to this principle."[26]

The agency took a step in this direction in January 1950 when it made area coverage a requirement to obtain loan contracts for distribution systems.[27] This was necessary for a variety of reasons, as indicated by a report of a field conference in mid-1950.

> It was brought out that some of the co-op managers are lax in signing up members because they do not want members "getting on their necks" for service. Too, some of them are anxious or pessimistic about economic conditions in their particular co-op areas; the managers and boards of directors are becoming too conservative, viewing the program on a

short-term basis. Here again, it is a job of "selling" area coverage to the co-op people, pointing out that electricity will be the last thing farmers will do without, spending more time with the managers on this question.

How should field representatives handle co-op boards? "If you are dealing with conservative boards who hesitate to proceed, or are under the impression that they have reached saturation, the best plan is to have them certify as to the percent of saturation. In many cases this will break down the board's hesitancy in applying for additional loans to complete area coverage."[28]

The controversial capital credits program asked co-ops to record the capital each member paid into the cooperative as part of their monthly electric bill. In large part this was another attempt to forestall co-ops selling out to power companies by educating members that they received their electricity from a consumer-owned cooperative, not a power company. Co-ops resisted the plan because of the expense, time, and vast changes in bookkeeping involved. The REA persisted and reported in 1952 than two-thirds of its co-ops had adopted the plan after a six-year struggle. That year, two co-ops in Indiana and Louisiana even refunded capital credits. Although the average refund was only fourteen dollars per member in both cases, the refund checks were the highlight of the annual meetings and provided another means to combat member apathy.[29]

It is difficult to judge how well the REA addressed another prewar problem: too few women as co-op leaders. Wickard did not push the issue as hard as John Carmody had in the 1930s; only once in his tenure, for example, did the *Rural Electrification News* publicize the prewar policy of joint membership for husband and wife.[30] But some women attained high positions. In 1952, five women sat on co-op boards of directors in Texas (four of whom were officers), and nine on boards in Illinois (three of whom were officers). Amelia McNeill was secretary of the board of the Monroe County (Iowa) Rural Electric Cooperative in 1949. Although Ruth Stevenson, the first woman president of a co-op, stayed on her board in Illinois as a regular member until 1950, I have found evidence of only one woman president elected after the war: Anna Boe Dahl, who became president of the Sheridan Electric Cooperative in Montana in 1956 (compare table A.10).[31]

New to the program were women managers. In the late 1940s, Delta Scales managed the Upshur Rural Electric Cooperative in Gilmer, Texas; Auby Snow managed the Halifax Electric Cooperative in Brattleboro, Vermont; and Madge Robertson managed the Washakie Rural Electric Cooperative in Worland, Wyoming. Following the path taken by many women who succeeded in business and politics in this period, Scales, apparently a widow, had followed her husband, Samuel, in the job. The REA's attitude toward women managers is unclear. The agency seems to have eased Scales out of her position around 1951 (she ended up doing well in the real estate business). Field representative Bernard Krug showed more sympathy for Snow in Vermont in 1951 but still said (probably paternalistically), "It is somewhat pathetic to see this lone female, battling against almost incalculable odds to put over the idea of cooperative rural electrification."[32]

Rebuilding Rural Telephony

The REA's success in electrifying rural America contrasted sharply with the situation in telephony. The percentage of farmsteads having electricity climbed rapidly after 1935, but the percentage having a telephone dropped during the Great Depression, reaching a low point of one-fourth of farm homes in 1940. More farmers installed phones during the agricultural prosperity induced by World War II, but the figure had risen only to about a third of farm households in 1945, probably because of a shortage of equipment during the conflict (see fig. 2). A few years later, the *Prairie Farmer* contrasted the dismal rural figures with conditions in urban areas.

> Farmers want to live like civilized people. They know what even ordinary folks in town have and they think they have earned the same conveniences.
>
> Yes, we have made some progress. Electricity has put the old kerosene lamp on the shelf for good in almost 90 percent of the homes in *Prairie Farmer* Land. On many farms the weather-beaten old privy out back has now been replaced by shiny new indoor facilities. That's progress.
>
> Did someone mention telephones? Ouch! That's a horse of another color.[33]

One reason was that farm people preferred other consumer goods to the telephone. When deciding what expenses to cut during the hard times of the depression, many farmers chose to give up their phone but keep the car, even though automobiles cost more to operate.[34] In addition to providing transportation, the car, as we have seen, could easily be converted into a stationary source of power or into a small truck to haul produce to town. For large numbers of farm families living on the margins, the telephone seemed like a luxury compared with the multipurpose automobile. Radios were also more popular than phones. Because their operating costs were generally comparable—the annual cost of replacing batteries was about the same as paying the phone bill for a year—farm people seem to have preferred the entertainment and information value of the radio over that of the telephone, especially in the 1930s.

Technical difficulties also plagued the telephone. As we have seen, a hum often could be heard on telephones hooked up to lines near (or often on the same poles with) power lines. This electrical interference was an acute problem for telephone cooperatives that built one-wire systems, which used the earth as the return wire, instead of the more expensive system of a twisted pair of wires. The REA won most of the lawsuits brought by small telephone companies in the 1930s and did not pay to modernize their lines, thus contributing to the decline in telephone service.

Yet telephone cooperatives held up surprisingly well under the circumstances. The U.S. Bureau of Labor Statistics obtained reports from sixteen hundred telephone cooperatives in 1936 indicating that the economic crisis had not decimated these hardy associations. Most had fewer than fifty members, but most had also been in business for more than a quarter of a century. Minnesota had the largest number by far (1,653 known cooperatives, with 765 responding to the survey). Iowa—usually regarded as the home of independent telephony—came in a distant second in regard to the number of groups (272 known cooperatives, with 137 responses), but the larger Iowa co-ops reported a number of subscribers comparable to that in Minnesota (26,161 to 33,347). A more comprehensive survey made by the U.S. Census Bureau in 1937 found almost 33,000 telephone co-ops across the country. The average mutual with a switchboard, of which there were about two thousand, served 146 patrons at a cost of less than one dollar a month. The average mutual without a switchboard, the vast majority of mutuals, served only 12 subscribers. Combined, the mutuals accounted for about 40 percent of the farm telephones in the country at the time. Nine midwestern and southern states (Iowa, Missouri, Illinois, Kansas, Minnesota, Indiana, Nebraska, Texas, and Oklahoma) had about two-thirds of all mutuals. Iowa was far in the lead with about one-third of them. A 1951 study noted that 37 telephone cooperatives in Indiana traced their origins to the early part of the century.[35]

The cooperatives survived two world wars and the Great Depression, but most had not updated their equipment nor maintained it very well during these years. A study conducted by Purdue University in 1948 estimated that probably one-fifth of telephones in rural Indiana gave unsatisfactory service.

The long list of problems included grounded (one-wire) lines, inadequate maintenance (resulting in missing insulators, missing poles, vines growing over the wires, trees untrimmed, and so forth, which made the lines practically useless in wet weather), obsolete equipment (i.e., magneto, hand-cranked sets of World War I vintage), a large number of subscribers per line, inadequate volume of business, inadequate financing (no provisions for depreciation, and reluctance on the part of members to furnish additional capital), and management difficulties. The study attributed some of these effects to changes in rural life after World War II. One was the coming of the REA. "Before electric lines became common, these grounded lines were reasonably satisfactory. But with REMC [Rural Electric Membership Cooperatives] or other electric lines near, the 'hum' on grounded lines reduces or destroys their usefulness."[36]

Another change was the growth of commercialized farming. "Inadequate maintenance often is associated with farmers' decreased willingness to perform miscellaneous specialized services. In recent years farmers are inclined to work a larger proportion of their own time with crops and livestock. There they can utilize their own equipment better than they could early in the 1900s when mutual telephone companies were started."[37] Ironically, the Purdue Extension Service, as part of the USDA, had promoted this type of specialization, which led to poor maintenance of mutual lines. As more and more farmers became "modern," they increasingly bought services, rather than performed them to support such neighborhood agencies as telephone cooperatives.

AT&T was not insensitive to solving what researchers in the Extension Service and the Bell system called the "rural telephone problem." Bell's chief statistician wrote an internal memo in 1941, saying, "We have had occasion in the past to call attention to the problems presented to the Bell system by the low telephone development in much of the rural area of the country and to point out that lack of progress in making telephone service available (with all that 'available' implies) might at some time have political repercussions." That occurred when the 1940 census reported dismally low figures for telephones on farms and influential papers like *Wallaces' Farmer*, whose editor was "on leave of absence as Vice President of the United States," began to ask why.[38]

The REA also became interested in telephony at this time. Its leaders may have felt partially responsible for the decline of rural telephony because many financially strapped mutuals could not afford to metallicize their lines to avoid electrical interference from nearby REA lines. Electric co-op members could also report outages and other line problems much more quickly by telephone, a growing concern as the REA expanded.[39] Whatever the reasons, the REA cooperated with AT&T's renowned Bell Laboratories in 1940 to experiment on ways to use a co-op's power lines for telephony. In this "power-line carrier" technique, developed by Western Electric and Bell Labs before the war, electronic equipment impressed an audio signal representing a telephone conversation at a sending station upon the alternating-current "carrier" running through the power line, then recovered the hitchhiking signal at a receiving station. Engineers from Bell Labs and the REA worked together from 1940 to 1942 to test this approach on rural power lines in Indiana and Maryland, and Bell Labs developed electronic telephone sets for this application. Normally, AT&T would have proceeded to the next stage of development, but Bell Labs, a major contractor during World War II, announced in early 1943 that it would "lay aside the work for the duration."[40]

In 1944 several groups began to make plans to revive rural telephony after the war. That November, the United States Independent Telephone Association appointed a "rural telephone service committee" composed of Bell and independent leaders to work out approaches to solving the problem. AT&T firmly supported the effort, which included developing its power-line carrier technology.[41] But some southern progressives in Congress, all Democrats whose states were far behind in rural telephony, took a different tack. Alabama senator Lister Hill and representative Luther Patrick introduced the first rural telephone bills in the Senate and House in late 1944 and 1945. Only 5 percent of Alabama's farms had telephones, compared with 79 percent in Iowa. The bills called for a Rural Telephone Administration to grant low-cost government loans (1.75 percent amortized over thirty-five years) to cover unserved areas, much like the REA had for electrification, except borrowers had to provide at least 15 percent of the capital. Texas representative W. R. Poage, another Democrat, introduced a separate telephone bill in 1945 that called for the REA to administer the program under its principle of area coverage. The REA, its co-ops, and the National Rural Electric Cooperative Association supported the bill. Private companies, who considered the REA's cooperatives to be a harbinger of socialism, opposed it.[42]

AT&T took several actions against the proposed legislation. Shortly after Hill introduced his bill in the Senate in late 1944, Keith McHugh, vice president of AT&T, who cochaired the joint committee on rural telephony with the independents, informed Bell operating companies, "It is needless to say that I believe the Rural Telephone Administration proposed in the Hill bill is wholly unnecessary." Contrary to what the senator had claimed, McHugh said, the number of rural telephones had increased since 1935, mainly because of a "widespread rural telephone program" begun by AT&T at that time. Bell Labs was also developing power-line carrier technology with the REA. AT&T took this argument to a wider audience in April 1945 by publishing a lead story on rural telephony in a company magazine. The company promised to spend $100 million to reach 1 million more farms and update existing systems (the same amount of money requested by Hill) and to continue developing new technology for rural lines. The bills, however, never made it out of committee.

That December, AT&T brought the REA, which had supported the bills, fully into its publicity campaign by issuing a press release on the first (experimental) call made over an REA line, in Arkansas, using Bell Labs equipment. AT&T emphasized its cooperation with the REA to convince Congress of its ability to meet the challenge of rural telephony after the war.[43] In early 1946, AT&T announced that it had extended telephone service to "about seventy-five thousand additional rural families in Bell territory in 1945." To further develop the rural market, Bell told the U.S. Independent Telephone Association in February, it would license its power-line carrier royalty-free to all manufacturers of telephone equipment. AT&T also announced that it would conduct experiments in the summer with a radiotelephone system for eight ranchers in Colorado, to develop yet another method of serving geographically isolated farm families.[44] Despite Bell's well-publicized efforts to improve rural telephony, Representative Poage cooperated with the REA in 1947 to introduce a revised telephone bill that favored private companies. But it died for lack of support from the Truman administration.[45]

The REA and AT&T cooperated in other ways during this period. In 1947 the two organizations, which for about two years had discussed the legal and financial aspects of using power lines and poles for telephone purposes, announced a standard contract that gave electric cooperatives two options regarding "joint use of facilities." Co-ops could permit Bell companies to string telephone lines on REA power poles, or they could use the power-line carrier technique. By the fall of 1948, nearly one hundred REA co-ops in seventeen states had signed agreements with Bell to provide power-line carrier service. Sixteen co-ops in ten states had signed agreements allowing telephone companies to string lines on REA poles in a way that avoided electrical interference.[46]

Neither system became widespread. Although Bell had employed power-line carriers for several years and had expended considerable effort to design the M1 carrier system for rural applications, an AT&T engineer noted in 1948 that the rural situation "presented numerous problems not previously encountered." Ironically, when REA power lines carried telephone signals, the lines exhibited "some of the characteristics of a ground-return circuitry, a type of line which the Bell System abandoned many years ago for voice transmission." The electrical properties severely weakened the telephone signals. Just as troubling, "carrier currents impressed on the line radiate electromagnetic waves that may interfere with radio reception, and similarly waves such as radio signals or static will be picked up by the line and may interfere with the carrier signals."[47] Although engineers could fix the technical problems by rebuilding lines (at considerable expense), social issues peculiar to the practice of rural telephony proved to be just as problematic.

Meanwhile, the progress of rural telephony seemed uncertain. In December 1948, AT&T celebrated its one-millionth rural subscriber added since the war, with a telephone call between the lucky subscriber, a North Carolina farmer, and President Truman. An REA researcher conceded in 1948 that the Bell companies were "largely responsible" for the increase in rural telephony since 1940 but concluded that AT&T had not taken action until faced with the threat of legislation and estimated that only one-half of all farms would have telephones by 1950, even if the company met all of its objectives. E. C. Weitzel, an REA program analyst, reported that AT&T's "publicity does not claim a million new *farm* subscribers. Their claims are limited to a million rural subscribers. Of course, the U.S. Independent Telephone Association is widely misinterpreting this figure in favor of farm subscribers," an indication that the independents were still cooperating with Bell on the rural telephone issue. Weitzel covered this and other matters in material he was "preparing for those interested in current telephone legislation."[48]

The people he probably had in mind were Representative Poage and Senator Hill, who had reintroduced their telephone bills a few weeks earlier, in early January 1949. Those testifying against the bills—which, as in 1947, aimed to make rural telephony universal by amending the REA Act rather than creating a Rural Telephone Administration—included AT&T, the U.S. Independent Telephone Association, who wanted the Reconstruction Finance Corporation to administer the program, and electrical utilities, which were fighting a rearguard action against the REA. On the other side, Administrator Claude Wickard helped Poage revise his bill to be more acceptable to the White House by giving priority to existing companies, and he later testified in favor of the bill. Wickard wrote in his diary, "I look upon the Bell [System] with mixed feelings." Bell had added rural lines in densely populated areas and had worked with the REA to develop joint-use programs, including the power-line carrier system. But "apparently Bell does not intend to go into the sparsely settled areas or any other which does not offer a pretty high return. The same thing can be said for the large independents." He also complained that Bell and independent companies had worked up the fears of small companies that the REA would take over their businesses.[49]

Bell's opposition was to no avail. Supported by all major farm groups, the NRECA, and the REA, a revised bill finally passed in the summer of 1949. The strongest support came from southern and border states, which were most in need of rural telephony. President Truman signed the bill into law as an amendment to the REA Act in October. Clyde Bailey, of the U.S. Independent Telephone Association, told Wickard, "Maybe some of our people are still pretty well dissatisfied [with the bill] but we feel it is a measure that we can live with. . . . As good American citizens, believers of the democratic way of life, we want to do what we can now by way of cooperation with you to see that the full purpose and effect and intent of Congress in that legislation is carried out."[50]

The Hill-Poage Act, as the legislation came to be known, reflected the cold-war political debate about public and private enterprise. Cooperatives were eligible for loans, but other public agencies, such as the public utility districts, which had electrified rural Nebraska and Washington state, were excluded. For the first year of the program, the act required the REA to give preference to borrowers already providing telephone service, that is, to existing commercial companies and mutuals. The REA would allocate thirty-five-year loans at 2 percent interest for financing improvements, expansion, new construction, and the acquisition or operation of telephone lines, facilities, and systems. Borrowers agreed to provide area coverage and put up equity capital ranging from 10 to 50 percent of the loan. Recalling that many farm people had given up their telephones during the depression, the REA required equity to meet three financial hazards: the "possibility of subscribers giving up service when economic conditions are unfavorable;" damage due to sleet and snow storms; and low earnings in rural areas.[51]

Although both the telephone and electrification efforts were government loan programs administered by the REA to provide a utility service to farmers, there were important differences. The agency required substantial equity for telephone loans, often twenty-five to fifty dollars from subscribers, rather than the nominal membership fee of five to ten dollars for those signing up for electricity. Newly formed cooperatives did not dominate the telephone program as they had in electrification, mainly because of the existence of many small telephone companies and mutuals. Yet the REA had learned enough from its experience in working with co-ops to mandate several loan provisions aimed at controlling their abuses. Borrowers could not sell their systems without approval by the REA (a reflection of the ongoing struggles to prevent co-ops from selling out to electric utility companies). The REA approved all co-op managers and could remove them, directors of a co-op could not receive a salary, and co-ops had to establish a system of capital credits.[52] By the end of the first year of the program, the REA's Policy Committee had prepared forty administrative bulletins covering every conceivable subject, from area coverage to insurance requirements to the standardization of system designs. After taking into account suggestions from staff in the field, the agency released the bulletins to borrowers in early 1951.[53]

The program proved to be popular. The REA received 475 applications for loans during the first six months of the program, most of them from established companies. The first allocations, in fact, went to the Florala Telephone Company in Alabama (Senator Hill's state) and the New Lisbon Telephone Company in Indiana, both for improving the service and extending the area covered by these privately owned companies. The first loan to a cooperative formed by the REA went to the Emery County Farmers' Union Telephone Cooperative (later called the Castle Valley Telephone Association) in Orangeville, Utah, which had offered to construct 177 miles of line and install an up-to-date dial system. A mutual company serving two towns in the valley with magneto phones was to be abandoned. The Fredericksburg (Virginia) and Wilderness Telephone Company installed the first REA-funded system (an upgrade to dial) in September 1950. President Truman came down from Washington to make the first call at a ceremony celebrating the "cutover" to the new system.[54]

By June 30, 1950, eight months into the program, the REA had allocated more than $3.5 million to 17 borrowers. They expected to serve about ten thousand new subscribers and improve service to seven thousand others. Fifteen of the loans went to private companies, only two to cooperatives. The figures improved dramatically in the program's second year. By the fall of 1951, the REA had allocated $45 million to 122 borrowers: 82 of them were existing companies, and 40 were new co-ops. The number of subscribers shot up to 167,000.[55] The influence of the REA in reviving rural telephony—directly, by loaning money to build and improve systems, and indirectly, by stimulating Bell and large independents to add rural lines—can be seen in the Census Bureau data in figure 2. The percentage of farm households with a telephone began to increase steadily after the passage of the Telephone Act in 1949.

Yet reviving rural telephony was more difficult, in many ways, than electrifying the farm. As the *Rural Electrification News* noted,

It is apparent for one thing, that the telephone problems of most rural communities are quite complex—more complex, generally speaking, than the problems that arise in rural electrification. A telephone system must be designed so that people can communicate readily with whom they have business and social contacts. Moreover, nearly every rural community has some existing telephone facilities that must be fitted into any plan for improvement and expansion.

Another problem arises because so many of the rural telephone companies in need of financing are extremely small. It is obvious that many of these systems are going to have to be combined to form larger and more economical units if the majority of American farms are to have telephone service.[56]

The difference between the social geography of telephony and electrification had been noted as early as 1945, when Gordon Persons, the head of the Alabama Public Service Commission and a former REA engineer, testified in support of Senator Hill's bill. "It makes no difference," Persons said, "whether electricity is brought into a farm from East, North, South, or West. However, every farmer has some certain community in which his business and personal interests lie. This means that existing telephone companies are in a very good position to do a great part of the postwar extension of rural lines."[57] The REA made a similar observation in its annual report for 1950.

One of the obstacles to this [joint-use] procedure is that telephone systems must be developed in accordance with the *community interests of the subscribers*. Power consumers, on the other hand, are not concerned about the geographical design of their lines. In other words, telephone subscribers want to talk to their community centers and to a particular pattern of local subscribers, and systems must be designed to meet this requirement; hence, it is usually impossible to make full use of electric facilities for telephone service.[58]

In the REA's first loan allocation, for example, the Florala Telephone Company contracted to build 130 miles of new telephone lines but only 80 miles of joint-use lines (probably on shared poles) with a nearby REA co-op.[59]

Faced with the persistence of community-based communication patterns in the postwar era, in addition to the technical problems with power-line carrier systems noted earlier, the REA decided to discontinue this state-of-the-art technology. The response was similar to the case of the mutual construction of technology and society discussed in chapter 1, when Bell engineers at the turn-of-the century came up against the entrenched habit of listening in on party lines and altered the design of magneto systems to accommodate this social practice.

The REA's experience in organizing telephone co-ops illustrates these complexities. In one case early in the program, a group of farmers near Lawton, Iowa, who owned their own lines and were being switched by a private company in town organized the Lawton Cooperative Telephone Association before the Telephone Act was passed, then applied for an REA loan. The co-op began negotiating the purchase of at least three small telephone companies in the fall of 1950 in order to meet the REA's requirement for area coverage. According to field representative T. E. Orman, the owner of the town system was "definitely opposed to anything in which the Government has any part." Orman noted that the mutuals opposed the co-op, as well. "Resistance is building up fast against them [the co-op] in their effort to get a telephone system underway here. Selfishness is the big factor as displayed by those holding the existing properties."[60]

Four months later, in April 1951, Orman blamed the slow pace of the project on differences between the negotiating practices in farm communities and those in urban societies.

> Mr. Olson [the owner of the town system] asked $72,300 for the two exchanges. After some three hours of bartering, as only farmers can do it, with their offer of $60,000 for the two exchanges, a compromise of $65,000 was reached. . . . Some local sentimental feelings enter into this sort of deal which would not be noticed in a large company dealing with minority stockholders. Nobody wants to step on anyone's toes too hard if they can help it. All concerned will [have to] live with their neighbors for a long time.[61]

Yet matters did not end there, because the organization of the entire system—which interconnected three mutuals, the town, and new farm subscribers by buying out companies, making plans to rebuild lines and replace obsolete equipment, and signing up subscribers—was not completed until December 1951, more than a year after the first negotiations began.[62]

As it had in the early days of the electrification program, the REA experienced many problems with farm families refusing to join a government-run co-op. In early 1951, the Lawton cooperative hired a right-of-way man from a neighboring electric co-op to conduct a sign-up campaign. By March, he had brought in 436 telephone subscribers at twenty-five dollars each, but many farmers doubted if the co-op would succeed.[63] In August, field representative W. B. Bridgforth reported that 551 of the 800 farms in the area had joined the co-op and that at least 600 could be signed. "This is all that the local Board feels that they can count on to take service immediately, feeling rather sure that many of the remaining 200 will connect up eventually." Some families seem to have been influenced by editorials in local newspapers commenting on an article in *Reader's Digest* that accused the REA of requiring an Iowa co-op to borrow twenty thousand dollars more than it requested and pay the difference to an REA engineer.[64] In December, Bridgforth thought that most of the 175 farmers who came to an information meeting were "satisfied with their present service and did not favor Cooperatives." He was able to convince some people to change their minds and help finish the sign-up campaign.[65]

Many farmers objected to the high membership fee needed to meet REA's equity requirements. A field representative in Alabama said a local leader "claimed that he was having trouble signing up the residents of the [rural] community of Cotaco because the people told him that if they needed a tel. they would go down the road aways and use the neighbor's tel., and that most of the men sat around the store except when they were actually in the crop and then claimed not to have enough money for anything." Owen Lynch, head of a rural telephone committee in Indiana, wrote his congressional representative to complain about a fifty-dollar membership fee. He thought the REA Telephone Act "is just a great big joke. . . . If it is going to cost ten times as much to get phones as it did to get lights, the average community cannot afford it. If they could stand that figure, private companies would have served it before now." Writing to the representative, Claude Wickard replied that the REA required more equity for the telephone program because farmers would more likely give up the telephone than electricity in hard times because they had invested much more in wiring and electrical appliances than in a telephone set.[66]

In the summer of 1951, consultant Daniel Corman recommended reducing the sign-up fee to five dollars and requiring an installation charge of ten dollars. "Once service is established, I believe the problem of selling the rural resident on telephone service will have ended and that the problem will change to one of meeting the demand for service."[67] The REA addressed the issue in late 1952 by publishing "Tips for Your Final Sign-Up Campaign" in *Rural Electrification News*. The canvassing techniques included placing radio and newspaper advertisements and distributing equity certificates. The agency reduced the amount of equity required by taking into account the number of customers already served by telephone companies applying for a loan, and it relaxed the survey requirements. Recognizing that "people often hesitate at first to sign up in a new organization before it has proved that it can do the job," the REA replaced its farm-to-farm sign-up method with an area survey of probable subscribers, a technique common in the telephone industry. Surveyors determined which farms would probably take service "by using certain objective standards . . . such as, the appearance of the buildings, type of farming operations, extent of farm mechanization, and the like." The REA thought this would give a "more accurate picture of the market" than the number of farms signed up.[68] The change in policy indicates that, just as in the electrification program, the agency adopted the sales techniques of its competitors—the private utility companies.

One technique the REA copied from its own electrification program was the energization ceremony. In 1952, the agency announced that at least nineteen rural communities in eleven states would celebrate "REA Telephone Day," marking the cutover from an old telephone system to REA lines. President Truman attended one of these celebrations in Virginia, as we have seen. Taking a page from the ceremonies organized by electric co-ops, the Williston Telephone Company in South Carolina "launched its new [dial] system in 1952 with a mock funeral for an old hand-crank telephone. This event attracted a great deal of attention and has been copied elsewhere."[69] This particular modernizing ceremony of burying an old-fashioned artifact differed from the mock funeral of a kerosene lamp described in the last chapter, in that it was more believable that a farm family would discard a hand-crank telephone than a kerosene lamp. Farmers lit their houses by kerosene lamps when the electricity went out (as it often did at this time), but they could not talk over a magneto set on a network designed for a dial telephone.

Independent telephone companies also resisted the REA. Corman noted in early 1952 that the Iowa Telephone Association "has been quite hostile to the REA telephone program and Mr. Miller [its executive secretary] has spearheaded the opposition. The *Reader's Digest* gathered much of the information they used in the article 'Phony Business' in the October 1951 issue from Mr. Miller." Corman talked to Miller at length and thought he had "accomplished quite a bit in breaking down his resistance to the telephone program." He suggested that Miller show the new REA movie, the *Telephone and the Farmer*, to all independents in Iowa. Miller agreed, provided he could "do a very limited amount of editing which would better adapt the picture to [a] commercial company application."[70] Field representatives in New York state reported that owners of private companies feared the REA would take them over. When a field representative explained to one company that Congress had mandated area coverage, the owner's "brother Louis (who carries no weight) said we are not socialists, we want to run our own business."[71]

In view of the fact that the REA's electric cooperatives had achieved a good deal of success in meeting similar charges and overcoming the opposition of farmers and private companies, it is not surprising that many of them got into the telephone business. The agency officially encouraged them to assist the new program by conducting surveys of possible subscribers and organizing telephone co-ops.[72] Before the Telephone Act was passed, an Iowa co-op had organized the Winnebago Cooperative Telephone Association, which received an REA loan to merge forty-three mutuals to serve more than two thousand subscribers. A big reason for its success was that the social geography of telephone service matched that of electric service in this area, enabling about 85 percent of the members of the telephone co-op to be members of the electric co-op (through a joint-use arrangement). During the telephone program's first year, electric cooperatives helped organize the South Plains Telephone Corporation in Lubbock, Texas, the Poka-Lambro Rural Telephone Cooperative in Tahoka, Texas, and the Eastern New Mexico Rural Telephone Corporation in Clovis, New Mexico.[73] Other co-ops did not fare as well. A field representative reported in 1951 that the manager of the Steuben Rural Electric Cooperative in New York "has washed his hands of the whole affair" of trying to convince a group of farmers to work with a mutual company and apply for an REA loan.[74]

The growth of the telephone program affected the organization of the REA. The number of applications for telephone loans grew rapidly, whereas the building of electric lines slowed considerably to accomplish the goal of area coverage. Because cooperatives were also calling for more local control over their operations, the REA reorganized itself in July 1952 to reflect these shifts in its program. Enough people were transferred from electrical divisions to the telephone side of the house to increase its staffing by 50 percent. The ten regional field offices, which had been in place since the late 1930s, were consolidated into five electrical distribution area offices. Many divisions were abolished in order to create a streamlined organization.[75]

The REA had barely had time to readjust itself when Dwight Eisenhower returned the presidency to the Republicans in November 1952 for the first time in twenty years. In January, Eisenhower appointed Ezra Taft Benson secretary of agriculture. Benson, a critic of New Deal price supports for agriculture, asked Claude Wickard, one of the few New Dealers left in gov-

ernment, for his resignation in March, with two years remaining in his supposedly nonpartisan, ten-year appointment as administrator of the REA. Udo
Rall, one of the original New Dealers in the agency, wrote Wickard to comment, "That you were not able to complete your term is tragic for the REA
program." The new administrator was Ancher Nelsen, newly elected Republican lieutenant governor of Minnesota and former vice president of a rural
electric statewide. Nelsen proceeded to cut more staff and reorganize an
agency that lost more autonomy as part of the USDA's Division of Agricultural Credit Services.[76] When Nelsen resigned in 1956 to run for governor of
Minnesota, Eisenhower replaced him with Dave Hamil. A former Colorado
legislator who had a good deal of experience building an REA co-op in his
home state, Hamil's REA pushed the percentage of farms with phones in 1959
to the high figure of 65 percent shown in figure 2.[77]

CHAPTER 9

(Re)forming Rural Life

How did farm people use electricity and telephony in the postwar era? There is little doubt that those who farmed in the increasingly depopulated countryside after the war created a material culture in the home and on the highway that was not much different from that established by millions of men and women who commuted from sprawling automobile suburbs to work in large towns and cities. The exposure of rural youth to urban ways of life in World War II created a good deal of this consumer demand. The owner of a supply business in North Carolina lamented, "We'll have to have electricity to get tenants—good tenants. They won't come otherwise; they want their electric refrigerator, radios, and washing machines and all that."[1] Despite the high demand for these goods, the postwar attempt to urbanize the family farm (which was often seen as a process of suburbanization after the war) was still a contested process; farm people used the material culture of electrification and the telephone to create their own rural modernities.

Increased prosperity did not eliminate the work of rural reformers and other promoters of consumer technology. The high level of farm income between 1945 and 1964, despite a slight decline in the 1950s, meant the Rural Electrification Administration had less trouble persuading farmers to sign up for electricity and wire their houses. But consumer resistance impelled the agency to continue its prewar efforts to convince farmers to use more electricity once they were hooked up. Farm men and women may have besieged the co-op to build lines to their homes during the material shortages after the war, but they did not break down the doors of the co-op or those of local merchants to buy appliances for the house and barn. Electrical modernization still had to be carefully nurtured and vigorously promoted in the (presumably) consumerist, postwar era. Consumer resistance thus allowed farm families to use the new technologies to create their own versions of modernity, to (re)form their lives with electrification and the revitalized rural telephone.

Pushing Electricity after the War

In mid-1946, about a year after Claude Wickard took the helm of the REA, the agency reorganized its consumer activities. Wickard abolished the Cooperatives' Operations Division, transferring most its members to a new Management Division. Utilization field representatives, some newly hired, joined wiring and plumbing specialists in the Applications and Loans (A&L) Division.[2] In the fall of 1946, Wickard outlined the tasks of various units regarding assistance to borrowers, in a statement that set goals for what soon became know as the Power Use and Member Education Program. One of the A&L Division's functions was to train the co-ops' electrification advisers to "insure that members' use of electricity in their household and farm operations will be adequate to provide financial soundness to the borrower."[3]

Although this sounds like the traditional load-building activities of private power companies and the early REA, an internal study of the program, conducted in 1950, claimed the agency had "gradually shifted from 'merchandising' to 'load building' to the very recent emphasis on 'power use and member education.'" Instead of encouraging higher levels of consumption, the new program "aimed at efficient, effective use of power that is available" (power shortages may have influenced the goals of the program). The approach would increase consumption in the long run, but it supposedly put the "interests of the consumer" first rather than the desire to immediately raise kilowatt averages. The REA argued that adding member education to the program would show co-op members that its work differed from the load-building activities of utility companies. If the two elements of the program were not integrated, there was also the "danger that, under the guise of assisting the co-op in its power use activities, outside groups unfriendly to the co-op's objectives might attempt to undermine the members' confidence in their co-op."[4] This warning alluded to the attempted takeover of Virginia's Craig-Botetourt cooperative by a power company, discussed in the previous chapter, in which the utility company had used neighborhood meetings about the electric range to criticize the co-op's rates.

The structure of the Power Use and Member Education Program differed from George Munger's and Clara Nale's work before the war, primarily because the main function of the A&L Division was to make loans, not to educate consumers. But some of their prewar colleagues did take leading roles.

At the Washington office in 1950, Louisan Mamer and Oneta Liter headed the effort in home electrification, Daniel Teare in farm electrification. Other specialists ran programs in education of cooperative members, wiring, plumbing, installation loans, rural industries, irrigation, and liaison with the U.S. Department of Agriculture's Cooperative Extension Service. Below them were seven agricultural engineers and ten home economists, who covered ten regions of the country. The latter group included Agnes Wilson and Mary Alice Willis of the original home economics staff hired in 1937. These agents worked with committees formed by thirty-one rural electrification statewide organizations in 1950. They supervised a co-op program, which was typically implemented by a co-op's electrification adviser.[5]

The electrification advisers were the heart of the program. Performing functions similar to those of the prewar project-utilization specialists, they wrote the co-op's newsletter, lobbied the board of directors to institute capital credits, arranged the annual meeting, and tried to persuade farm families to buy more appliances—activities the REA considered vital for ensuring the long-term health of co-ops. After a slow start in 1947, the A&L Division convinced more than two hundred co-ops to hire advisers two years later, almost three times the number of co-ops that hired utilization specialists before the war.[6] The number of co-ops employing electrification advisers steadily grew to 467 (about one-half the number of co-ops) in June 1952, when they employed 528 advisers. Many more women entered these ranks after the war, but women still constituted only one-fourth of the advisers (139) in 1952 (and in the two previous years), despite the greater number of women employed as regional field representatives. Contradicting the view that professional women would fare better in the North than in the South in this period, the only regions that approached gender equality in 1952 were the Southeast (60 men to 50 women) and the Southwest (85 men to 57 women). The most lopsided were the Northeast (81 men to 15 women) and the North-central region (94 men to 6 women). The states having more women than men advisers in 1950 were all in the Southeast and Southwest: Arkansas (9 women to 2 men), Georgia (10 to 4), New Mexico (4 to 3), and Texas (17 to 10).[7] Part of the reason may have been that the REA pushed commercial agricultural uses (e.g., electrification of dairying, poultry raising, and the machine shed) in the more prosperous North, which also had more dairy farms, and tried to catch up with household appliances in the poorer South, although poultry was becoming a big business in that region.[8]

At a regional conference held at St. Louis in 1947, co-op leaders and REA officials discussed the question of whether the electrification adviser should be

"a man or a woman" (i.e., a home economist or an agricultural engineer, following a gendered division of labor that had not changed much since World
War I). Oneta Liter reported, "In Oklahoma and Arkansas the need appears
to be one in which a woman would be the most desirable type of employee,
while in Missouri the tendency is toward employing agricultural engineers to
meet the farm electrification needs of the state." New co-op leaders debated
this topic at their meeting in 1950, concluding that "where a co-op is large
enough to afford two electrification advisers, it is good to have a man and a
woman," to cover both farm and home promotions. Although most electricity was consumed in the house, the co-op leaders thought it was important to
promote agricultural uses (a view shared by Wickard, a former secretary of
agriculture). They concluded that "each co-op must answer this question for
itself."[9]

Although gender relations changed for the better in this period, it was
still difficult for women to achieve equal pay for equal work in REA co-ops.
In late 1948, experienced field representative Dora Haines reported that Delta
Scales, the woman manager of a Texas co-op we met in the previous chapter,
had noted that "it was not easy to sell her board on the idea of paying a
woman a good salary to do a good job in the field of adult education." Some
board members had the idea "that no woman can earn more than $150, $200
at the outside, a month. That is the reason she [the electrification adviser] gets
such a low salary." Male agricultural and electrical engineering graduates
started at $200 and $225 a month at this time at co-ops in Florida and
Kansas.[10] During the Korean War, W. P. Nixon, head of a region in the A&L
Division, told a co-op manager in Kansas, "In view of the need for men in the
armed services, it might be somewhat easier to find a qualified home economist for the job." But in the spring of 1953, as the war was drawing to a close,
male electrification advisers advanced into higher paying positions, such as
co-op managers, while women stayed in home economics jobs. In 1955, the
REA came out in favor of hiring female advisers, stating that "steadily increasing electric loads reported by REA borrowers with trained home economists are a good answer to the question posed by some managers, 'Are women
paying their way as power use specialists?'"[11]

There was a large amount of resistance to the power-use program,
whether it was conducted by men or women. Slightly more than one-half of
the co-ops did not employ electrification advisers in 1952, and several discontinued the program after a year or so. This varied from two-thirds of co-ops
not participating in the program in the West to one-half not complying in the
Southwest.[12] In the early 1950s, co-op managers in South Carolina, Iowa, and

Illinois convinced their boards not to hire advisers, mainly because of the expense.[13] Twenty-three co-ops discontinued the program between 1950 and 1952. Field representatives reported that the manager of a Florida co-op "had not been able to see any advantage in having an E.A. and didn't know of any increased kWh usage which was attributable to his activities," that "appliances were being placed on the line [of a Virginia co-op] at such a rapid rate that the [board of] directors could not see the need for a program," and that the firing of women advisers by three Oklahoma co-ops "without warning . . . has greatly affected the morale of the remainder of the Home Economists working for REA co-ops" in the area.[14] Advisers commonly complained that managers and boards took little interest in their work.[15]

The REA combated this problem by using its leverage of approving loans to urge co-ops to hire electrification advisers. Nixon told the manager of a Kansas co-op applying for an extension loan in 1951 that new members "may need more time to utilize electric energy more fully [as shown by historical statistics]. However, these new members need considerable guidance and assistance in achieving satisfactory results in all applications of electrical energy." Because the co-op's average consumption per farm was below the REA's postwar expectation of 250 kilowatt-hours per month, Nixon explained, it needed a power-use program.[16] Many in the agency seem to have shared Nixon's view, but others maintained that the postwar focus should not be on load building. Sheridan Maitland said in 1950, "Many REA officials connected with member education will frankly admit that a distressingly large number of co-ops have hired Electrical Advisers for the sole purpose of 'building load.'"[17] George Dillon, head of the Power Use and Member Education Program, wrote on the margins of a field report, which criticized the firing of an adviser for not increasing a co-op's kilowatt average, that load building "is not the only phase of an Educ. Program." At least two of Dillon's staff expressed this view in their reports.[18]

The tension between load building and power use was similar to the tension between sales and education experienced by home economists before the war. In 1951, field representative Betty Williams attended a professional home economics meeting in New Mexico with five women co-op advisers. She noted with pride that "Crickett [Taylor, one of the advisers] (may the Lord bless her!) refused pleasantly and politely to permit us to be called commercial Home Economists!"[19]

Williams's colleagues also tried to educate utility company representatives in the educational approach of the REA. Eleanor Delany convinced "commercial home economists" at a university short course on frozen foods—another new technology—to use USDA research in their training, rather than company information. Elizabeth O'Kelley "made it a point to tactfully explain [to two male General Electric representatives at a short course on frozen foods in Virginia] that ours was an educational program on parts, features, and principles of operation of home freezers and that we therefore could not afford to mention one manufacturer's product in comparison with another." She was pleased with the result: "Never have I seen such perfect cooperation from commercial sales promotional personnel." Plumbing specialist Earl Arnold was not as satisfied with an advisers' training school in Georgia. He noted that another General Electric (GE) representative "scrupulously avoided promoting his particular brand, but his whole discussion was a sales talk for electric water heaters."[20] Oneta Liter expressed this criticism well when she applied for a promotion in 1950.

> This broad program for the use of electricity by farm people, must if it is to yield the most desirable results, reach far beyond connecting houses to distribution lines and piling into those houses, loads of power-consuming equipment (satisfactory end results will be obtained if a sound program is established). Farm people should buy the electrical equipment they need and can pay for and then learn to use it efficiently and productively. A new household refrigerator or learning the practical economics of a farm freezer is still, even with high farm prices, a considerable business venture for a large group of farm people in this country, not to mention the purchase of a plumbing system and bath room equipment.[21]

Liter's formal training may have inclined her toward these views, which were widely held by educators in home economics; she may also have argued the REA line to be considered for the position.[22] In any event, promoters of rural electrification not on the REA staff, such as co-op personnel, had difficulty making what looked to them like fine distinctions between "education" and "sales." The REA's educational efforts often looked like sales promotions. The agency's statewide training schools taught co-op advisers to promote new appliances such as home freezers and television sets, which used plenty of electricity. The REA stated in its annual report for 1950, "The success and effectiveness of these organizations [co-ops] depend on the extent to which all members accept and meet their responsibilities and privileges they enjoy as consumers. . . . Rural consumers need advice and assistance in selecting and effectively using various types of electrical equipment."[23]

As we have seen, many co-ops ignored the "educational" part of the

Power Use and Member Education Program and unabashedly interpreted it as old-fashioned "load building," even to the extent of using this politically incorrect term. The directors of a New York co-op reported in 1947, "The Manager gave a brief report on his conference with the load-building representative from REA who was here during the past week and he informed the Board that Miss Wood would return again in the fall to help formulate plans for a load-building program." Other co-ops criticized the REA on its own score. A field representative reported in 1953, "The power-use adviser at East River Power Cooperative [in South Dakota] stated that he would like to have REA furnish him some radio scripts covering various uses of electricity such as appliances, water systems, etc., rather than the propaganda type now being received. See what you can do for him."[24]

The charge of propaganda is not surprising, because the REA used the same merchandising and load-building methods it had developed before the war. Co-op newsletters proliferated in the late 1940s and early 1950s as the Information Division taught electrification advisers layout, typesetting, and how to write personal-interest stories at special training schools.[25] Newsletters in Indiana, Montana, and Iowa still carried "honor rolls" of members who used the most electricity each month. A Georgia newsletter proudly printed the names of families who had purchased major appliances. One in Colorado listed those who had bought irrigation pumps.[26] The slogan of the first issue of *The Light Line,* published in 1953 by the Socorro Electric Cooperative in New Mexico, was "Use More Electricity." Veteran co-op watcher Udo Rall objected to the flagrant load-building message, saying the slogan "may seem all right to co-op management, but will not impress the members. They need to be given reasons why it is to their advantage to do so." Rall suggested "Use More Electricity—To Save More Time and Labor," yet another attempt to turn prewar "load building" into postwar "power use and member education." Rall apparently had no problem with the slogan of a 1953 newsletter for a Mississippi generating co-op, "Go All Electric Today—The Modern Way," even though the first part was a common utility-company slogan, presumably because "modernization" encapsulated the REA's educational message. Reinforcing the image were full-page ads for GE refrigerators.[27] Willie Wirehand, the National Rural Electric Cooperative Association's cowboy-farmhand version of the suburban stick figure, Ready Kilowatt, greeted readers from the pages of many newsletters.[28] Newsletters also promoted high school essay contests on the value of rural electrification, combated the antitax image of co-ops, and offered useful advice on such matters as how to report outages and how to avoid the dangers of ungrounded washing machines.

Home economists still gave cooking demonstrations at annual meetings and county fairs. Crowds as large as seven hundred watched Peggy Bailey bake cakes, pies, rolls, and cookies using an electric range at a North Carolina fair.[29] Kitchen parties were in vogue more than ever, and remained so through the 1950s, now that rising farm income made electric ranges more afford-able.[30] The REA recommended that electrification advisers use the following techniques to promote the sales of small appliances: gift promotions, holiday demonstrations, floor demonstrations, combined demonstrations with deal-ers, club programs, youth programs like 4-H Club and Future Farmers of America, special household activities programs, community usage (e.g., school lunches and buffet suppers), clinics on the use and repair of appliances (to "promote getting appliances out of cupboards and pantries and putting them into active use"), special promotional activities with dealers, study courses, educational discounts, and appliance loans.[31] Considerations of co-op load and diversity factors (i.e., the power-use patterns of different customers), rate schedules, farm income, and the saturation levels of household appliances were also important factors in designing promotions.[32]

A field representative reported that an adviser in Michigan was "toying with the idea of carrying a supply of 100-watt lightbulbs with him at all times. When he makes a visit to a home and notices that a room such as a kitchen or bathroom is being served with a 60-watt bulb, he will remove it and insert the higher wattage bulb, perhaps at no cost to the member for the initial bulb."[33] A familiar means of reaching parents through their children was a 1951 Lighting the Study Center Contest, in which high school students in-vestigated the lighting at home (usually inadequate) and suggested improve-ments. Field representative Bernard Krug tried to help Auby Snow's strug-gling co-op in Vermont by writing letters to the members. "One of them is aimed directly at the young people on the co-op lines—the teenagers—its purpose being to take advantage of the influence that the youngsters wield when it comes to buying new things for the home."[34]

There were some changes in postwar techniques. The Information Divi-sion listed more than one hundred movies, filmstrips, and slide sets available from manufacturers and the REA to promote all aspects of the program. But Pare Lorentz's award-winning movies, *The Plow That Broke the Plains* and *The River,* were not listed. Joris Ivens's *Power and the Land* was, and in late 1949, before the McCarthy era got under way, the REA recommended showing the critically acclaimed documentary by the leftist director. On the twentieth an-niversary of the film in 1959, the REA estimated that more than a million people had seen *Power and the Land,* which was still occasionally being shown. In 1946, the REA made its own, politically conservative film, *Bob Marshall Comes Home*, about a returning war veteran, and recommended General Electric's prewar film, *Bill Howard, RFD*.[35] Co-ops got into radio broadcasting near the end of the war and immediately thereafter as a new way to reach farm families. *Rural Electrification News* praised this method by describing co-op radio programs in Texas, South Dakota, Indiana, Ohio, Missouri, Iowa, Georgia, and New York from 1944 to 1947.[36] The Information Division also spent a good deal of effort helping co-ops develop radio advertising to complement their newsletters. Co-ops took up the new medium of television when it became popular in the 1950s.[37]

More men did home economics work after the war. Regional home economist Elizabeth O'Kelley marveled at how well GE's Frank Lambert demonstrated scalding and packaging vegetables for home freezing in 1951. "His demonstration was far more effective than if it had been done by a woman, for he did it with such ease and simplicity that it was made evident that freezing of food is the simplest as well as the most nutritious and flavor-saving method of food preservation." The implication was that freezing food could not be as difficult as old-fashioned canning if a man could do it. Another field representative reported that "Oscar Johnson, the president [of a co-op in South Dakota], is doing part-time power-use work, and currently has a range program going."[38] Some male electrification advisers demonstrated home appliances when the co-op did not employ a woman adviser.[39]

Another postwar change was that home economists geared their promotions to their interpretation of the expressed needs of women rather than proscribing those needs. The manager of the Lighthouse Electric Cooperative in Texas explained in 1955, "Our home economics program is based on giving the rural housewife what she wants. We used to figure we were the best judge of her needs. Nancy [Morckel] plans her program to suit popular demand." Although this ideal reflected a general trend in home economics in this period, Morckel's promotional methods for the electric range seem almost identical to prewar techniques of trying to sell an unpopular, expensive appliance that was also a great "load builder."[40]

Although the REA did not revive the circus atmosphere of the popular Farm Tour, several groups put together abbreviated regional shows for annual meetings (much like those given in the early part of 1941) after the war. Organized primarily to increase attendance, these "rotational annual meetings" started in 1949 and were popular in the South. The statewide organization in Kentucky published a booklet for prospective exhibitors, stating that one hundred thousand had attended annual meetings and Electric Farm Shows at seventeen Kentucky co-ops in 1949; they expected double that number in 1950. Alabama, Virginia, Kentucky, Tennessee, and Mississippi arranged Home and Farm Electrical Expositions to visit several co-ops in the summer of 1950. A Tennessee co-op reported an attendance of eight thousand at its annual meeting in 1951. Entertainment in Kentucky for 1950 included hillbilly music, a comedian, fireworks, a beauty contest, the popular 4-H Club tractor-driving contest, and more than three hundred appliance exhibits. Entertainment for the 1952 show, whose theme was "Go All Electric," featured cowboys, cowgirls, and a trapeze act. Alabama's Electrical Fair had a night for "local townspeople" that included a talent show and Frigidaire's "Previews of Progress, and electrical science show."[41] In the northern states, similar caravans that traveled to co-ops in Minnesota were called "farm circuses." Twenty-one South Dakota cooperatives contracted for traveling shows in 1953.[42]

Cooperatives continued granting consumer loans in the postwar period. The Cullman County (Alabama) Membership Corporation contracted with a local bank in 1946 to provide consumer loans for appliances sold by area dealers. But most consumer loans were made under the REA Act; the Electric Home and Farm Authority was dissolved in 1942 because of wartime restrictions on credit and the manufacture of home appliances.[43] In 1948, about 1 percent of all REA loans financed wiring, plumbing, and the purchase of appliances. The agency provided money at 2 percent interest to co-ops, which lent it to members at 4 percent interest for five years (the other terms were 20 percent down and a five-hundred-dollar limit on the loan). The USDA Farmers Home Administration estimated that under the Federal Housing Act of 1949, 135,000 farmers would be able to borrow an average of twenty-five hundred dollars each during the four-year program (at an interest rate of 4 percent with terms running from five to thirty-three years). Preference was given to veterans. As with the case of housing loans in the New Deal, analyzed well by historian Ronald Tobey, the money would enable electrical modernization of the farm home.[44] In order to build load, many co-ops also paid, in part or in full, for the installation of ranges and water heaters that used a lot of electricity.[45]

The issue of co-op merchandising took on a new look during the anticommunist fervor of the early cold war, when appliance dealers criticized it as government competition. When a Missouri dealer wrote Vice President Harry Truman about the matter in 1945, the REA's deputy administrator, Vincent Nicholson, replied that "it has always been a policy of REA to discourage our borrowers from engaging in the business of selling electric appliances except to the extent of obtaining for their members certain items which are not readily available from established local dealers." The appliance co-ops, though, were separate from electric co-ops, and, besides, some co-ops thought they should follow the lead of private utility companies and sell appliances. The next year, Wickard spoke out against the practice because it often harmed relations with dealers. A field representative reported that an Illinois dealer thought "government competition in the electrical appliance field was entirely unfair and that he is definitely opposed to this practice." The co-op continued merchandising "regardless of suggestions made by REA field personnel that the activity be curtailed or eliminated." The agency blamed dealer apathy near a Kentucky co-op on its merchandising; North Carolina dealers opposed a co-op's selling plumbing fixtures. Although another North Carolina co-op stopped merchandising because it had "incurred the ill will of established dealers," an Illinois co-op enjoyed the goodwill of dealers, probably because it did not cut prices below retail. The co-op saw itself as a pacesetter for the dealers.[46]

That sentiment helps explain the persistence of co-op merchandising. In 1950, the REA counted seventeen co-ops that merchandised in ten states in all sections of the country except the Far West.[47] The REA seems to have been ambivalent about the practice (as implied in Nicholson's response to the Missouri dealer). In at least one instance, a field representative used the threat of a co-op selling appliances to discipline reluctant dealers. E. C. Collier reported in 1953 that he had "advised dealers [near an Illinois co-op] to get busy and assist in load building or [the] co-op will merchandise next year." Elwood Olver, head of the Iowa Rural Electric Cooperative's educational program, exhorted the state's electrification advisers to "whip the dealers into shape. Get them to sell appliances so that our cooperatives will not be forced into the appliance business. . . . Do something about these dealers selling bottled-gas equipment over the electric."[48] In Wisconsin and Vermont, REA field representatives thought co-op merchandising would help fight bottled-gas competition and promote the sales of electric ranges.[49]

I have found little evidence that co-ops resisted these programs in the vigorous manner they had before the war. Olver reported in 1952 that an Iowa co-op opposed the promotion of electric ranges because "selling something you do not have results in poor service; also, [electric] ranges hinder peaks [peak usage periods]." An REA field representative said in 1951 that the electrification adviser of the Central Electric Power Association in Mississippi "is an old time employee . . . [who] is not interested in having any of his members get up into the four-mill rate. He wants to have all of them below that part of the rate structure." George Dillon wrote on the margins of this report, which was written to Walter Rich of the A&L Division: "Rich—Looks like this Mgr. needs continuous working over or eventually replacement. Ha!"[50]

There is also less evidence that farm people resisted the REA's sales efforts as they had before the war, probably because the farm population now consisted of a smaller number of wealthier families. A field representative noted some resistance to installment buying in Maine in 1951. "Some of these people [on Swan's Island] are very conservative and will not take advantage of the Section 5 [REA] loan, but will wait until they have enough money to do the job [of wiring their houses], as we see no amount of talking will change this."[51] Home economists in the Extension Service faced some opposition as well. Iowa women tuned in to local women "neighboring" as radio homemakers more often than to broadcasts of expert advice from the college. A survey conducted in the late 1950s reported that a large number of farm women preferred to hear their home economics lessons from local leaders than from college-educated agents.[52] The home economists of the REA probably faced similar responses to their sales techniques, but postwar prosperity and the high value placed on consumerism undoubtedly made their lot easier than that of their sisters before the war.

Much more resistance came from Republican politicians, who had criticized the REA's "extension" activities in the Hoover Commission report. When Ancher Nelsen took over the REA in early 1953, one of his first acts was to abolish the Power Use and Member Education Program at the staff level in Washington. Reporting on a short course she gave at the Virginia Polytechnic Institute that summer, which had been scheduled before the program was abolished, home economist Louisan Mamer remarked, "It was heartwarming to make my last stand in REA power-use work at VPI, where our work is so highly regarded. . . . Most pleasing of all was Maude Wallace's reaction—she feels it is a shame to abolish the staff and the fine collection of materials and teaching aids, which have been assembled. She said it was impossible for a state to render this type of service and regretted REA's decision to discontinue such a program on a national scale."[53] Mamer continued to work for the REA for many years, but the new leadership left the promotion of electrical modernity more in the hands of power-use specialists in local agencies—statewides and co-ops.[54]

Patterns of Electrical Consumption

As in the prewar period, farm families interacted with the REA and the Extension Service to fashion their own versions of electrical modernity. The outlines of these consumption patterns were drawn by promoters—toward increased use of electricity and purchase of a full complement of electrical appliances—but farm people controlled to a large extent the details of the pattern and when, or whether, it was completed. This can be seen at the aggregate level in figures for the consumption of electricity. At the end of the war, the average consumption for households hooked up to the REA had increased to a respectable level of 84 kilowatt-hours per month (see table A.15). But not until 1947, a full decade after the division was established, did farms reach the goal of 100 kilowatt-hours per month set by George Munger's Utilization Division. Consumption grew at a fairly rapid rate thereafter, especially in the late 1940s, and passed the 350 kilowatt-hour-per-month mark in 1960. Although the record was impressive, the REA had expected more. In 1946, Daniel Teare estimated the usage of a full-fledged electrical household at 524 kilowatt-hours per month, which he thought farmers could afford by using more electricity in agricultural operations.[55] Teare's figures assumed that farm people would cook meals and heat water with electricity, which, however, only a small number chose to do.

Table A.15 masks enormous regional differences in the usage of electricity, which were brought to light by an exhaustive survey of randomly selected farms conducted by the USDA's Bureau of Agricultural Economics in Georgia, Iowa, Kansas, Tennessee, and Washington in 1947-48. The farms were supplied by power from private companies and municipalities, in addition to the REA. The bureau found the increase in consumption to vary widely by region. More than five hundred farms in the cotton-producing Upper Piedmont area of Georgia increased their consumption from 50 kilowatt-hours per month in 1938 to 88 kilowatt-hours in 1947. The figures for a similar number of farms in the general livestock-producing area of eastern Iowa were more impressive (almost tripling, from 67 to 183 kilowatt-hours per month in this period), and those for a dairying area in northwestern Washington were even more so (nearly quadrupling, from 92 to 358 hours). The bureau suggested "that in a given area the use of electricity is principally a function of three major factors—[the length of] time [a farm has been electrified], income, and size of specific farm enterprises, particularly dairy and poultry enterprises." Statistics on the use of electricity on farms in north-central North Dakota from 1943 to 1949 and on REA farms in Washington and Arkansas from 1947 to 1951 support these conclusions.[56]

Income was a crucial factor, but the bureau noted in its study of Georgia, Iowa, and Washington that "at all income levels there was a wide variation in the amount of electricity used by individual farms." In the groups having elec-

tricity for five years, this ranged from 500 to 16,000 kilowatt-hours per month
for those with an income in excess of twenty-five thousand dollars per year
and from 200 to 2,600 kilowatt-hours for those with an annual income of less
than twenty-five hundred dollars. The bureau could find no obvious reasons
for these enormous variations within income brackets but did point out some
considerations, which we observed for the prewar period in chapter 7. Some
families "have delayed the installation of a water system until a new home is
built. Some continue to use a coal-burning range because there is no central
heating in the home. Some will continue to use old equipment until it is worn
out and must be replaced. Still others cling to traditional methods because established habits are difficult to break." Furthermore, many farmers interviewed "have talked about the times when the power went off and what they
had to do to cook meals or to pump water or to keep baby chicks from chill-
ing. [Thus] outages have militated against the greater use of electricity on
farms."[57] The prospect of losing a winter's supply of meat if the electricity
failed was a large concern for those who contemplated buying a home freezer.
The instructions by home economists on how to save the situation may have
simply drawn attention to the problem.[58]

These consumer choices, entrenched habits, and concerns about outages
indicate that postwar consumerism did not suddenly displace prewar ideas of
rural modernities based on the selective use of electricity. The investigation
conducted by the Bureau of Agricultural Economics indicates, instead, that
farm people in the late 1940s fashioned individual electrical modernities of
their own choosing.

Regional variations were also apparent in the ownership of appliances.
Another study done by the Bureau of Agricultural Economics, of eighty-four
hundred randomly selected electrified farms in 1947, shows some of the same
economic and social factors noted in chapter 7 (see table A.16). The poorer
South had much lower "saturation" levels than the wealthier North and West.
But differences in climate played a similar role as they had before the war. The
South favored refrigerators over washing machines, and the North bought
more washing machines than refrigerators. Water systems, vacuum cleaners,
and ranges were still unpopular for the reasons noted previously. The bureau's
study of farms in Georgia, Iowa, and Washington showed similar results, as
did investigations of mostly prosperous farms in Oregon and Missouri in 1945
and 1948.[59]

Regional differences in the percentage of washing machines, water systems, and water heaters continued with the increase in prosperity and the
growth of consumerism, as shown by the 1950 Census Bureau figures for these

appliances on electrified farms in table A.17. But, as would be expected, the percentage of all farms having running water in 1950 is lower than that of farms having water systems and washing machines in that year (see table A.8). Figures were not published for refrigerators, presumably because the Census Bureau was more interested in the new technology of home freezers. The regional buying patterns of freezers followed income rather than environmental lines.

Although sales of washing machines and refrigerators climbed right after the war, farm people were reluctant to buy other appliances. Referring to the poor showing of small appliances in the Oregon survey, veteran home economist Maud Wilson predicted that for appliances "now owned by a considerable number of families the saturation point promises to be considerably short of 100 percent; clocks, coffee makers, portable fans, hot pads, hot plates, razors, grills, and sewing machines are in this category."[60] The other studies bear this out, probably because most of these appliances still carried an aura of urban luxury in the countryside.

The slow adoption of bathrooms still puzzled promoters. In a study of forty prosperous farms in north-central Ohio in 1946, the USDA's Bureau of Human Nutrition and Home Economics found that these "families were at various stages along the road to the goal of a completely modern house." All of them had a kitchen sink, but only twenty-four had "completely equipped bathrooms," partially because plumbing equipment on order had been delayed. Farmers in eastern Iowa told researchers in 1948 they would not install water systems because it was difficult to prevent pipes from freezing in the region's severe winters; they would wait until they built a new house. The age of families was also a factor. A study of seventy-three farms in Mississippi in 1948 found that older "homemakers" were more likely to have running water in the house, but those in the younger group "had more often added modern bathroom facilities, such as flush toilets and bathtubs with running water." The researchers speculated on a cultural factor: younger women "may have had more exposure to modern plumbing equipment and hence a greater desire for it."[61]

A recent study of southern Illinois by Jane Adams supports this view. Several farm women interviewed "added indoor bathrooms after their husbands died. Many men (but not, apparently, women) resisted bringing toilets into the house; it violated their notions of sanitation."[62] Many other farm men undoubtedly carried over the common idea that human wastes should be discharged outside the house, which we noted in chapter 7, from the prewar to the postwar period.

Demand for bathrooms was sluggish in the consumerist 1950s because of high cost, entrenched habits, and poor water supplies. The REA's Earl Arnold noted in 1951 that because a complete plumbing system cost about as much as a second-hand car, farmers purchased the system in stages. "The farmer goes to his dealer looking for a pump. In order to help the farmer in his selection, the dealer asks where water is going to be used and is assured it is to be used only for a kitchen sink. That is all!" Later, a bathroom would be added. A postwar study of 108 farms showed an average cost of about six hundred dollars for a complete water system. Louisan Mamer stated that "long-established habits of conserving water" on the farm—habits that probably date to the droughts of the 1930s—mitigated against using a bathroom in a middle-class urban manner.[63] In 1955, only 14 percent of farms connected to a large co-op in southern Kentucky had running water. The co-op's power-use adviser explained: "Farmers had their own reasons for putting off getting water systems. Some of them blamed it on the shortage of water. But we explained that every farm has some source of water and that an electric system could be used to tap it. There was also the problem of financing. . . . There was just so much money to go around and water systems were last on the farmer's buying list." Consumer resistance was not restricted to the South. The Nodak Rural Electric Cooperative in North Dakota staged two "Plumbaramas" in 1956, involving local dealers, plumbers, and Extension Service agents, to sell water systems to farmers reluctant to buy on the installment plan. Co-ops in Arkansas, Wisconsin, Minnesota, and Virginia put on similar demonstrations in 1958; by that date, only 60 percent of electrified farms had automatic-pressure water systems.[64] Declining farm prices and income in the 1950s were probably factors.

Selling electric ranges was even more difficult. One factor was the greater cooking requirements in the countryside. In Maud Wilson's 1945 study of Oregon farms, more than a third of the one hundred families with ranges "also used wood or coal stoves for cooking." Sixteen percent of the range owners "reported that they were inadequate for the preparation of family meals, 53 percent for company meals, and 46 percent for canning. Only 22 percent of the group reported their ranges adequate for all of these uses. The need for more units [burners] accounted for 50 percent; larger units, 21 percent; smaller units, 4 percent; and more space around the units, 25 percent of the inadequacies listed." These families obviously thought the coal or wood stove was better suited than the smaller electric ranges for the kitchen practices of farm life in Oregon. The availability of wood was a factor in some regions. A Mississippi farm woman told researchers in 1948, "It's there; it seems as if we ought to use it."[65]

Liquefied petroleum (LP, or bottled gas) provided the main competition for the electric range, as well as the electric water heater and the electric refrigerator. As early as 1944, the president of the Homegas company in Wichita, Kansas, published an article in a gas trade journal entitled, "Who's Afraid of the REA?" The article cited a government study that LP gas was much cheaper than electricity for cooking (eleven dollars per year, compared with fifty-nine dollars per year for electricity) and for heating water (fourteen dollars compared with one hundred dollars per year). The difference for refrigeration was not as much (twelve dollars compared with nineteen dollars per year).[66] Despite the rural electrification boom in the late 1940s, the REA had good reason to be concerned about the inroads of bottled gas. The Bureau of Agricultural Economics found in 1951 that 43 percent of all farms in Oklahoma and Texas used LP gas, 37 percent in the Great Plains, 27 percent in the Corn Belt, and 25 percent in both the Mountain states and the Pacific states.[67] A survey of an REA cooperative in New Mexico that same year found that 68 percent of its rural customers used LP gas, compared with only 36 percent in 1936. The investigator concluded that "the failure of these consumers to attain their expected [electrical] saturation is believed to be principally due to the competition provided by liquid petroleum gas for cooking, house heating, water heating, and refrigeration." All of the LP gas customers cooked with gas, three-fourths heated their houses with it, two-thirds used gas refrigerators, and one-half heated water with gas. The report attributed the situation to the delay in establishing REA co-ops in the state and to the fact that co-ops gave a discount rate only for water heaters.[68]

A systemwide problem was that appliance dealers often pushed LP gas instead of electric ranges. An open letter from the Pioneer Rural Electric Cooperative in Piqua, Ohio, to manufacturers of electrical appliances summed up many of the complaints on this score in 1950. The letter acknowledges that the "electrical appliance industry has done a great job of creating the desire for modern appliances and does draw customers into the dealer. [But] there you will find the root of the problem. The dealer turns them to bottled gas" because of its low operating cost, even to the point of selling equipment at cost to get the replacement-bottle business. "There was a time back in 1936 when we first went into operation that [the dealers] might have been able to point to frequent power failures on our lines, as well as the power companies, as a reason for bottled gas serving major appliances. We, as all other power companies in recent years, have brought our rural lines up to the point where our power failures are not any more frequent than forgetting to order another bottle of gas."[69] Co-op leaders noted that cooperatives often gave a list of their members to dealers who carried both gas and electrical appliances, a practice that sometimes boomeranged because the dealers sold gas appliances to co-op families.[70]

Neither dealers nor farm families saw this as a backward step, because bottled gas was a newer technology than electricity. Ironically, the progressive ideology of technology, which had served rural electrification well for half a century, now created a dilemma for the REA. Kermit Overby, head of the agency's Information Services Division, was alarmed about the competition from bottled gas in 1947. But he then observed, "No REA co-op, of course, wants to be in the position of advising its members against any home or farm improvement solely on the ground that the improvement involves the use of a fuel other than electricity." Elwood Olver told Iowa's electrification advisers in 1951, "Your first reaction [to the competition from bottled gas] may be that you cannot stop progress, and, hence, we should just let bottled gas take over!"[71] Rather than caving in, Overby and Olver exhorted co-ops to sell electric ranges, water heaters, and house heaters on the time-honored virtues of electricity, namely that it was clean, safe, fast, automatic, and economical (if enough electricity was used to put the consumer in the lowest rate schedule). Olver told his advisers to remind farmers that "gas explodes."

But many REA agents disliked these tactics. Field representative Phillip Voltz complained in 1952 that "LP gas continues to be a problem in the area and unfortunately the co-op [in Illinois] seems to fight gas in both letters and word-of-mouth by the use of the 'gas-is-dangerous' technique." Home economist Louisan Mamer had observed a similar situation in a training program about ranges that she conducted for electrification advisers in Kentucky in 1949. "In general, the attitude of the Kentucky group seems to be to put on fighting gloves in relation to the LP gas opposition. Kerr, Mary Alice, and I all tried to pour oil on troubled waters, but we felt we did not succeed too well." Emphasizing their profession's preference for education to sales, they recommended taking a positive approach to promoting electrical cookery rather than the negative one of criticizing LP gas.[72] One way co-ops did this was to pay the installation charges for the 220-volt service needed to operate an electric range.[73] Another method was to pay for the installation and buy back old appliances. An "Old Range Round-Up" conducted by the Nodak Rural Electric Cooperative in North Dakota "corralled hundreds of old appliances" in 1958. The co-op estimated that the promotion cost twenty thousand dollars but increased revenue by fifty thousand dollars a year in additional electrical consumption.[74]

Despite the innovative promotional efforts, the problem of low usage did not disappear in the late 1950s. The Nebraska Interindustry Electric Council investigated consumption patterns within the Southwest Power District in Palisade in 1957 by interviewing a sample of 190 farm and 250 town members who used less than 100 kilowatt-hours per month. Although average farm consumption on this project was a respectable 331 kilowatt-hours per month and the town homes averaged 216 kilowatt-hours, 20 to 30 percent of rural and small-town consumers on REA systems still averaged less than 100 kilowatt-hours per month at this time. The Nebraska council found that the age of consumers was a big factor. "The low users on farms, while not as old a group as the town [where 68 percent of the low consumers were over sixty], nearly all are over forty." Four years of drought had also reduced the incomes of these farmers, mostly livestock and feed producers. Furthermore, the "saturation of LP gas equipment is high, even in refrigeration." To improve the situation, the council recommended that the project purchase 150-watt lamp kits, electric frying pans, and roasters; sell these door-to-door; find a dealer who was willing to recondition electrical appliances for sale, and work with Extension Service agents to improve farm income in the area.[75]

Using the Telephone, the Radio, and Television

Farmers used communication technologies in prewar patterns, as well. This is especially true for the telephone. In 1945, the Northwestern Bell Telephone Company investigated the possibility of rehabilitating rural lines in Iowa by surveying about one thousand subscribers on company-owned lines and another one thousand service-station subscribers (i.e., farmers who formed mutual companies to own and maintain their lines, which were switched through Bell exchanges). When asked the most important reason for having a telephone, a little more than half of those on company lines and about the same percentage of those on mutual lines marked "emergency reasons." Business purposes came in a close second, with the two groups at 42 percent, while "general and misc. uses" came in last, at 10 to 13 percent. The breakdowns in usage, for business ("operating your place"), domestic ("running the household"), and social ("general purposes"), show the types of activities discussed in chapter 1. Calls made to town for spare parts, the veterinarian, and making appointments were high on the list. But so were local calls for "trading help" and "visiting with friends or relations," which illustrates the

persistence of older agrarian customs in the mid-twentieth century. The minuscule 7 percent of calls to the county agent marks the continued rural resistance and distrust of these reformers.

Many of the questions asked about party lines. About one-third of families on company lines and those on mutual lines marked "often" to the question, "Do other people listen in on your conversation?" About one-half of
those on company lines and about the same percentage of those on mutual
lines said they would like to have a device installed to prevent this practice.
But they were not that concerned about it. When asked, "Do you object to
other people listening in on your calls," more than one-half of those on company lines and nearly two-thirds of those on mutual lines said "no." This response, and the fact that three-fourths of the subscribers said the amount of
ringing on the line seldom or never bothered them, supports the argument
made in chapter 1 that farm people welcomed listening in much more than
commentators, then and now, maintain. Iowa farm families in 1945 would
have liked the telephone company to provide more privacy, but they had woven the party line into the fabric of rural life in ways that were acceptable and
(probably) enjoyable to them.

The survey also asked rural folk to compare the importance of the telephone with that of electricity and the automobile. It is not surprising that
nearly two-thirds of both groups of subscribers considered the car to be an
"absolute necessity," but only a third to one-half ranked the telephone in that
manner. Electricity rated a little higher than the telephone, even though both
groups thought it gave more value for the money than the phone. Farmers on
mutual lines ranked the telephone significantly higher in the "absolute necessity" category than did rural subscribers on company-owned lines, probably
because when they originally built the lines they had connected to specific
neighborhoods and trading areas with which they wished to communicate.
Subscribers on mutual lines most likely ranked electricity lower than those on
company lines because a higher percentage of those on mutual lines did not
have electricity (20 percent to 13 percent). The fact that hardly anyone in either group called the telephone, electricity, or automobile a "luxury" indicates
that these consumer goods had lost an urban image prevalent since the
1910s.[76]

Other evidence supports many of these observations. A study conducted
by Purdue University in 1948 found that rural subscribers in Parke County,
Indiana, made two-thirds of their calls for business purposes and one-third
for "nonbusiness or social purposes." The results differed somewhat from the
Iowa study in that "business calls to other farmers were slightly more frequent

than business calls to town." Farmers in Washington County, Indiana, said that 36 percent of their calls were business calls to other farmers and 35 percent were social calls. The researchers noted that "only 29 percent of the calls were reported as business calls to town. It appears that trading work and other neighborhood ties are still very important among farmers of Washington County," an older, primarily agricultural area. The average number of calls per week in both counties was about fifteen, similar to that in the Iowa survey, and farmers in both counties made fewer long-distance calls than did city residents.

An important question was whether the farm phone should be considered a business or residential telephone. Rural reformers and farmers themselves had called farming a business since the early part of the century, although they viewed it as a unique business in which the "employees" were a family that lived and worked on the premises. But business rates were higher than residential rates in all telephone systems, including those funded by the REA. The agency advised its borrowers in 1953 to classify the farm telephone as a business phone only if the farmer wanted to list the number under a business name, such as Sunnyvale Dairy. The REA's concern about the fragile economics of rural telephony, which had declined rapidly during the Great Depression, enabled it to sidestep the thorny issue of declaring whether a farmstead was a business or a home and whether the farm telephone should be used primarily for business or social purposes.[77]

The radio grew rapidly in popularity after the war. Sixty percent of all farms had radios in 1940, 83 percent in 1947. Although the figures varied by region (from 60 percent, in Alabama, to 96 percent, in Washington state), 80 percent of farms in most states had radios in working order in 1947.[78] The 1948 study of Mississippi mentioned earlier found that far more of the farm families surveyed owned table radios than the more expensive consoles and that about one-fourth of the families owned more than one radio.[79] Radio interference caused by power lines and electric appliances, discussed in chapter 6, was still a problem after the war. Among the offenders were electric drills, sanders, razors, food mixers, and fluorescent lighting. Loose connections between the motor and frame of a refrigerator and in wall plugs were also culprits. Poor line construction caused much interference, as did the move to higher voltages around 1950 until the REA convinced manufacturers to modify their equipment. The problem was worse in the countryside than in the city because of the low strength of signals in rural areas and the fact that the popular table radios were not designed to pick up such weak signals. The REA recommended that farmers string up an external antenna if they lived

more than fifty miles from a broadcast station.[80] A copper wire running between two trees with another wire leading to the house thus became a symbol
of rural radio in the 1950s.

Determining what farm men and women listened to was just as difficult
after the war as it had been before the conflict. The Federal Communications
Commission asked the Bureau of Agricultural Economics to conduct a survey in 1945. The bureau interviewed about twenty-five hundred households in
116 counties throughout the country, usually interviewing men and women
separately. When asked, "What kind of program would you miss most if your
radio gave out?" farm and rural nonfarm people overwhelmingly said the
news (about 80 percent for each group). Every other category fell below 20
percent. Farm people preferred market reports and farm talks and had less interest in sports broadcasts, but the differences with the nonfarm group in the
remaining categories were slight. Both men and women ranked news the
highest, but (constructed) gender differences were evident in men's preference
for market reports and farm talks and women's for serial stories. The only regional difference was the much higher preference for religious programming
in the South.[81]

Television quickly became popular in the countryside, despite the fact
that most stations could be received well only at a distance of fifty to one hundred miles, depending on the strength of the station and weather conditions.
The Census Bureau reported that 36 percent of all farms had TV sets in 1954,
up from 2.4 percent in 1950. Regional ownership patterns were similar to
those for other large appliances. Farms in the South Atlantic, East South-
central, and West South-central states fell under the 30 percent figure,
whereas more than 50 percent of farms in the Far West and most of the Midwest owned TVs. The Mountain states were at the 25 percent level, probably
because few stations served the sparsely populated region. Television was
prone to the same type of interference that afflicted radio, but quality sets suffered less from power-line problems. Poor reception in rural areas was the
biggest problem for many years.[82]

Modernizers saw the new medium as yet another way to send their messages to farm people. In 1949, the REA predicted that "just as rural families
need radio to give them weather reports and market information, indications
are that they eventually will use television to learn the latest and most effective methods of farming and farm homemaking." Noting that most stations
tended to "slant their programs" to the much larger urban audience, the REA
described a popular rural program begun by an eastern station. The program
included a "typical rural church Christmas service," a 4-H Club project

demonstration, a "barn dance including features such as corn-husking, nail-driving, and pie-eating contests," home economics lessons, and demonstrations of soil-heating cables.[83] Later indications are that farm people preferred to listen to more urbane fare, like the soap operas and comedy shows that migrated from radio to television, but also to country music. The REA's predictions were out of tune with the type of rural modernity that already had been achieved with the radio in rural America. Louisan Mamer was closer to the mark when she suggested in 1952 that farm women could use a TV set to earn money by baby-sitting neighborhood children and that television would keep bedridden patients quiet and children busy.[84]

Social Changes

Imbued with the progressive ideology of electricity shown in the scissors picture in the introduction (fig. 1), the REA predicted sweeping social changes for rural electrification. Before the war, Administrator John Carmody had told a staff conference, "Few other activities carried on in communities, whether by private industry or by public bodies, have so—if I may use the word here—electrified the thinking of the people as the coming of electricity in the community, introducing new ways of thought, putting them in the position to be receptive to new ideas within a year, that they would have resisted throughout their lives, as their parents resisted them."[85] One of the new ideas was racial equality. The REA reported in 1937 that "prospective users [in the South] sometimes waived the segregation of races, to say nothing of the willingness of political parties to work together." At an electrification meeting held in a Louisiana courthouse, which was normally divided into sections for whites, Negroes, and Indians, the "entire room was jammed with listeners of all three races." Another social change was the fostering of a community spirit. Farm men and women "get more than electricity," Carmody told Congress in 1937. "They get a sense of community responsibility and neighborliness. . . . This new spirit [of cooperation] spreads from the farms themselves to the nearby towns whose merchants are brought still closer."[86]

Louisan Mamer filled out the prewar utopian picture in a talk she gave to the Electrical Women's Round Table in New York City in 1952, a conference of home economists working in the electrical industry. Drawing on the themes of earlier promoters and rural reformers, Mamer listed "some results of home electrification on living in rural areas" that "homemakers and educational leaders" had given her. Rural electrification was "changing women's ways of doing things," increasing their income, encouraging "young farm

women to stay in rural areas and marry farm men, to keep youth on the farm,"
decentralizing industry, filling labor shortages, stimulating the renovation of
houses, improving rural health, and uplifting farm life. "Rural living is becoming a way of life, throughout life, for more rural people, rather than just a
means of making a living to retire to town."[87]

Although historians have made many of these same claims,[88] the evidence to support them is mixed. In regard to race, REA agents often worked
with African American co-op members. Although this occurred in a segregated manner, it was better treatment than blacks received from many New
Deal agencies, especially the Agricultural Adjustment Administration, with
its pro-landlord policies enacted against sharecroppers. In 1951, Elizabeth
O'Kelley reported that an electrical adviser gave demonstrations of home
freezers in Virginia to "those who need it greatly—our colored members. . . .
When some seemed surprised that she was having this for the colored group,
she told them she was doing so because they wanted help and asked for it and
she would do the same for any of their groups whenever they wanted it." Walter Rich reported on two all-black community meetings held on Hilton Head
Island in South Carolina that year.

> They were the leaders who had come to get the "message," as they ex-
> pressed it, and carry it back to the others. . . . This was my first experi-
> ence with an all-Negro audience. Their faces and their responsiveness
> plainly showed their appreciation of our meeting with them and for the
> information we had brought them. There is no question but that sound
> relationships have been established with these people and our sincere
> hope is that [co-op] Manager Perry [who was at the meeting] will carry
> on the educational and power-use activities which we have started
> here.[89]

The situation was not common in the South. An REA field representative reported that one co-op's electrification adviser made "range calls to both their
Negro and White members, which is unusual for Mississippi."[90]

At the annual meeting of another South Carolina co-op in 1951, Eleanor
Delaney reported that "there were almost as many colored members as white
members and they occupied half of the seating space." But there were racial
tensions. Following the elections for a board of directors, a "Negro member
got up and asked why they couldn't have a Negro director. Since the voting
had already taken place, it was explained that it was too late for such a suggestion. Needless to say, this suggestion by a Negro caused quite a bit of com-

ment after the meeting, because in this section of the South they aren't ready to have Negro directors." After the meeting, Delaney made a home visit to a black family that had won a refrigerator as a door prize. "It was a pleasure to see how much this piece of equipment meant to this family. It was worth more than everything else they had put together and they were very happy about it." Delaney, however, did not seem concerned about whether they could afford to run it.[91]

The issue of changes in community with the coming of electricity is difficult to analyze, mainly because of numerous other factors like the depopulation of the countryside in the 1950s and 1960s. Before the war, the REA proudly reported that co-ops were strengthening rural communities, as was electrification itself. Community refrigerators were "becoming solidly established" in the Tennessee Valley in 1937. The REA lit traditional meeting places and schools. In 1938, the agency thought the electrification of Grange halls was particularly significant, "in view of the long battle waged by the Grange and other farm organizations to bring central-station service to rural areas." By 1941, the REA served about twelve thousand rural schools, almost seven thousand of which were one-room buildings—a sign of the resiliency of this institution.[92] After the war, electrification enabled teachers to use slide projectors, movie projectors and sound equipment, intercoms, microphones, and the radio news report. Electric lights, running water, and refrigerators served school lunch programs and after-school meetings. The REA vigorously supported the gendered activities of 4-H Clubs, Future Farmers of America, and Future Homemakers of America, which continued the prewar strategy of educating young people to convince parents to modernize the farmstead. Rural clinics, hospitals, gyms, and baseball diamonds were established near REA co-ops.[93]

More attention has been paid to the relationship between rural electrification and the work patterns of farm women. Many women thought electrical appliances saved effort and time. In 1937, *Wallaces' Farmer* solicited letters about improvements in the farm home brought about by electricity. A woman from Tama County, Iowa, wrote, "The good fairy, electricity, has waved her magic wand across my path and now I lead a charmed life. . . . No water to be carried uphill; no waste water to be carried out; no kerosene lamps to be cleaned and filled; no hand-scorching sad-irons to be used; no fuel to clutter up my kitchen in pails and boxes; no ashes to be swept up and carried out. . . . It seems too good to be true."[94] Gertrude Dieken, woman's editor of the *Farm Journal,* spoke for many farm women in an address before a national electrification conference in 1946. "In my opinion, asking a farm woman whether she wants electricity flowing through the walls of her home and making it come alive is like asking her 'Would you care to go to heaven?'" Dieken cited many letters from farm women about their long days and blamed farm men for modernizing the barn ahead of the house. "Until it is electrified, the farm home is sadly out of step with the society in which it exists." Mrs. Joseph Zikmund, wife of a co-op director in Nebraska, speaking at a conference in 1948, claimed that electricity "goes right inside the home and revolutionizes the homemaker's physical routine, her mental attitudes, enhances her value to society, and relieves her of hard labor."[95]

Other observers questioned the utopian effects of electricity and thus inadvertently provided evidence for the ironies of household technology discussed in chapter 3. Although the REA generally sold electrical appliances as laborsaving devices that would relieve the lot of the overworked farm woman, the agency sometimes considered whose work was actually being saved. Chief home economist Clara Nale said in 1937, "In using an electric range, the farm wife is greatly reducing the amount of work necessary for her husband in felling trees and sawing and splitting the logs, as well as eliminating work for herself." Nale argued that this was a selling point, not a detriment to her female clients, because she thought the range freed both sexes for other farmwork. A farm woman reported in 1938 that a new electric water pump saved the labor of her husband in maintaining a gas engine and that of her boys in pumping water for the house. In 1952, Louisan Mamer quoted an Iowa survey that "some farmers said electrification had meant saving in labor and benefits from timeliness in doing some jobs, because those jobs had been taken over, wholly or partly, by the women and children of the family."[96]

An Indiana farm woman, Helen Musselman, recently recalled that a bathroom created more work for mother. "Grandmother always laughed when she would come to visit me, though, after we had our bathroom. She said, 'My goodness, you spend more time cleaning that bathroom, in spite of all the convenience. . . .' She said she used to do her washing on Monday and she'd take the wash water out to the old-fashioned privy and scrub it out. It was done once a week, and that was the chore. And she said, 'You scrub your bathroom every day—sometimes twice a day.'"[97]

Social scientists thought the question of whether electrical technology saved labor time would be settled by the time studies of farm women, which home economists had conducted since the 1920s. Before World War II, researchers could easily compare farmhouses with and without electricity in order to correlate time spent on housework with electrification, but these distinctions became less noticeable after the war. Although many of the early researchers had not noticed any time savings with electricity, their colleagues expected that the average time spent on housework after the war would decline because of the modernization of the farmhouse. They were surprised, then, when their data showed that farm women still spent fifty-two to fifty-four hours a week on housework, the same amount of time as before the war. Although it will be recalled from chapter 3 that the time studies were biased toward well-educated, middle- and upper-class, native-born, white women, the data do inform us about the work patterns at least of this class of farm women.[98]

The time-use researchers responded in a variety of ways to data on the irony of laborsaving household technology. Marianne Muse of Vermont was surprised that she found no correlation between time spent on preparing meals and the availability of running water and that her data showed a negative correlation between time spent on sweeping and the ownership of hand sweepers and vacuum cleaners. But rather than use the prewar studies as a guide to investigate a general relationship between technology and time spent on housework, she explained the apparently paradoxical data on the basis of not being able to isolate one cause from another.[99] Elizabeth Wiegand, a student of Jean Warren's at Cornell University, followed up her mentor's work in 1936 by investigating the same area of New York state in 1954. Although Warren had deemphasized the irony of household technology, Wiegand stated it clearly. "Modern equipment may make it possible for the homemaker to do better home-making work and save her some energy, but it does not necessarily save her time." She noted that some technologies (running water and electric or gas stoves) saved time, but others (freezers and electric mixers) did not. Comparing her results with Warren's, she noted that farm women seemed to have shifted their time from food preparation to care of clothes and marketing, a point Maud Wilson had made in 1929.[100] In a 1956 study of Wisconsin farm women, May Cowles thought it "remarkable" that the time spent on housework had not changed since the mid-1920s, despite the many changes in household technology. Like Wiegand, she noticed a shift from time spent on food preparation to that spent on purchasing, management, and family care.[101] But Sarah Manning's 1968 survey of Indiana farm women only noted that the ownership of dishwashers, washing machines, and ironers did not correlate with less time spent on meal preparation. She did not analyze the phenomenon nor note any shifting patterns of time usage.[102]

How were the postwar studies received? The earliest reference I have found in the REA literature and archives to any of the time studies is the paper Louisan Mamer presented to the Electrical Women's Round Table in 1952. Mamer cited Maud Wilson's 1929 study finding that women, when asked about the benefits of electrification, placed the savings of energy first, time second, and productivity third. Commenting on the long hours documented in the studies (fifty-two hours per week on homemaking, plus nine on gardening and farmwork), Mamer said, "I now know why my mother [who lived on an unelectrified farm] had to retire to town when she was fifty." Although Mamer believed that electricity generally saved time, she acknowledged that the time studies often showed the opposite effect. "Frequently, however, no time is saved at all by having electric equipment, a woman just does more work in the same length of time," thus raising her standards of housework. Unlike many home economists, Mamer critiqued this type of productivity. "In some cases she is unwise to do this; she might better continue leaving certain things unironed and start doing a little more resting, reading, or playing with the children." Mamer also observed that variations in family size, arrangement of kitchens, and attitudes of researchers and those being studied created difficulties in interpreting the data. She seemed ambivalent about the middle-class, urban domestic ideal, quoting a male electrification specialist that "it does no good to save her time if [the homemaker] has to go out and do work on the farmstead." But Mamer reported that female advisers thought one of the benefits of electricity was to "permit women to help with farmwork."[103]

Other home economists addressed the time studies more extensively. In each new edition of their popular textbook on home management first published in 1942, Paulena Nickell and Jean Dorsey analyzed the studies at length. By the time of the third edition in 1963, they observed that the amount of time spent on housework had not changed since the 1920s. They noted changes in the pattern of time use, such as less time spent on preparing food and more time spent caring for clothes. They explained the latter phenomenon by saying that washing machines enable women to wash more often. In all four editions, Nickell and Dorsey deemphasized the irony of laborsaving devices, which Dorsey had clearly stated in 1932, by shifting the emphasis from the savings of the home worker's time to the savings of her energy.[104]

This situation changed after sociologist Joann Vanek published her analysis of the time-use studies in the early 1970s.[105] Before Vanek, every researcher had held household technology as an independent variable. Although home economists differed about how to identify and interpret data that usually showed no correlation between this variable and time spent on housework, they included the data in their studies. After Vanek showed in an exhaustive manner that time spent on housework had remained remarkably constant from the mid-1920s to the mid-1960s, despite the obvious technological changes that had occurred, at least one major study dropped technology as an independent variable. In 1976, Kathryn Walker, third in a line of time-use researchers at Cornell, noted that "contrary to the opinion of many, average time used by wives for household work has not been drastically reduced because of technological developments in automatic equipment such as dishwashers, washers, and garbage disposers." Like Maud Wilson in 1929, Walker observed that women did more work with up-to-date technology and shifted their work patterns toward consumption and child care.[106]

What do the time studies say about nonhousehold work? The amount of time full-time home workers spent on farmwork did not decline dramatically during the increased consumerism following World War II, remaining at a level of about eight to twelve hours a week. In a 1943 study of fifty Illinois farm women selected by home economists, thirty-four of the women used the kitchen to prepare eggs for market, sixteen to wash separator or milk containers, eleven to handle produce for sale, and seven to prepare cream and milk for market.[107]

This study and the constant level of farmwork noted in the time-use studies for middle-class farm women complements the historian Katherine Jellison's modification of the thesis that household technology always led to more work for mother. Jellison persuasively argues that many postwar farm women, those who had not adopted the urban domestic ideal to such a large extent, used the time savings possible with some technologies not to do more housework in the same amount of time but to work off the farm or to work more in the field, especially driving tractors and running to town for spare parts for farm machinery.[108]

In regard to the question of technological determinism, the time-use studies reinforce my argument that household technologies and other supposedly urbanizing inventions like the telephone and automobile did not transform rural life in a wholesale manner. Instead, farm people assimilated these technologies into existing social patterns, which expanded to some extent as a result. The time-use surveys show that a large number of middle- and upper-class farm women employed some "laborsaving" household technologies to do more work during the culturally defined, standard workweek of about fifty-two hours in the house. Indeed, social norms seem to have played as large a role as the purchase of new technology in determining what constituted the length of a "typical" workweek for these women.

Other areas of gender structure and identity did not change as much. Technical competence as an element of masculine gender identity, discussed in chapter 2 in relation to the early automobile, seems to have prevailed for electrical appliances after World War II. In a survey of more than two thousand farm operators (all men) and seven hundred "professional workers" (male and female agricultural professionals) in Missouri in 1948, a majority of both groups listed in order of importance twenty-one items family members should know about electricity and electrical appliances. "It was believed that girls should know items 1 to 7 [selection, purchase, and operation of equipment, plus replacement of fuses], that women should know items 1 to 10 [which included making simple repairs to appliances, locating the cause of blown fuses, and determining the profitability of electrical equipment], and that these items and the remaining ones [concerned with "nonhousehold equipment," rigging up a portable motor, and wiring] should be known by boys and men." What electrical knowledge and skill farm youth, women, and men actually possessed and used in Missouri or elsewhere in this period is unknown. In a 1946 study of one hundred urban women—in Syracuse, New York—home economists found that only "27 percent of the women turned equipment which refused to operate over to the man of the house to look over before tinkering with it themselves or calling a service man. About one-fourth of the women could repair electric cords and replace blown fuses."[109]

Gender mattered for the telephone in the postwar era as much as it had before the war. Researchers from Purdue University observed in 1949 that "men and women [in Parke County, Indiana] both agreed that women used the phone most frequently. Many men said they did not like to use the phone, so they had the women call for them."[110] These gendered perceptions of telephone use have been analyzed well by anthropologist Lana Rakow, who conducted a study of Prospect, Illinois, in the 1980s. Addressing the party line, operator functions, and telephone neighborliness, Rakow's oral histories "illustrate how the telephone is subtly woven through the lives of women and the life of the community and must be understood within those broader contexts." But men (farmers and telephone employees) often viewed women's work done over the telephone as trivial gossip, much like telephone men did before the war. Rakow concludes that "women in Prospect use the telephone, then, for maintaining relationships with family and friends, for accomplishing the ongoing structure of the community, for giving and receiving care and emotional support, and for doing the business work that enters the private sphere." This was both gender work and gendered work, in that it (re)created and expressed gender relations through the medium of the telephone.[111]

By 1960, the REA, Congress, manufacturers, and farm people had reached most of their goals in rural electrification and telephony. Their combined efforts electrified nine out of ten farmhouses and brought telephone service to about two-thirds of them. Postwar prosperity had enabled them to create a basis for further attempts to (sub)urbanize the family farm. Yet this process was contested, just as it was in the early years of the telephone, the automobile, the radio, and electricity on the farm, but without the violence of attacking automobiles or cutting down telephone and power lines. Farm men and women selectively used the material culture of new technologies to create their own versions of rural modernity, many of them individual modernities, within the networked services of electrification and telephony. These new forms of farm life were not predetermined by any internal "logic" of these technologies nor the social goals of the REA and other promoters. They represented the results of negotiations worked out between farm people and modernizers in the first decades following the war.

CONCLUSION

Consumers All?

How well did reformers accomplish their mission of using new communication, transportation, and household technologies to make rural life more like urban life in the twentieth century? A comparison of two yearbooks of the United States Department of Agriculture indicates that that agency, at least, thought the job was completed by the mid-1960s. Figure 10 shows the production-oriented cover of the 1940 yearbook. A county agent checks the crop patterns of a prosperous farmer to see if he is in compliance with the acreage allotment of the New Deal's agricultural adjustment and soil conservation programs.[1] The farm man's striped overalls, although more stylish than the traditional denim variety, mark him as both a man and a farmer. The "absent other"—the farm woman—is presumed to be working in the house in the background, and absent children in the house, barn, or field. In this new era of "farmers in a changing world," a farm family cooperates with a modernizer (the county agent) to implement a government plan to control production and bring them out of an economic depression caused, in part, by increased agricultural productivity.

Figure 11 shows the consumption-oriented cover of the 1965 yearbook. According to this view, the desired urbanization—now interpreted as a suburbanization—of the family farm, the goal of reformers since the beginning of the century, has been achieved. Indeed, it is difficult to tell this television-age, "father knows best" family from a suburban family. The man wears a fashionable bow tie and cardigan sweater. The equally fashionable woman is now present in the picture, purse in hand, as a modern consumer. The children play and study, rather than work. The absent others are now the modernizers in town. The couple and their children signify "consumers all" in two respects: farm families, just like urban and suburban families, are consumers,

272

and all members of the family are individual consumers, even the dog, who eats out of a dogfood can instead of hunting rabbits or eating scraps from the table.[2]

The contents of the two yearbooks show similar contrasts. *Farmers in a Changing World*, the 1940 yearbook, contains lengthy articles on the history of agriculture, analyses of the seminal farm policies of the New Deal, impressive statistical studies of the current state of agriculture, and portents for the future. Historians and social scientists have long considered the book to be a classic piece of scholarship.[3]

Consumers All, the 1965 yearbook, is more superficial. Short articles by home economists, agricultural engineers, forestry experts, and social scientists inform consumers about houses, household equipment and furnishings, finances, safeguards, activities, food, and clothing. The matter-of-fact articles primarily address the needs of suburbanites; the specific concerns of farm families are briefly mentioned by only a few authors. A USDA house plan includes a farm office and a "convenient area for Dad and the kids to hang their clothes and wash up when coming in from the field and play," which shows the staying power of the department's gendered domestic ideal. In purchasing a home freezer, the homemaker should consider that a farm family will need five to six cubic feet of space per person, whereas an "urban or rural family" that shops once a week will need only three to four cubic feet per person. Instead of discussing the prewar concern about saving the overworked farm woman, the article on leisure tells city families about "vacation farms" and "dude ranches." On the other hand, the expert on sanitation addresses the needs of many farmers in the mid-1960s by saying, "The earth pit privy, properly located, constructed, and maintained, is a satisfactory facility for the home without indoor plumbing." The author gives detailed instructions on where to locate and how to build and maintain a modern outhouse.[4]

The USDA, of course, had good reason to question whether average farm families in 1965 looked like the foursome on the cover of *Consumers All*. Other yearbooks in the 1960s show farm people hard at work in the field, barn, and house, usually dressed in overalls, jeans, or conservative town clothes. *Consumers All* depicts the (purported) achievement of the long-standing urbanization ideal of the USDA. Artistic license may have been taken to portray all consumers, including farmers, as stereotypical suburbanites on the yearbook's cover, but the articles assume that farm families have essentially the same housing, financial, food, and clothing needs as suburban and urban consumers. The only difference, from the USDA's point of view, seems to have been that farmers lived where they worked, in a house that was as modern as

Figure 10. Farmers in a Changing World. A Nebraska farm man shows a county agent his crop patterns in order to qualify for government assistance programs
Source: U.S. Department of Agriculture, *Yearbook of Agriculture, 1940* (Washington, D.C.: GPO, 1940), cover.

Figure 11. Consumers All. The USDA's depiction of the alleged urbanization (or suburbanization) of the American farm in 1965, the goal of the Country Life movement since the beginning of the century

Source: U.S. Department of Agriculture, *Yearbook of Agriculture, 1965* (Washington, D.C.: GPO, 1965), title page.

the 1960s mechanized farmstead. The sharp drop in farm population, from 30.5 million in 1940 to 13 million in 1964 (see table A.1), also helps explain why USDA home economists continued to shift their focus from rural to town and urban consumers in this period, a process that appears to have begun with nutritional programs during the Great Depression.[5]

Yes, but perhaps the creators of *Consumers All* were not that far off base in depicting farm families finally entering the (sub)urban consumerist era? More and more farm families became consumers, in that they purchased more and more of their food, clothing, shelter, and transportation rather than producing it on the farm. Perhaps agricultural reformers and modernizers succeeded more completely than my account suggests?

Although I have stressed the contested aspects of the attempts to urbanize rural culture, there is little doubt that the introduction of new communication and household technologies into the countryside helped to transform farm life in the first six decades of the twentieth century. Farm people used the telephone, the automobile, the radio, and electric appliances to create a material culture that resembled that of urban and suburban areas. Yet I have argued that this process was much more fractured, and the outcomes much more uncertain, than the images usually associated with the terms *transformation, urbanization,* and *modernization.* Promoters disagreed about how to introduce so-called urban technologies into rural America; many of them saw it as a losing proposition. A material culture derived from the city did not completely urbanize farm life. Farm people were not passive consumers who accepted new technology on the terms of the reformers. Instead, they resisted, modified, and selectively used these technologies to create new ways of rural life. They followed their own paths to modernity.

This was particularly true before 1930, when networks of modernity were rather limited in scope. Farm families often cooperated to build and operate their own telephone systems, connecting up their neighbors in existing community patterns and recreating the rural institutions of musical gatherings and "visiting" on the party line. Farmers exercised a good deal of power in the social geography of access competition in rural telephony. AT&T acquiesced to the social practice of eavesdropping by designing a system that permitted it. Farm people tamed the "devil wagon" roaring from the city into the country by actively resisting its inroads, then modifying it to wash the clothes, plow the fields, and carry produce to town. Existing gender relations enabled farm men to make these alterations, a process that usually reinforced gender identities, symbols, and structures. Farm women and men traveled farther and more frequently to town with the auto in the 1920s and 1930s but did not become urbanites in the process. Manufacturers brought out gas-engine washers, tractors, and pickup trucks to capitalize on the consumer innovations of farm men and women. In a similar manner, farm people tuned in the country, not the city, with the radio, and modernized their houses "off the grid" before World War II, with an innovative assortment of nonelectric appliances. Those who made their own electricity were bitten by the bug of electrical modernization and wanted more.

By the 1930s, farm people had interacted with the commercial mediating networks of manufacturers and dealers to create rural modernities using household technologies and communication networks of telephone lines, automobile roads, and radio broadcasting stations. The expanding agricultural complex came into play more extensively with the founding of the Rural Electrification Administration in 1935, a government body that used the techniques of commercial companies and rural organizations to build distribution and mediating networks to electrify America's farms. Its field agents organized cooperatives, signed up reluctant farmers, and pushed electrical appliances to relieve the overworked farm woman and transform the farm home into a rural version of (sub)urban middle-class life. Farm people resisted the process by not signing up for the cooperative, running the co-op in accordance with local values, using less electricity than the REA desired, and not buying a full complement of appliances. Full-time home workers used electrical appliances in culturally established ways by doing more work in the same amount of time or using the time saved to work off the farm or to work more in the field.

Despite the supporting postwar climate of farm prosperity and a new wave of consumerism, the REA still had to work hard to mediate between producer and consumer, to sell farm people on the virtues of electrical modernization. Farmers bought more washing machines and refrigerators but were slow to install bathrooms because of the high cost, poor water supplies, and entrenched cultural habits (especially among older farm men). By focusing on the resistance to electrical modernization, I do not mean to leave the impression that farm people did not value electrification. They did so from the beginning of the REA program. But up to the 1960s, they wove the new technology into rural life by valuing certain items—lights, irons, and radios, then washing machines and refrigerators—more highly than "urban luxuries" like coffeemakers, vacuum cleaners, and flush toilets.

Before the war, electrification was a novelty. Several farm families, who had wired their houses for electricity and had accidently left the switches in the "on" position before they went to town or into the fields, thought the house was on fire when, returning home, they approached a house that was fully lit when the REA turned on the power. A Virginia schoolteacher reported that her first- and second-graders, who had never seen electric lights, were greatly excited by the new technology. "Some of them were so startled that they could not talk. . . . The youngsters were entranced. I had to snap the lights on and off many times before they were satisfied. Later the sky cleared, but the children begged so hard that I kept the lights on all day."[6] By the 1950s, electrification in farm country was commonplace. One measure of how thoroughly farmers had woven it into the fabric of their lives is the volume of complaints raised when the electricity went out. An electrification adviser on an Illinois cooperative told the REA in 1952 that electricity "is definitely paying its way in rural homes and if one does not believe it, they should be here in our office on one of the rare outages. You'd think the world was coming to an end if service is interrupted for any length of time."[7] Like urban and suburban folk, farm people probably felt the same way about the telephone, automobile, and radio.

By 1965, all four technologies had become a taken-for-granted part of modern life in the city, suburbs, and the farm, as the USDA's phrase, "Consumers All," implies. It is the convergence of these technologies, in fact, that provides the strongest evidence for the yearbook's point of view. Although the telephone, the automobile, the radio, and electrification have their separate histories—which is how I have told most of my story—promoters and farm people combined them early and often. Telephone companies and the REA used automobiles—and later, two-way radios—to build and maintain their lines. Home economists used the telephone and automobile to schedule and make home visits. Rural Electrification Administration crews patrolled the lines, listening to their car radios to check for radio interference caused by power-line faults. Farm people often substituted one technology for another, when they switched from rural free delivery to the telephone to the radio to television to obtain market and weather reports, for example. But over time, they used these technologies as an ensemble. Farmers telephoned to see if implement dealers had spare parts for the tractor or combine, then drove to town to pick them up. Electrification allowed radio listening to become more popular and provided the much greater power needed by television sets. Electricity pumped water for the bathroom and kitchen. By the mid-1960s, electricity did more and more of the chores depicted in the 1930s scissors picture (fig. 1).

Yet *Consumers All* is wrong to imply that these technologies were irresistible, powerful social forces. As we have seen, most farm families chose to purchase them selectively, to create *individual* modernities. Before World War II, many farmers owned an automobile and a telephone but did not have electricity. Others had electricity and an automobile but no telephone. After the war, a large number of families enjoyed all four technologies but still trod the well-worn path to the outhouse. New technology did not bring about the consumerist utopia predicted by its promoters. The telephone, the automobile, the radio, and electrification—individually or as a group—did not stop the migration to the city, reduce the hours of full-time home workers, or eliminate urban-rural differences.

Although modernizers wanted to reform rural life by urbanizing it, the REA, for one, became concerned about an unintended consequence: the effect of the growing number of suburbs on REA cooperatives. In 1959, the Anoka Electric Cooperative in Minnesota found itself facing what the REA called "the Big New Problem." Simply stated, it was "no less than the problem of pleasing a new type of rural consumer—the citybred commuter family. It is also the knotty problem of making the ex-city families on the lines members of the co-op in spirit as well as name." The transplanted urbanites thought of the co-op as just another power company and did not attend annual meetings. Much to the surprise of the co-op, they did not know how to make simple repairs such as replacing blown fuses nor how to use electric dryers and automatic washing machines. In sum, the co-op manager said nonfarm people were harder to please than farmers because they were used to more service in the city. The co-op planned to hire a full-time home economist to handle the situation. The Prince William Electric Cooperative in Manassas, Virginia, faced similar problems in 1960 in dealing with the "invasion" of federal employees who commuted to Washington.[8] These reactions reveal that the USDA's (sub)urbanization ideal was not always a welcome outcome for the farm-oriented REA. In 1959, the REA—an agency founded by a patrician Philadelphia reformer—praised a telephone co-op in Texas for being "strictly rural and strictly cooperative."[9]

The staying power of rural ways of life in the face of social change—evident in the urban-rural tensions in REA co-ops and a major theme of this book—was recognized in the late 1970s by rural sociologists in a collection of articles entitled *Rural U.S.A.: Persistence and Change.* In his overview to the volume, Thomas Ford observed that

> the most intriguing feature of these ordered patterns of population characteristics, social and economic behavior, values, beliefs, and attitudes associated with community size is not their development but their persistence. The graded differences [in urban and rural social characteristics] have continued long after the dissipation of social forces that originally were thought to account for them (such as the dissimilar occupations of rural and urban residences) and after the development of new influences (such as rapid transportation and the mass media) that many rural social scientists thought would eliminate rural-cultural distinctions.[10]

In one of the articles, Olaf Larson questioned a claim he and his colleague Everett Rogers had made in 1964 that differences between urban and rural values were fast disappearing—the theme of the *Consumers All* yearbook published a year later. A survey of Gallup polls and rural studies done since then led Larson to "challenge any assumption that all the important rural-urban differences in values and beliefs are rapidly vanishing, if they have not already been obliterated by the forces of mass society." Differences had narrowed regarding equity issues in race and gender but had widened with respect to questions of sexual freedom and the importance of religion. Cultural pluralism was evident in rural America, especially between nonfarm and farm populations. Farm men and women generally agreed with the values of nonfarm rural adults, but their values were more at odds with those of big city people, and they were slower to change than other rural folk.[11]

Ford explained the persistence of rural social structures and culture on the basis of an ecological approach. Social forms arose from and were modified by "human cultural adaptation to environmental circumstances. Taking into account all of its aspects, including human population density, the environmental milieu of the city is appreciably different from that of smaller towns, villages, and rural areas. So long as these environmental differences persist, differences in social life will also persist."[12] Although neither Ford nor the other authors in the volume discussed consumer technology, Ford's conclusion supports the view of historian Lizabeth Cohen that interwar Chicago workers drew on ethnic community resources—their environment—to resist the modernizing influences of consumer technologies like chain stores and mass media.[13] I have also argued that the structure and culture of rural society were resources that enabled farm men, women, and youth to resist the inroads of (supposedly) urbanizing technologies and weave them into existing patterns of rural life.

My main goal, however, has been to understand the reciprocal relations between technology and social change. I have emphasized how farm people interacted with promoters to modify and selectively use communication, transportation, and household technologies to change their environment, to create rural forms of modernity that differed from those envisioned by the modernizers. My story about technology and the contested transformation of rural America thus argues that the surprising persistence of rural culture in the face of its near disappearance amid rapid technological and social change is best understood as a mutual construction of technology and society. It forms part of a larger story historians and sociologists of technology have begun to tell, in which cultural beliefs and structures shape and are shaped by sociotechnical networks composed of artifacts, knowledge, and practices.

The main symbols of middle-class farm life in the United States when *Consumers All* was published in 1965 were not cardigan sweaters and dogfood cans. The outward signs of that life were a pickup truck in the driveway, a long propane tank nestling against a clapboard farmhouse, and farm machinery parked everywhere. The material culture in the farmhouse resembled that in urban and suburban homes, but work patterns and leisure activities were still those of farm men and women. Double-wide mobile homes and satellite dishes for cable television were added to the mix in the late twentieth century.

At the same time, city and suburban folk drove imported pickup trucks and, in the 1990s, adopted sports-utility vehicles as a sign of (sub)urban independence and ruggedness.

The fact that mobile homes were originally developed for suburbia, that satellite dishes represented the high point of information technology in the 1980s, which is usually seen as an urban or suburban phenomenon, and that rural people initially used the technology to subvert the corporate monopoly of TV programming reinforces the argument made in this book that the attempts to urbanize rural life were part of a complex, contested process.[14] Even in the late twentieth century, when middle-class consumer culture has pervaded all aspects of American life, farm people still use new technology in innovative ways to create their own forms of modernity. Midwestern farmers driving huge air-conditioned wheat combines, for example, use cellular phones to communicate with their spouses about when to pick up the next load in the grain truck and where to store it.[15]

On the other side of the issue, the migration of pickup trucks out of the countryside, like that of country music in the 1920s, shows a continued ruralization of the city, town, and suburbs. Both processes illustrate that technology has not been an autonomous social force at the everyday level. Although the Old Order Amish in Pennsylvania demonstrate this in an extreme manner, middle-class farm people have also been able to choose, to a large extent, how to weave modified communication, transportation, and household technologies into daily life. We are beginning to understand how they, like users of consumer technologies in the town and city, employed resistance and selective adaptation to help produce technological and social change.

Appendix

Table A.1
GENERAL FARM STATISTICS, UNITED STATES, 1890–1969

Year	Number of Farms (millions	Farm Population (millions)	Farm Population as Percentage of Total Population	Net Income of Farm Operators from Farming, per farm (in dollars)	Personal Income of Farm Population, per farm (in dollars)
1890	4.565	24.771	42	—	—
1900	5.740	29.875	42	—	—
1910	6.366	32.007	35	652	—
1915	6.458	32.440	32	667	—
1920	6.454	31.974	30	1,196	—
1925	6.372	31.190	27	1,041	—
1930	6.295	30.529	25	651	—
1935	6.812	32.161	25	775	1,135
1940	6.102	30.547	23	706	1,245
1945	5.859	24.420	18	2,063	2,937
1950	5.388	23.048	15	2,421	3,780
1954	4.782	19.019	12	2,606	3,857
1959	3.711	16.592	9	2,795	4,866
1964	3.158	12.954	7	3,564	6,535
1969	2.730	10.307	5	5,674	10,093

Source: U.S. Department of Commerce, Bureau of the Census, *The Statistical History of the United States from Colonial Times to the Present* (New York: Basic Books, 1976), pp. 457, 483.
Note: Throughout the tables in this appendix, percentages are rounded. The five-year periods often mask year-to-year swings. Net income of farm operators from farming, per farm, fell in the following years:

1911 (from $652 to $525)	1931 (from $651 to $506)	1954 (from $2,626 to $2,606)
1913 (from $693 to $581)	1932 (from $506 to $304)	1955 (from $2,606 to $2,463)
1920 (from $1,395 to $1,196)	1936 (from $775 to $639)	1959 (from $3,189 to $2,795)
1921 (from $1,196 to $517)	1938 (from $905 to $668)	1964 (from $3,708 to $3,564)
1926 (from $1,041 to $919)	1949 (from $3,044 to $2,233)	1967 (from $5,019 to $4,730)
1927 (from $919 to $883)	1952 (from $2,946 to $2,896)	
1930 (from $945 to $651)	1953 (from $2,896 to $2,626)	

Table A.2
U.S. FARMS REPORTING OWNERSHIP OF TELEPHONES, 1920-1959
(PERCENTAGE)

Year	United States	Northeast	Midwest	Far West	South
1920	39	46	66	32	20
1930	34	49	62	35	14
1940	25	40	43	28	9
1950	38	65	61	51	16
1954	49	77	69	67	26
1959	65	87	80	81	45

Source : U.S. Department of Commerce, Bureau of the Census, *U.S. Census of Agriculture, 1959: General Report,* vol. 2, *Statistics by Subjects* (Washington, D.C.: GPO, 1960), p. 210.
Note: The table combines Census Bureau geographical divisions into four regions as follows:
Northeast: New England (Maine, New Hampshire, Vermont, Massachusetts, Rhode Island, Connecticut) and Middle Atlantic (New York, New Jersey, Pennsylvania) states.
Midwest: East North-central (Ohio, Indiana, Illinois, Michigan, Wisconsin) and West North-central (Minnesota, Iowa, Missouri, North Dakota, South Dakota, Nebraska, Kansas) states.
Far West: Mountain (Montana, Idaho, Wyoming, Colorado, New Mexico, Arizona, Utah, Nevada) and Pacific (Washington, Oregon, California) states.
South: South Atlantic: (Delaware, Maryland, D.C., Virginia, West Virginia, North Carolina, South Carolina, Georgia, Florida), East South-central (Kentucky, Tennessee, Alabama, Mississippi), and West South-central (Arkansas, Louisiana, Oklahoma, Texas) states.

Table A.3
U.S. FARMS REPORTING OWNERSHIP OF AUTOMOBILES, 1920-1959
(PERCENTAGE)

Year	United States	Northeast	Midwest	Far West	South
1920	31	33	53	42	14
1930	58	68	81	72	39
1940	58	71	82	73	37
1950	63	75	82	77	45
1954	71	82	87	84	55
1959	80	87	91	88	67

Source : U.S. Department of Commerce, Bureau of the Census, *U.S. Census of Agriculture, 1959: General Report,* vol. 2, *Statistics by Subjects* (Washington, D.C.: GPO, 1960), p. 220.
Note: For the states that make up these regions, see the notes to table A.2.

Table A.4
U.S. FARMS REPORTING OWNERSHIP OF RADIOS, 1925–1950 (PERCENTAGE)

Year	United States	Northeast	Midwest	Far West	South
1925	5	11	7	5	1
1930	21	43	38	31	5
1940	60	—	—	—	—
1950	92	—	—	—	—

Source: U.S. Department of Commerce, Bureau of the Census, *Statistical Abstract* (Washington, D.C.: GPO, 1930), p. 648; and *U.S. Census of Housing: 1950*, vol. 2, *General Characteristics* (Washington, D.C.: GPO, 1953), pp. 1–9.
Note: For the states that make up these regions, see the notes to table A.2.

Table A.5
U.S. FARMS REPORTING ELECTRIFICATION, 1920–1954 (PERCENTAGE)

Year	United States	Northeast	Midwest	Far West	South
1920[a]	7	15	10	15	3
1930	13	35	17	37	4
1940	33	66	40	60	19
1950	78	92	84	87	71
1954	93	97	95	95	90

Source: U.S. Department of Commerce, Bureau of the Census, *U.S. Census of Agriculture, 1954: General Report*, vol. 2, *Statistics by Subjects* (Washington, D.C.: GPO, 1956–57), p. 199.
Note: For the states that make up these regions, see the notes to table A.2. Figures include all sources of electricity: central-station service and home plants.
[a] Includes gas or electric light.

Table A.6
U.S. FARM HOUSEHOLDS REPORTING OWNERSHIP OF SELECTED CONSUMER
GOODS, 1919–1936 (PERCENTAGE)

| Year | 1919 | 1925 | 1935–36 | |
			Low Income	High Income
Telephone	72	—	29	52
Automobile	62	80	71	80
Radio	—	22	54	81
Washing machine	57	43	47	64
Refrigerator	—	7	4[a]	20[a]
Running water	32	47	16	43
Stationary bathtub	20	21	—	—
Indoor toilet	10	17	10	33
Electric lighting	—	—	19	44
Electric or gas lights	—	30	—	—

Source : Florence E. Ward, "The Farm Woman's Problems," *Journal of Home Economics,* 12 (1920): 437–57 (1919 survey); Ellis L. Kirkpatrick, *The Farmer's Standard of Living* (New York: Century, 1929), pp. 136–38, 151–52, 155; and Mary Sherman, "What Women Want in Their Homes," *Woman's Home Companion,* Nov. 1925, pp. 28, 97, 98 (1925 survey); Day Monroe, "Patterns of Living of Farm Families," in U.S. Department of Agriculture, *Yearbook of Agriculture, 1940* (Washington, D.C.: GPO, 1940), pp. 848–69 (1935–36 survey).
Note: 1919: survey of about 10,000 farms in 33 northern and western states by U.S. Department of Agriculture home extension agents. 1925: survey of about 40,000 farms in 28 states by the General Federation of Women's Clubs, as part of a national survey of about 450,000 urban and rural households in 35 states financed by *Woman's Home Companion.* 1935–36: Consumer Purchases Study of 66 farm counties in 21 states by the Bureau of Home Economics (except for automobile figures); income groups are nonrelief families making $1,000–1,249 (low) and $2,500–2,999 (high) per year, which included in-kind products of the farm (e.g., fuel and food).
[a] Mechanical refrigerator.

Table A.7
U.S. FARM HOUSEHOLDS REPORTING OWNERSHIP OF SELECTED CONSUMER
GOODS, 1915–1928 (PERCENTAGE)

Consumer Good	Iowa-1 1915–16	Michigan 1917	Iowa-2 1918	Iowa-3 1920–21	Wisconsin 1927–28	New York 1928
Telephone	93	49	87	90	77	65
Vacuum cleaner	54	43	28	21	—	37
Automobile	53	—	58	91	87	67
Washing machine	48	67	40	51	—	30
Running water	40	11	18	28	13	30

Refrigerator	39	15	8	24	—	21
Stationary bathtub	33	12	18	24	—	16
Gas lighting	33	2	15	16	—	7
Electric or gas iron	25	14	16	33	—	26
Indoor toilets	24	10	18	21	—	24
Electric lighting	11	1	6	24	21	12
Radio	—	—	—	—	34	39

Source : George H. von Tungeln, "A Rural Social Survey of Orange Township, Black Hawk County, Iowa," Iowa State College Agricultural Experiment Station, *Bulletin,* no. 184, Dec. 1918 (Iowa-1); Ilena M. Bailey and Melissa F. Snyder, "A Survey of Farm Homes," *Journal of Home Economics,* 13 (1917): 346–56 (Michigan): George H. von Tungeln, "A Rural Social Survey of Lone Tree Township," Iowa State College Agricultural Experiment Station, *Bulletin,* no. 193, Mar. 1920 (Iowa-2); George H. von Tungeln and Harry L. Eells, "A Rural Social Survey of Hudson, Orange, and Jesup Consolidated School Districts, Blackhawk and Buchanan Counties," Iowa State College Agricultural Experiment Station, *Bulletin,* no. 224, Nov. 1924 (Iowa-3); E. L. Kirkpatrick, J. H. Kolb, Creagh Inge, and A. F. Wileden, "Rural Organizations and the Farm Family," University of Wisconsin Agricultural Experiment Station, *Research Bulletin,* no. 96, Nov. 1929 (Wisconsin); Helen Canon, "The Family Finances of 195 Farm Families in Tompkins County, New York, 1927–1928," Cornell University Agricultural Experiment Station, *Bulletin,* no. 522, May 1931 (New York).
Note: Iowa-1: survey of 142 Iowa farms; Michigan: survey of 91 southern Michigan farms; Iowa-2: survey of 85 Iowa farms; Iowa-3: survey of 386 farms; Wisconsin: survey of 282 Wisconsin farms; New York: survey of 195 New York farms.

Table A.8
U.S. FARMS REPORTING RUNNING WATER, 1920–1954 (PERCENTAGE)

Year	United States	Northeast	Midwest	Far West	South
1920	10	30	11	26	3
1930	16	56	19	41	6
1940	22	56	20	46	8
1950	38	60	51	53	22
1954	59	86	64	86	46

Source : U.S. Department of Commerce, Bureau of the Census, *U.S. Census of Agriculture, 1954: General Report,* vol. 2, *Statistics by Subjects* (Washington, D.C: GPO, 1956–57), pp. 199–200.
Note: For the states that make up these regions, see the notes to table A.2.

Table A.9
REA ANNUAL STATUS, 1935–1960

Year[a]	Number of Systems in Operation	Number of Co-op Borrowers	Lines Energized (in miles)	Total Number of Consumers	Number of Farm Consumers[b]	Electrified Farms Served by REA (%)
1935	2	—	0	0	—	—
1936	29	—	3,000	7,508	—	—
1937	126	249[c]	16,500	43,878	—	—
1938	350	319[c,d]	67,317	176,382	—	—
1939	548	551[c,d]	181,359	435,566	—	—
1940	685	617[c]	267,846	674,495	—	—
1941	773	660[c,d]	348,062	902,266	—	—
1942	803	778[c]	378,015	1,012,284	—	—
1943	811	800	390,058	1,087,801	837,000	—
1944	826	831	410,471	1,216,798	—	—
1945	848	884	449,579	1,408,918	1,080,000	38[e]
1946	869	930	506,838	1,683,901	1,300,000	—
1947	911	947	603,064	2,046,095	1,580,000	—
1948	952	955	759,494	2,518,450	1,960,000	49[c]
1949	995	976	943,385	3,040,425	2,370,000	—
1950	1,007	986	1,088,777	3,413,407	2,250,000[f]	53[e]
1951	1,016	986	1,178,515	3,665,966	2,460,000	52[c]
1952	1,020	986	1,244,645	3,858,396	2,590,000	54[c]
1953	1,022	954	1,297,264	4,024,826	2,700,000	—
1954	1,024	936	1,332,937	4,174,346	2,710,000	61[e]
1955	1,026	936	1,361,605	4,251,250	2,470,000	—
1956	1,026	927	1,382,737	4,361,896	2,470,000	—
1957	1,030	928	1,405,327	4,466,444	2,550,000	—
1958	1,030	926	1,424,441	4,596,270	2,620,000	—
1959	1,032	927	1,445,835	4,721,630	2,550,000	72[c,g]
1960	1,038	928	1,465,315	4,825,802	—	—

Source : Rural Electrification Administration, *Annual Statistical Reports, 1943–1960* (Washington, D.C.: U.S. Department of Agriculture, 1943–60). See also Series S 147–159, in U.S. Department of Commerce, Bureau of the Census, *The Statistical History of the United States from Colonial Times to the Present* (New York: Basic Books, 1976), pp. 829–30. The June 30 figures for the number of co-op borrowers are from Rural Electrification Administration, *Report of the Administrator* (Washington, D.C.: GPO), for that year.

"Number of Co-op Borrowers" for 1943–60 understates the number of co-ops because some co-ops had paid off their loans. The figure is sometimes greater than "Systems in Operation" because not all co-ops had begun operations during the reporting year.

[a] As of December 31.

[b] Calculated from REA's estimates of the percentage of farm consumers for each year, rounded to three significant figures.

^c As of June 30.

^d Calculated from percentage of borrowers that were co-ops.

^e Calculated from census data on number of farms reporting electricity on April 1 of census year, rounded to two significant figures.

^f The Census Bureau changed its definition of what constituted a farm for the 1950 census, which lowered the number of farms and, thus, the number of "farm consumers" for 1950 and thereafter.

^g Calculation based on 1956 figure of 96% farms reporting electricity.

Table A.10

WOMEN ON REA CO-OP BOARDS OF DIRECTORS, 1938–1945

State	Co-op	Earliest Known Date	Name of Director
Alabama	AL-30	1940	Mrs. Gordon DeRamus
Arkansas	AK-18	1940	Mrs. Nannie Formen
			Mrs. D. K. McCurry
			Mrs. Oscar Sears
	AK-24	1941	Mrs. Crane
Illinois	IL-32	1938	Mrs. Edith Serven
			Mrs. Blanche Noper
			Mrs. Helen Wicks
	IL-33	1938	Mrs. Ruby Hurst
			Mrs. Ruth Stevenson
Kansas	KS-?	1940	Mrs. Fred Shaull
	KS-?	1940	Mrs. M. E. Baxter
	KS-?	1941	Mrs. Madge Addington
			Mrs. Owen Smith
	KS-14	1938	Mrs. Paul Bossi
			Mrs. H. R. Wenrich
	KS-25	1938	Mrs. Alta Hollinsworth
			Mrs. Zona Smith
	KS-29	1938	Mrs. Rosa Weber
Louisiana	LA-12	1938?	Mrs. G. A. Newcomer
			Mrs. T. A. Woodbridge
	LA-17	1940	Mrs. M. D. Wren
Michigan	MI-37	1938	Ruth Brandmair
Montana	MT-?	1941	Anna Boe Dahl
	MT-13	1939	Mrs. O. E. Bolon
			Mrs. Ellen Logan
			Unknown
Nebraska	NE-44	1938	Mrs. Betz
			Mrs. Mabel Gillespie
New York	NY-19	1945	Mrs. Mary Kabosius
	NY-20	1943	Mrs. Archie Campbell

State	Co-op	Earliest Known Date	Name of Director
	NY-21	1941	Mrs. Katie Darmstadt
			Bernice Shautt
Ohio	OH-42	1940	Mrs. Olive M. Folkerth
			Mrs. Guy Kreider
Pennsylvania	PA-4	1941	Mrs. Mabel Conroe
			Mrs. Thomas Schlosser
Texas	TX-40	1940	Mabel Bryan Morriss
	TX-86	1940	Mrs. George Black

Source : Addie Lee Collier to Dora Haines, Aug. 27, 1940 (AL-30) and M. Rempel to Haines, Nov. 28, 1940 (AR-18), Records of the Rural Electrification Administration, entry 89, box 1 (letterheads); Dora Haines, field report, July 18, 1941 (AR-24), Records of the Rural Electrification Administration, entry 19, box 2; Harold Severson, *Architects of Rural Progress: A Dynamic Story of the Electric Cooperatives as Service Organizations in Illinois* (n.p.: Association of Illinois Electric Cooperatives, [1966?]), pp. 139, 246–55; Kenneth E. Merrill, *Kansas Rural Electric Cooperatives: Twenty Years with the REA* (Lawrence: University of Kansas Center for Research in Business, 1960), pp. 140, 144, 147, 170, 185, 188; Gary A. Donaldson, "A History of Louisiana's Rural Electric Cooperatives, 1937–1983" (Ph.D. diss., Louisiana State University, 1983), pp. 106–7, 128; Frank Wilson to Franklin Roosevelt, May 11, 1938 (MI-37), Franklin Delano Roosevelt Papers, FDR office files 1570, Box 3 (letterhead); David Long, "'We're Not Isolated Now!': Anna Boe Dahl and the REA," *Montana,* 39 (spring 1989): 18–23; O. E. Bolon to John Carmody, Mar. 16, 1939, (MT-13), Eleanor Roosevelt Papers, Box 1489; *Rural Lines,* July 1959, p. 23; Boyd Fisher, field report, July 26, 1938 (NE-44), Records of the Rural Electrification Administration, entry 13, box 14; Board of Directors Minutes, Feb. 8, 1945 (NY-19), and Dec. 6, 1944 (NY-20), Records of the Rural Electrification Administration, entry 77, box 94; *Steuben Advocate,* June 13, 1941 (NY-21); Weltha North to Udo Rall, Oct. 14, 1940 (OH-42), Records of the Rural Electrification Administration, entry 90, box 14 (letterhead); Elva Bohannan, field report, Jan. 25–28, 1941 (PA-4), Records of the Rural Electrification Administration, entry 89, box 21; D. B. Lancaster to George Munger, Sept. 10, 1940 (TX-40), Records of the Rural Electrification Administration, entry 86, box 21 (letterhead); Mrs. A. C. Thomas to Haines, Feb. 20, 1940 (TX-86), Records of the Rural Electrification Administration, entry 90, box 5 (letterhead).

Note: The list is incomplete because it is drawn from a variety of sources rather than from a systematic search of documents from the Rural Electrification Administration's 1,000 co-ops.

Table A.11

REA-SERVICED FARMS REPORTING OWNERSHIP OF SELECTED APPLIANCES,
1938–1941 (PERCENTAGE)

Appliance	1938	1939	1940	1941
Radio	86	82	88	90
Iron	81	84	84	85
Washing machine	47	59	55	55
Refrigerator	26	32	33	42
Toaster	24	31	29	32
Water pump	17	19	15	18
Vacuum cleaner	16	21	21	21
Hot plate	12	19	15	15
Motors of less than				
1 horsepower	9	18	15	15
Coffeemaker	6	6	8	9
Range	5	3	4	4
Cream separator	5	14	8	8
Milking machine	2	4	2	3
Chicken brooder	1	3	4	7
Water closet	—	6	6	6

Source : Rural Electrification News, July 1938, pp. 4–10; *Rural Electrification News*, Jan. 1940, pp. 6–8; *Rural Electrification News,* Oct. 1940, pp. 10–11; and J. Stewart Wilson to Robert Craig, et al., Aug. 8, 1941, Records of the Rural Electrification Administration, entry 17, box 4.

Note: The number of projects in each survey varied from 46 (1938) to 123 (1939), the average length of service from 8.4 months (1938) to 19.3 months (1941), the number of customers responding from 17,100 (1938) to 70,893 (1941), and the percentage responding from 64% (1938) to 69% (1939).

Table A.12

ECONOMIC COMPARISON OF ELECTRICAL APPLIANCES, 1936-1940

Appliance	Average Price (in dollars) (1936)	Price Range (in dollars) (1936)	Average Usage Per Month (in kWh) (1940)	REA Farm Ownership (%) (1938)
Refrigerator	164	75–595	35	26
Range	130	72–430	120	5
Water heater	73	—	240	—
Washing machine	66	39–525	3	47
Vacuum cleaner			2	16
Floor type	54	—		
Hand type	14	—		
Water pump		37–265	8–10	17
Radio			8	86
Table model	41	—		
Console	79	—		
Toaster	6	—	3	24
Iron	4	—	5	81
Percolator (metal)	4	—	5	6
Hot plate	3	—		12
Milking machine	135	—	1.5–2.5 per cow	2
Cream separator	—	—	0.5 per 1,000 lbs. milk	5

Source : Electrical Merchandising and Radio Retailing, *Appliance Specifications and Directory, Including Refrigerators and Radio Sets, 1936* (New York: McGraw-Hill, 1936), pp. 82–88, 90, 110, 112, 114, 116–17; and *Electrical Merchandising,* 44 (Jan. 1938): 10–11. The estimated kilowatt-hours (kWh) per month comes from Rural Electrification Administration, *A Guide for Members of REA Cooperatives* (Washington, D.C.: USDA, 1940), pp. 35–36, copy in Franklin Delano Roosevelt Papers, Department of Agriculture Records, box 1. For similar cost and kWh figures, see Sears, Roebuck, and Co., "Rural Electrification Comparative Cost Data," 1938, University Archives, University of Illinois, Urbana. The price for the milking machine comes from the Sears booklet.

Table A.13
NATIONAL VERSUS REA OWNERSHIP OF SELECTED APPLIANCES, 1940
(PERCENTAGE)

Appliance	U.S. Wired Homes	REA Customers
Iron	95	84
Radio	81[a]	88
Washing machine	60	55
Refrigerator	56	33
Toaster	56	29
Vacuum cleaner	48	21
Coffeemaker	33	8
Hot plate	17	15
Range	10	4

Source : Electrical Merchandising, 63 (Jan. 1940): 10, 14–15; and *Rural Electrification News*, Oct. 1940, pp. 10–11.

Note: In 1940, 22.7 million urban and rural nonfarm homes, and 1.8 million farms, were wired; the Rural Electrification Administration had 43,000 new customers in the same year.

[a] Percentage of total households (wired and nonwired), calculated from U.S. Department of Commerce, Bureau of Census, *Historical Statistics of the United States from Colonial Times to the Present* (New York: Basic Books, 1976), Series A 320–334 and 335–349 (p. 42) and Series R 93–105 (p. 796).

Table A.14
GROWTH OF THE REA, 1936–1960

Year	Lines Added		Consumers Added		Co-ops Added	
	Miles	Growth Rate (%)	Number	Growth Rate (%)	Number	Growth Rate (%)
1936	3,000	—	7,508	—	0	0
1937	13,500	450	36,370	480	0	0
1938	50,909	310	132,504	330	70	28[a]
1939	113,858	170	259,184	150	232	73[a]
1940	86,579	48	238,929	55	66	12[a]
1941	80,216	30	227,771	34	43	7.0[a]
1942	29,953	8.6	110,018	12	118	18.0[a]
1943	12,043	3.2	75,517	7.5	22	2.8
1944	20,413	5.2	128,997	12	31	3.9
1945	39,108	9.5	192,120	16	53	6.4
1946	57,259	13	274,983	20	46	5.2
1947	96,226	19	362,194	22	17	1.8
1948	156,430	26	472,355	23	8	0.8
1949	183,891	24	521,975	21	21	2.2
1950	145,392	15	372,982	12	10	1.0
1951	89,738	8.2	252,559	7.4	0	0
1952	66,130	5.6	192,430	5.2	0	0
1953	52,619	4.2	166,430	4.3	−32	−3.4
1954	35,673	2.7	149,520	3.7	−18	−1.9
1955	28,668	2.2	76,904	1.8	0	0
1956	21,132	1.6	110,646	2.6	−9	−1.0
1957	22,590	1.6	104,548	2.4	1	0.1
1958	19,114	1.4	129,826	2.9	−2	−0.2
1959	21,394	1.5	125,360	2.7	1	0.1
1960	19,480	1.3	104,172	2.2	1	0.1

Source : Calculated from table A.9.
Note: Unless otherwise noted, figures are for December 31 of each year. "Growth rate" is the percentage over previous accumulated total.
[a] Fiscal year.

Table A.15
AVERAGE MONTHLY CONSUMPTION OF ELECTRICITY BY REA RESIDENTIAL
CONSUMERS, 1943–1959

Year[a]	KWh/Month	Growth Rate (%)
1943	71	
1944	75	5.6
1945	84	12
1946	90	7.1
1947	105	17
1948	121	15
1949	134	11
1950	147	9.7
1951	166	13
1952	182	9.6
1953	201	10
1954	223	11
1955	242	8.5
1956	263	8.7
1957	283	7.6
1958	311	9.9
1959	334	7.4
1960	357	6.9

Source : Rural Electrification Administration, *Annual Statistical Reports, 1943–1960* (Washington, D.C.: U.S. Department of Agriculture, 1943–60). Figures are from the annual reports for 1943–47 and from a summary table for 1948–60. Annual report figures for 1948–51 differ slightly from the summary figures, which are also given in U.S. Department of Commerce, Bureau of the Census, *The Statistical History of the United States from Colonial Times to the Present* (New York: Basic Books, 1976), Series S 147–159, pp. 829–30.
Note: "Growth" is the percentage over previous accumulated total.
[a] As of December 31.

Table A.16

ELECTRIFIED U.S. FARMS REPORTING OWNERSHIP OF SELECTED
ELECTRICAL APPLIANCES, 1947 (PERCENTAGE)

Region	Washing Machine	Refrigerator	Water System	Vacuum Cleaner	Range	Water Heater	Sewing Machine
United States	43	38	27	25	12	10	7
New England states	70	58	45	43	14	8	17
Mid-Atlantic states	76	59	53	58	20	17	13
East north central states	75	58	47	52	26	15	12
West north central states	44	35	21	26	10	8	6
South Atlantic states	25	29	20	9	6	6	4
East south central sates	18	22	10	6	6	4	2
West south central states	23	22	14	9	2	3	5
Mountain states	53	49	32	34	22	16	13
Pacific states	83	69	55	52	34	36	26

Source: U.S. Department of Agriculture, Bureau of Agricultural Economics, "Farm Homes Use Wide Variety of Electrical Appliances," Apr. 20, 1948, attached to E. C. Weitzell to Mr. Haggard, Nov. 29, 1948, Records of the Rural Electrification Administration, entry 27, box 3.

Table A.17

ELECTRIFIED U.S. FARMS REPORTING OWNERSHIP OF SELECTED
ELECTRICAL APPLIANCES, 1950 (PERCENTAGE)

Region	Washing Machine	Water System	Water Heater	Home Freezer
United States	75	48	22	15
New England states	85	57	19	20
Mid-Atlantic states	91	67	31	25
East north central states	92	68	30	23
West north central states	87	51	21	15
South Atlantic states	58	37	15	10
East south central states	58	24	13	7.7
West south central states	54	32	11	13
Mountain states	85	52	33	18
Pacific states	89	68	50	20

Source: U.S. Department of Commerce, Bureau of the Census, U.S. Census of Agriculture, 1950: General Report, vol. 2, Statistics by Subjects (Washington, D.C.: GPO, 1952), p. 213.

Abbreviations

AES	Agricultural Experiment Station
AH	*Agricultural History*
AT&T	American Telephone and Telegraph
ATTA	American Telephone and Telegraph Company Archives, Warren, New Jersey
BHE	Bureau of Home Economics Papers, Record Group 176, National Archives, Washington, D.C.
BLR	*Bell Laboratories Record*
CBS	Columbia Broadcasting System
CEL	Carl E. Ladd Papers, Division of Rare Books and Manuscript Collections, Cornell University Library, Ithaca, New York
CHE	College of Home Economics Papers, Division of Rare Books and Manuscript Collections, Cornell University Library, Ithaca, New York
CL	*Country Life*
CNT	Clesson Nathan Turner Papers, Division of Rare Books and Manuscript Collections, Cornell University Library, Ithaca, New York
CREA	Committee on the Relation of Electricity to Agriculture
CW	Claude Wickard Papers, Franklin Delano Roosevelt Library, Hyde Park, New York
EM	*Electrical Merchandising*
ER	Eleanor Roosevelt Papers, Franklin Delano Roosevelt Library, Hyde Park, New York
EWL	Emil W. Lehmann Papers, University Archives, University of Illinois, Urbana, Illinois
FDR	Franklin Delano Roosevelt Papers, Franklin Delano Roosevelt Library, Hyde Park, New York
FJ	*Farm Journal*

FMCA	Ford Motor Company Archives, Henry Ford Museum and Greenfield Village, Dearborn, Michigan
FT	*Ford Times*
Grange Proc.	*Journal of Proceedings of the National Grange of the Patrons of Husbandry*
HB	*Harper's Bazaar*
HS	Harry Slattery Papers, William R. Perkins Library, Duke University, Durham, North Carolina
IEEE	Institute of Electrical and Electronics Engineers
JEH	*Journal of Economic History*
JHE	*Journal of Home Economics*
JMC	John M. Carmody Papers, Franklin Delano Roosevelt Library, Hyde Park, New York
KJTE	Karl J. T. Ekblaw Papers, University Archives, University of Illinois, Urbana, Illinois
LBJ	Lyndon Baines Johnson Papers, Lyndon B. Johnson Library, Austin, Texas
LD	*Literary Digest*
LHB	Liberty Hyde Bailey Papers, Division of Rare Books and Manuscript Collections, Cornell University Library, Ithaca, New York
MA	*Motor Age*
MH	*Manitoba History*
MLC	Morris L. Cooke Papers, Franklin Delano Roosevelt Library, Hyde Park, New York
MVHS	Mound Valley Historical Society, Mound Valley, Kansas
MW	*Motor World*
NAL	National Agricultural Library, Beltsville, Maryland
NCHR	*North Carolina Historical Review*
NBC	National Broadcasting Company
NELA	National Electric Light Association
NRECA	National Rural Electric Cooperative Association
NYT	*New York Times*
PF	*Progressive Farmer*
REA	Rural Electrification Administration
REAA	Records of the Rural Electrification Administration, Record Group 221, National Archives, Washington, D.C.
REN	*Rural Electrification News*
RL	*Rural Lines*
RNY	*Rural New Yorker*
SA	*Scientific American*
SMI	Suzanne Moon Oral History Interviews, conducted in 1994–95, in the possession of the author, Cornell University, Ithaca, New York

TC	*Technology and Culture*
UIA	University Archives, University of Illinois, Urbana, Illinois
USDA	United States Department of Agriculture
WF	*Wallaces' Farmer*

For the records of the Rural Electrification Administration (REAA), I have followed the REA system of designating a cooperative by the standard postal (two-letter) abbreviation of the state it is in followed by the co-op number, for example, NY-19. I have also adopted the REA system of using a double zero (oo) following the state's abbreviation to indicate issues of concern for more than one co-op in a state. A question mark following a state's abbreviation means that I have not been able to identify that cooperative's number.

Notes

Introduction

1. *WF,* Nov. 9, 1935, p. 653; partially reprinted in *REN,* Dec. 1935, p. 26.

2. USDA, Office of the Secretary, *Domestic Needs of Farm Women,* Report 104 (Washington, D.C.: GPO, 1915), p. 43.

3. John Steinbeck, *Grapes of Wrath* (1939; reprint, New York: Penguin, 1977), p. 180.

4. On the extensive literature on this transformation, see the section entitled "Agriculture and Rural Life" in the bibliographical note at the end of this book.

5. On this paradox, see William L. Bowers, *The Country Life Movement in America, 1900–1920* (Port Washington, N.Y.: Kennikat, 1974), pp. 61, 78–79; David B. Danbom, *The Resisted Revolution: Urban America and the Industrialization of Agriculture, 1900–1930* (Ames: Iowa State Univ. Press, 1979).

6. Jonathan Coopersmith, "The Electrification of Russia, 1880 to 1925" (Ph.D. diss., Oxford University, 1985), quotations on pp. 144, 168–76.

7. Quoted in *Report of Rural Electrification Administration, 1937* (Washington, D.C.: GPO, 1938), pp. v–viii, on p. viii.

8. Quoted in *REN,* Sept. 1935, p. 20.

9. Martha Bruère and her husband, Robert, had published a prewar series of articles on their survey of rural conditions; see "The Revolt of the Farmer's Wife," *HB,* Nov. 1912 to May 1913.

10. Claude S. Fischer, *America Calling: A Social History of the Telephone to 1940* (Berkeley: Univ. of California Press, 1992), pp. 93, 102.

11. Raymond Williams, *Keywords: A Vocabulary of Culture and Society,* rev. ed. (New York: Oxford Univ. Press, 1983), p. 22. For more documentation and discussion of the issues raised in this section and the next one, see the bibliographical note at the end of this book.

12. Pitirim A. Sorokin and Carle C. Zimmerman, *Principles of Rural-Urban Sociology* (New York: Henry Holt, 1929), chaps. 18, 27, quotation on p. 612.

13. Edmund de S. Brunner and J. H. Kolb, *A Study of Rural Society: Its Organization and Change* (Boston: Houghton Mifflin, 1935), p. 177.

14. Paul H. Johnstone, "Old Ideals versus New Ideas in Farm Life," in USDA, *Yearbook of Agriculture, 1940* (Washington, D.C.: GPO, 1940), pp. 111–70, quotations on pp. 159, 160, 162.

15. See, e.g., David B. Danbom, *Born in the Country: A History of Rural America* (Baltimore: Johns Hopkins Univ. Press, 1995), p. 233.

16. See, e.g., Michael K. Green, "A History of the Public Rural Electrification Movement in Washington to 1942" (Ph.D. diss., University of Idaho, 1967), p. 79; James H. Shideler, "Flappers and Philosophers and Farmers: Rural-Urban Tensions of the Twenties," *AH,* 47 (1973): 283–99, on pp. 296, 299; Bowers, *Country Life Movement in America,* p. 29; Paula M. Nelson, "Rural Life and Social Change in the Modern West," in *The Rural West since World War II,* ed. R. Douglas Hurt (Lawrence: Univ. Press of Kansas, 1998), pp. 38–57.

17. For a similar usage of *modern,* see Fischer, *America Calling,* p. 4.

18. Leo Marx, "The Idea of 'Technology' and Postmodern Pessimism," in *Does Technology Drive History?: The Dilemma of Technological Determinism,* ed. Merritt Roe Smith and Leo Marx (Cambridge: MIT Press, 1994), pp. 237–58; Ronald Kline, "Construing 'Technology' as 'Applied Science': Public Rhetoric of Scientists and Engineers in the United States, 1880–1945," *Isis,* 86 (June 1995): 194–221. For a similar usage of *technology,* see Judy Wajcman, *Feminism Confronts Technology* (University Park: Pennsylvania State Univ. Press, 1991), pp. 14–15.

19. Ruth Schwartz Cowan, "The Consumption Junction: A Proposal for Research Strategies in the Sociology of Technology," in *The Social Construction of Technological Systems: New Directions in the Sociology and History of Technology,* ed. Wiebe Bijker, Thomas Hughes, and Trevor Pinch (Cambridge: MIT Press, 1987), pp. 261–80.

20. Peter H. Argersinger and Jo Ann E. Argersinger, "The Machine Breakers: Farmworkers and Social Change in the Rural Midwest of the 1870s," *AH,* 58 (1984): 393–410; Karen J. Ferguson, "Caught in 'No-Man's Land': The Negro Cooperative Extension Service and the Ideology of Booker T. Washington, 1900–1918," *AH,* 72 (1998): 33–54; Danbom, *Resisted Revolution;* Melissa Walker, "Home Extension Work among African American Farm Women in East Tennessee, 1920–1939," *AH,* 70 (1996): 487–502; Jane Adams, *The Transformation of Rural Life: Southern Illinois, 1890–1990* (Chapel Hill: Univ. of North Carolina Press, 1994); Katherine Jellison, *Entitled to Power: Farm Women and Technology, 1913–1963* (Chapel Hill: Univ. of North Carolina Press, 1993); Mary Neth, *Preserving the Family Farm: Women, Community, and the Foundations of Agribusiness in the Midwest, 1900–1940* (Baltimore: Johns Hopkins Univ. Press, 1995); Hal S. Barron, *Mixed Harvest: The Second Great Transformation in the Rural North, 1870–1930* (Chapel Hill: Univ. of North Carolina Press, 1997); Donald B. Kraybill, *The Riddle of Amish Culture* (Baltimore: Johns Hopkins Univ. Press, 1989).

21. The *Oxford English Dictionary,* 2d ed., cites Frederick Allen, *Only Yesterday* (1931), for an early usage of the term *consumer resistance.*

22. On these traditions, see the bibliographical note. My story of the ways in which rural people wove new communications, transportation, and household technologies into rural ways of life in the twentieth century agrees with many conclusions reached by recent authors on the history of rural America. Neth, *Preserving the Family Farm,* chaps. 7–8, discusses this process as a selective adoption of consumer products to meet rural traditions; Barron, *Mixed Harvest,* p. 241, speaks of a "hybrid rural version" of consumer culture; Jellison, *Entitled to Power,* argues that farm women welcomed new technologies but rejected the urban domestic ideal of the modernizers; and Adams, *The Transformation of Rural Life,* places more emphasis on resistance. My story adds to these accounts by concentrating on the development of technological systems and mediating networks, the interactions among producers, mediators, and consumers, the strategies farm men and women used to resist, adapt, and modify technology, and the resultant changes in both technology and rural life.

23. For a comparison between the agricultural and military complexes, see Jim Hightower, *Hard Tomatoes, Hard Times: The Original Hightower Report, Unexpurgated, of the Agribusiness Accountability Project on the Failure of America's Land Grant College Complex* . . . (Rochester, Vt.: Schenkman, 1978), p. 91.

24. Roy V. Scott, *The Reluctant Farmer: The Rise of Agricultural Extension to 1914* (Urbana: Univ. of Illinois Press, 1970), pp. 108, 116–21.

25. Warren J. Gates, "Modernization as a Function of an Agricultural Fair: The Great Grangers' Picnic Exhibition at Williams Grove, Pennsylvania, 1873–1916," *AH,* 58 (1984): 262–79; Donald B. Marti, *Women of the Grange: Mutuality and Sisterhood in Rural America, 1866–1920* (Westport, Conn.: Greenwood, 1991), chap. 4.

26. Linda J. Borish, "Farm Females, Fitness, and the Ideology of Physical Health in Antebellum New England," *AH,* 64 (1990): 17–30; W. W. Hall, "Health of Farmers' Families," in *Report of the Commissioner of Agriculture, 1862* (Washington, D.C.: GPO, 1863), pp. 453–70, quotation on p. 453.

27. Marti, *Women of the Grange,* p. 78; Alan I. Marcus, *Agricultural Science and the Quest for Legitimacy: Farmers, Agricultural Colleges, and Experiment Stations, 1870–1890* (Ames: Iowa State Univ. Press, 1985), pp. 7–12, quotation on p. 7.

28. *LD,* 37 (1908): 235–36; and 38 (1909): 286; Clayton S. Ellsworth, "Theodore Roosevelt's Country Life Commission," *AH,* 34 (1960): 155–72; David B. Danbom, "Rural Education Reform and the Country Life Movement, 1900–1920," *AH,* 53 (1979): 462–74.

29. *Report of the Commission on Country Life* (1909; reprint, New York: Sturgis and Walton, 1911), quotations on pp. 114, 103.

30. *Des Moines Homestead* quoted in *LD,* 37 (1908): 965–66, on p. 965; Matthew Jansen to Theodore Roosevelt, Jan. 23, 1908, quoted in Bowers, *Country Life Movement in America,* p. 103; Bowers, *Country Life Movement in America,* chap. 7.

31. Olaf F. Larson and Thomas B. Jones, "The Unpublished Data from Roosevelt's Commission on Country Life," *AH,* 50 (1976): 583–99, on p. 597, emphasis in the original. For a history of urban-rural tensions in the United States and a traditional view of rural urbanization, see Johnstone, "Old Ideals versus New Ideas in Farm Life."

32. L. H. Bailey, *The Country Life Movement in the United States* (New York: Macmillan, 1911); Ellsworth, "Country Life Commission"; Lowry Nelson, *Rural Sociology: Its Origin and Growth in the United States* (Minneapolis: Univ. of Minnesota Press, 1969); Bowers, *Country Life Movement in America.*

33. Alfred C. True, *A History of Agricultural Extension Work in the United States, 1785–1923* (Washington, D.C.: GPO, 1928), pp. 100–115; Gladys Baker, *The County Agent* (Chicago: Univ. of Chicago Press, 1939); Wayne D. Rasmussen, *Taking the University to the People: Seventy-Five Years of Cooperative Extension* (Ames: Iowa State Univ. Press, 1989), chaps. 3–4; Cynthia Sturgis, "'How're You Gonna Keep 'Em Down on the Farm?': Rural Women and the Urban Model in Utah," *AH,* 60 (1986): 182–99; David B. Danbom, "The Agricultural Extension System and the First World War," *Historian,* 41 (1979): 315–33; Dorothy Schwieder, "The Iowa State College Cooperative Extension Service through Two World Wars," *AH,* 64 (1990): 219–30; Grant McConnell, *The Decline of Agrarian Democracy* (New York: Athenaeum, 1969), chap. 5.

34. Christiana McFadyen Campbell, *The Farm Bureau: A Study of the Making of National Farm Policy, 1933–1940* (Urbana: Univ. of Illinois Press, 1962).

35. Lizabeth Cohen, *Making a New Deal: Industrial Workers in Chicago* (Cambridge: Cambridge Univ. Press, 1990), especially pp. 119–24.

36. Jane Marie Pederson, "The Country Visitor: Patterns of Hospitality in Rural Wisconsin, 1880–1925," *AH,* 58 (1984): 347–64; Neth, *Preserving the Family Farm,* chap. 2.

37. On the growing literature on farm women, see the section entitled "Agriculture and Rural Life" in the bibliographical note.

CHAPTER 1: (RE)INVENTING THE TELEPHONE

1. *Telephony,* Aug. 1904, p. 168; May 1905, p. 454; and June 1904, p. 492.

2. Frederick Rice Jr., "Urbanizing Rural New England," *New England Magazine,* n.s., 33 (Jan. 1906): 528–41, on pp. 537, 540.

3. *WF,* Jan. 6, 1905, p. 17.

4. Jesse E. Pope, "Rural Communication," in *Cyclopedia of American Agriculture,* ed. Liberty H. Bailey, 4 vols. (New York: Macmillan, 1907–9), 4:312–20, on p. 312; T. N. Carver, "The Organization of a Rural Community," in USDA, *Yearbook of Agriculture, 1914* (Washington, D.C.: GPO, 1915), pp. 89–138, on p. 122.

5. Wayne E. Fuller, *RFD: The Changing Face of Rural America* (Bloomington: Indiana Univ. Press, 1964); Roy A. Atwood, "Routes of Rural Discontent: Cultural Contradictions of Rural Free Delivery in Southeastern Iowa," *Annals of Iowa,* 48 (1986): 264–73; Hal S. Barron, *Mixed Harvest: The Second Great Transformation in the Rural North, 1870–1930* (Chapel Hill: Univ. of North Carolina Press, 1997), chap. 5.

6. Milton Mueller, "Universal Service in Telephone History," *Telecommunication Policy,* 17 (1993): 352–69; David Gabel, "Competition in a Networked Industry: The Telephone Industry, 1894–1910," *JEH,* 54 (1994): 543–72; I. M. Spasoff and H. S. Beardsley, "Farmers Telephone Companies: Organization, Finance, and Management," *Farmers' Bulletin* (of the USDA), no. 1245, Dec. 1922, p. 4.

7. J. Leigh Walsh, *Connecticut Pioneers in Telephony: The Origin and Growth of the Telephone Industry in Connecticut* (New Haven: Telephone Pioneers of America, 1950), pp. 204–5.

8. *RNY,* Jan. 7, 1899, p. 10; *WF,* Jan. 6, 1905, p. 15; Jan. 13, 1905, p. 47; and Feb. 17, 1905, p. 223; Harry B. MacMeal, *The Story of Independent Telephony* (Chicago: Independent Pioneer Association, 1934), chap. 5; Charles A. Pleasance, *The Spirit of Independent Telephony* (Johnson City, Tenn.: Independent Telephone Books, 1989), chaps. 7, 13.

9. *SA,* 82 (1900): 196; *RNY,* May 26, 1900, p. 366; *Telephony,* Dec. 1902, p. 262.

10. *Telephony,* Mar. 1903, p. 176; and Nov. 1903, pp. 383–84; *Outlook,* 72 (1902): 631–32; U.S. Department of Commerce, Bureau of the Census, *Special Reports: Telephones and Telegraphs, 1902* (Washington, D.C.: GPO, 1906), p. 6; John Sabin to F. P. Fish, May 13, 1903, ATTA, box 1342; A.B.S., "Farmers Line Sets: Five Bar Generators," Apr. 27, 1906, ATTA, location 21-06-02-08; Spasoff and Beardsley, "Farmers Telephone Companies," pp. 4, 12; Michael L. Olsen, "But It Won't Milk the Cows: Farmers in Colfax County Debate the Merits of the Telephone," *New Mexico Historical Review,* 61 (1986): 1–13. Claude Fischer, "The Revolution in Rural Telephony, 1900–1920," *Journal of Social History,* 21 (1987): 5–26, on p. 7, mentions a barbed-wire line built in California in 1892.

11. *Telephony,* Aug. 1904, p. 168; Dec. 1904, p. 564; Mar. 1905, p. 277; and Feb. 1906, p. 138.

12. *Telephony,* Sept. 1904, p. 258; Apr. 1905, p. 370; Apr. 1907, p. 278; and Sept. 2, 1911, p. 290.

13. G. R. Johnson, "Some Aspects of Rural Telephony," *Telephony,* May 8, 1909, pp. 542–45, on p. 542; "In the Rural Line Districts," *Telephony,* May 11, 1912, p. 608.

14. *Telephony,* Feb. 25, 1911, p. 266; and Aug. 24, 1912, p. 266.

15. *RNY,* June 11, 1903, p. 437.

16. Kempster B. Miller, *American Telephone Practice,* 4th ed. (New York: McGraw-Hill, 1905); M. D. Fagen, ed., *A History of Engineering and Science in the Bell System: The Early Years (1875–1925)* (New York: Bell Telephone Laboratories, 1975), pp. 128–53, 497–502.

17. U.S. Department of Commerce, Bureau of the Census, *Special Reports: Telephones, 1907* (Washington, D.C.: GPO, 1910), pp. 23–27. The reported average of thirty-two telephones probably overstated the case somewhat, because "lines" also referred to systems having more than one circuit. See Bureau of the Census, *Telephones and Telegraphs, 1902,* p. 6.

18. Theodore Saloutos and John D. Hicks, *Twentieth-Century Populism: Agricultural Discontent in the Middle West, 1900–1939* (Lincoln: Univ. of Nebraska Press, 1951), chap. 3; Barron, *Mixed Harvest,* chap. 4.

19. Pleasance, *Spirit of Independent Telephony,* pp. 111–15.

20. Spasoff and Beardsley, "Farmers Telephone Companies," p. 9.

21. Roy A. Atwood, "Telephony and Its Cultural Meanings in Southeastern Iowa, 1900–1917" (Ph.D. diss., University of Iowa, 1984), pp. 159–60, 169–71; Fischer, "Revolution in Rural Telephony"; and the sources cited in note 25.

22. [Elmer R. Waite?] to F. P. Fish, May 31, 1903, ATTA, box 1342.

23. C. W. Thompson, "How the Department of Agriculture Promotes Organization in Rural Life," in USDA, *Yearbook of Agriculture, 1915* (Washington, D.C.: GPO, 1916), pp. 272a–272p; Spasoff and Beardsley, "Farmers Telephone Companies"; *RNY,* Feb. 8, 1908, p. 98; Deer Creek Cooperative Telephone Company, *Constitution and Bylaws,* adopted Dec. 31, 1909, CW, box 19; *Telephony,* Nov. 16, 1912, pp. 764–56.

24. *CL,* 24 (May 1913): 68; *Telephony,* May 11, 1912, p. 608 (Indiana quotation). Many other cooperatives are described in *RNY* and *Telephony* from 1898 to 1908.

25. *RNY,* Jan. 14, 1899, p. 25; June 10, 1899, p. 431; Aug. 31, 1901, p. 594; Aug. 12, 1905, p. 599; Feb. 8, 1908, p. 98; and Feb. 15, 1908, p. 118; *Telephony,* July 10, 1909, p. 44; Helen B. Gardner and Quenton T. Bowler, "The People's Progressive Telephone Company, 1912–1917: The Dream and the Reality," *Utah Historical Quarterly,* 61 (1993): 79–94.

26. *Telephony,* May 1906, p. 314.

27. W. O. Pennell, "A Study of the Mutual Rural Telephone Movement," May 12, 1906, ATTA, location 128-09-01-02, quotation on p. 8.

28. Atwood, "Telephony and Its Cultural Meanings," pp. 110–18, 139–44, 155, 170–71, 235–36, 422.

29. Spasoff and Beardsley, "Farmers Telephone Companies," p. 6; Bureau of the Census, *Telephones and Telegraphs, 1902,* pp. 6, 155, 163; Bureau of the Census, *Telephones, 1907,* pp. 11–12, 23–24, 125; Florence E. Parker, "Cooperative Telephone Associations, 1936," *Monthly Labor Review,* 46 (1938): 392–413, on p. 400.

30. Bureau of the Census, *Telephones and Telegraphs, 1902,* pp. 34–35; Bureau of the Census, *Telephones, 1907,* p. 41; L. D. H. Welch, "Statistics of Cooperation among Farmers in Minnesota, 1913," University of Minnesota AES, *Bulletin,* no. 146, Dec. 1914; B. H. Hibbard and Asher Hobson, "Cooperation in Wisconsin," University of Wisconsin AES, *Bulletin,* no. 282, May 1917.

31. Charles J. Galpin, *Rural Life* (New York: Century, 1918), chap. 4.

32. Pennell, "Mutual Rural Telephone Movement," pp. 3–4, 7–8, emphasis in the original.

33. *RNY,* Sept. 27, 1902, p. 655.

34. *Telephony,* May 1903, pp. 296–98; and June 1903, pp. 343–44, quotation on p. 343.

35. *Telephony,* June 19, 1909, p. 707.

36. Mueller, "Universal Service," p. 361.

37. *Telephony*, May 1907, pp. 304–5.

38. *Telephony*, Sept. 1903, p. 185; July 1904, pp. 34–35; Mar. 1905, pp. 255–58; May 1905, pp. 432–33; July 1905, p. 52; May 1907, p. 312; Sept. 25, 1909, p. 319; and Mar. 5, 1910, pp. 286–87.

39. *New York Telegram*, Sept. 1, 1908, reprinted in *Atlanta Constitution*, Feb. 15, 1909, attached to Evelyn Harris to J. D. Ellsworth, Feb. 17, 1909, ATTA, box 1363; *Country Gentleman*, June 14, 1919, p. 25; MacMeal, *Story of Independent Telephony*, pp. 24–26.

40. These contrasts often occurred in the same issue of a magazine. Compare, e.g., *Telephony*, July 1904, pp. 34–35, with p. 84 in the same issue.

41. Fischer, "Revolution in Rural Telephony," pp. 10–11.

42. The series, "In the Rural Line Districts," began in *Telephony* in June 1904 and ended on Nov. 23, 1912. The Alfalfa Granger cartoon series started on Dec. 7, 1912. For an example of sarcasm directed against farm people in a regular article, see *Telephony*, Nov. 1903, p. 339. These types of jokes were popular in other electrical journals; see Carolyn Marvin, *When Old Technologies Were New: Thinking about Electric Communication in the Late Nineteenth Century* (New York: Oxford Univ. Press, 1988), chap. 1.

43. *Telephony*, Apr. 1902, p. 117; Nov. 1903, p. 395; Jan. 1905, pp. 56–58; Nov. 6, 1909, pp. 478–79; and Oct. 28, 1911, pp. 523–25.

44. F. P. Fish to E. D. Trowbridge, Apr. 18, 1903, ATTA, box 1342; F. P. Fish to C. E. Yost, Aug. 30, 1902, ATTA, box 1214.

45. C. E. Yost to F. P. Fish, Apr. 14, 1902, ATTA, box 1214.

46. Angus Hibbard to John Sabin, Apr. 25, 1903; Maiden to John Sabin, Apr. 25, 1903; E. B. Smith to F. P. Fish, Apr. 29, 1903; and Charles Cutler to F. P. Fish, June 1, 1904, ATTA, box 1342; Mueller, "Universal Service," p. 361.

47. [Newland?] Carlton to F. P. Fish, May 16, 1903, ATTA, box 1342; F. A. Pickernell to F. P. Fish, Feb. 2, 1906, ATTA, box 1372.

48. See correspondence between Bell operating companies, American Bell, and F. P. Fish, 1899–1902, ATTA, box 1214; and between Bell operating companies and Fish, 1903–7, ATTA, boxes 1342 and 1372. On the Nebraska rates, see C. E. Yost to F. P. Fish, Apr. 14, 1902, ATTA, box 1214.

49. Morris F. Tyler to F. P. Fish, May 4, 1903; and [James Caldwell?] to F. P. Fish, Apr. 30, 1903, ATTA, box 1342.

50. Maiden to Sabin, Apr. 25, 1903; E. D. Trowbridge to F. P. Fish, Apr. 28, 1903; H. E. Hawley to F. P. Fish, Apr. 30, 1903; Morris F. Tyler to F. P. Fish, May 4, 1903; and F. A. Pickernell to F. P. Fish, Nov. 18, 1903, ATTA, box 1342; F. A. Pickernell to F. P. Fish, May 6, 1905, ATTA, box 1372; J. D. Ellsworth to F. P. Fish, Mar. 12, 1907, ATTA, box 1363; and Kenneth Lipartito, *The Bell System and Regional Business: The Telephone in the South, 1877–1920* (Baltimore: Johns Hopkins Univ. Press, 1989), p. 151.

51. George F. Durant to F. A. Pickernell, Dec. 4, 1905 (attachment); and [President of Northwestern] to F. P. Fish, Apr. 20, 1903, ATTA, box 1342; Pennell, "Mutual Rural Telephone Movement," p. 17.

52. *Telephony*, May 1907, p. 305.

53. John Sabin to F. P. Fish, May 13, 1903; C. B. Bush to J. D. Ellsworth, Dec. 13, 1907; and C. B. Bush to E. C. Bradley, Nov. 21, 1908 (quotation), ATTA, box 1363.

54. Lipartito, *Bell System and Regional Business*, pp. 152–53; F. A. Pickernell to E. J. Hall, Dec. 16, 1907, and attached reports, ATTA, box 1363.

55. "Memorandum Regarding Rural Telephone Service in the United States," July 8, 1912, attached to H. B. Thayer to Director of Posts and Telegraphs, Tunis, Algeria, July 23, 1912, ATTA, location 125-08-01-15.

56. *Telephony,* June 1903, pp. 384–85; Aug. 1906, p. 117; Apr. 1908, p. 269; and Feb. 6, 1909, p. 158; *Atlanta Constitution,* Feb. 15, 1909, attached to Evelyn Harris to J. D. Ellsworth, Feb. 17, 1909, ATTA, box 1363; *SA,* 104 (1911): 162.

57. F. A. Pickernell to Joseph Davis, Nov. 18, 1903, ATTA, box 1342; *Telephony,* Jan. 1905, p. 100; and Mar. 1908, p. 370 (quotation). One independent leader called the red-bull story "that classical old telephone joke—the hoary-whiskered, weather beaten, great-grandfather of them all"; see *Telephony,* Sept. 11, 1909, p. 255.

58. Donald B. Kraybill, *The Riddle of Amish Culture* (Baltimore: Johns Hopkins Univ. Press, 1989), pp. 143–48.

59. *RNY,* Dec. 20, 1902, p. 841.

60. *RNY,* Dec. 13, 1902, p. 837.

61. Claude S. Fischer, *America Calling: A Social History of the Telephone to 1940* (Berkeley: Univ. of California Press, 1992); Michele Martin, *'Hello, Central?': Gender, Technology, and Culture in the Formation of Telephone Systems* (Montreal: McGill–Queen's Univ. Press, 1991). Although Fischer studies rural telephony, he does not apply the convincing argument he makes—that urban and small-town people used the telephone to widen and deepen existing communication patterns—to the rural case.

62. Maiden to John Sabin, Apr. 25, 1903, ATTA, box 1342 (quotation); *Telephony,* May 1907, p. 309; and Mar. 5, 1910, p. 287; *RNY,* Feb. 15, 1908, p. 118.

63. K. W. Waterson to Joseph Davis, Feb. 17, 1904, ATTA, location 21-05-01-05; Thomas Wales to Joseph Davis, Mar. 25, 1904; H. D. McBride to Mr. Ford, Apr. 5, 1904; K. D. Waterman to Hammond Hayes, Jan. 4, 1905; and E. H. Bongs to Hammond Hayes, Jan. 13, 1905, ATTA, location 21-05-01-03. On the peg-count method, see Milton Mueller, "The Switchboard Problem: Scale, Signaling, and Organization in Manual Telephone Switching, 1877–1897," *TC,* 30 (1989): 534–60, on p. 550.

64. Chief Engineer, "Notes on Rural Lines," appendix I, p. 6, [May 1906], ATTA, location 21-05-01-04.

65. *Telephony,* May 8, 1909, pp. 542–45, on p. 542; and Feb. 6, 1909, p. 158.

66. See, e.g., *Telephony,* Oct. 1904, p. 347; Dec. 1904, p. 564; Aug. 26, 1911, p. 264; and Dec. 23, 1911, p. 796. This evidence contradicts the views of Fischer, "Revolution in Rural Telephony," p. 14, and Katherine Jellison, *Entitled to Power: Farm Women and Technology, 1913–1963* (Chapel Hill: Univ. of North Carolina Press, 1993), pp. 35–36, who maintain that (male) business use predominated on the farm in this period.

67. *Telephony,* Oct. 1901, pp. 150, 151; Marvin, *When Old Technologies Were New,* pp. 209–22; Fischer, *America Calling,* pp. 66–67.

68. *RNY,* Jan. 14, 1899, p. 25 (quotation); *Independent,* 54 (1902): 648–49; *Outlook,* 72 (1902): 631–32; *Telephony,* Apr. 1902, p. 117; Nov. 1903, pp. 383–84; Dec. 1903, p. 436 (quotation); Jan. 1906, p. 68; and Mar. 1906, p. 204; *SA,* 94 (1906): 500. The McLuhanite point is made by John Brooks, *Telephone: The First Hundred Years* (New York: Harper and Row, 1975), p. 116.

69. *Telephony,* Nov. 1903, pp. 383–84; June 10, 1911, p. 708; and July 22, 1911, p. 118.

70. USDA, *Yearbook of Agriculture, 1904* (Washington, D.C.: GPO, 1905), p. 16 (quotation); *Telephony,* Sept. 1903, pp. 185–86; July 1904, pp. 12, 32; Mar. 27, 1909, pp. 367–68; and Feb. 25, 1911, p. 266 (quotation); Atwood, "Telephony and Its Cultural Meanings," p. 312; *WF,* Jan. 23, 1925, p. 119.

71. *Telephony,* June 1904, p. 492 (quotation); May 1906, p. 348 (quotation); July 1904, p. 84; Dec. 1904, p. 564; Jan. 1905, p. 100; and Jan. 1907, p. 70,

72. Quoted in *PF,* May 14, 1910, p. 427.

73. *Telephony,* Sept. 1904, p. 258; Nov. 1904, p. 460; and Oct. 1907, p. 260.

74. "Rules and Regulations of the Mound Valley Telephone Company," in Mound Valley Telephone Books, ca. 1906 and ca. 1918, MVHS. For other examples of these rules, see *Telephony,* Aug. 1907, p. 103; Fischer, "Revolution in Rural Telephony," p. 258.

75. *Telephony,* Dec. 19, 1909, pp. 699–701; Western Electric Company, *How to Build Rural Telephone Lines,* [1910], ATTA, location 129-05-02. On the booklet's date, see *PF,* Sept. 17, 1910, which reprinted extracts from it.

76. Mitford M. Mathews, ed., *A Dictionary of Americanisms on Historical Principles,* 2 vols. (Chicago: Univ. of Chicago Press, 1951), 2:1425, lists two rather late usages (a 1920 novel and a 1948 folklore journal), but the term *rubbering* appeared frequently in *Telephony* from about 1902 onward.

77. [Elmer R. Waite?] to F. P. Fish, May 31, 1903, ATTA, box 1342; *Telephony,* Dec. 1907, p. 385.

78. *RNY,* Dec. 20, 1902, p. 841; *Telephony,* July 22, 1911, p. 118; *LD,* 49 (1914): 733.

79. Lana F. Rakow, *Gender on the Line: Women, the Telephone, and Community Life* (Urbana: Univ. of Illinois Press, 1992), p. 33.

80. *Telephony,* July 1905, p. 52; and Mar. 4, 1911, p. 294.

81. *Telephony,* Aug. 1907, p. 103; Deer Creek Cooperative Telephone Company, *Constitution and Bylaws;* "Rules and Regulations of the Mound Valley Telephone Company"; Spasoff and Beardsley, "Farmers Telephone Companies," p. 29 (recommended rules of service); Fischer, "Revolution in Rural Telephony," p. 258.

82. *Telephony,* Jan. 1906, p. 67; Apr. 10, 1909, p. 446; June 19, 1909, p. 707; Sept. 11, 1909, p. 254; July 22, 1911, p. 118; Sept. 16, 1911, p. 354; Nov. 11, 1911, p. 604; Dec. 2, 1911, p. 683; Dec. 23, 1911, p. 791; and Nov. 16, 1912, pp. 764–65; Atwood, "Telephony and Its Cultural Meanings," pp. 358–59.

83. *Telephony,* Mar. 1908, p. 230; and Oct. 14, 1911, p. 478; *WF,* Jan. 21, 1910, p. 96.

84. *Outlook,* 86 (1907): 767–68; *Telephony,* Apr. 10, 1909, p. 433; and Oct. 21, 1911, p. 500; *LD,* 49 (1914): 733; Angela E. Davis, "'Valiant Servants': Women and Technology on the Canadian Prairies, 1910–1940," *MH,* 25 (spring 1993): 33–42.

85. Suzanne Moon's interviews with Eva Watson, Feb. 21, 1995; George Woods, Mar. 18, 1995; and Lina Rossbach and Sonia De Frances, Jan. 21, 1995, SMI.

86. Eleanor Arnold, ed., *Party Lines, Pumps, and Privies: Memories of Hoosier Homemakers* ([Indianapolis]: Indiana Extension Homemakers Association, [1984]), pp. 150, 152; Eleanor Arnold, ed., *Voices of American Homemakers* (Hollis, N.H.: National Extension Homemakers Council, 1985), pp. 188–89.

87. Harriet P. Spofford, *The Elders People* (Boston: Houghton Mifflin, 1920), pp. 55–76, on p. 76.

88. Miller, *American Telephone Practice,* p. 423.

89. *Telephony,* May 1903, pp. 296–98; and Mar. 5, 1910, pp. 286–87, on p. 286.

90. *Telephony,* Apr. 1906, p. 280.

91. Miller, *American Telephone Practice,* chap. 24; Fagen, *Engineering and Science in the Bell System,* pp. 121–23.

92. Angus Hibbard to John Sabin, Apr. 25, 1903, ATTA, box 1342; Miller, *American Telephone Practice,* pp. 433–34; *Telephony,* Mar. 1905, p. 277; Apr. 10, 1909, p. 433; and Mar. 25, 1911, p. 382; *LD,* 49 (1914): 733 (reporting on a story in *Telephony,* Oct. 3, 1914).

93. P. L. Spalding to Joseph Davis, Nov, 5, 1903; Joseph Davis to P. L. Spalding, Nov. 9, 1903; and J. Fay, "Induction Coil for Farmers Lines," memorandum, Nov. 9, 1903, ATTA, location 21-06-02-08.

94. C. E. Paxson to Hammond Hayes, July 3, 1905; and Hammond Hayes to C. E. Paxson, July 6, 1905, ATTA, location 21-05-01-05.

95. See, e.g., Atwood, "Telephony and Its Cultural Meanings," pp. 278–315, 349–56; Fischer, "Revolution in Rural Telephony."

96. Fred W. Card, "Cooperative Fire Insurance and Telephones," in *Cyclopedia of American Agriculture,* ed. Liberty H. Bailey, 4 vols. (New York: Macmillan, 1907–9), 4:303–6, on p. 305.

97. Arnold, *Voices of American Homemakers,* p. 188. For similar views, see Atwood, "Telephony and Its Cultural Meanings," pp. 316–23.

98. "Directions for Classifying Records, in Study of Use of Time by Homemakers," Nov. 1927, categories I E 4, IV B 10, V C 1, BHE, entry 8, box 641.

99. Maud Wilson, "Use of Time by Oregon Farm Homemakers," Oregon State Agricultural College AES, *Bulletin,* no. 256, Nov. 1929, pp. 21–22, 59.

100. John H. Kolb, "Rural Primary Groups: A Study of Agricultural Neighborhoods," University of Wisconsin AES, *Research Bulletin,* no. 51, Dec. 1921, pp. 63–64.

101. Dwight Sanderson and Warren E. Thompson, "The Social Areas of Otsego County," Cornell University AES, *Bulletin,* no. 422, July 1923, p. 34.

102. *Telephony,* Oct. 1901, p. 150; June 1904, p. 492; May 14, 1910, p. 630; and Mar. 27, 1909, pp. 367–68, on p. 367.

103. On the reorganization of rural communities by RFD, see Fuller, *RFD*; Atwood, "Routes of Rural Discontent."

104. Fischer, *America Calling,* chap. 1, makes a similar point for the telephone in general.

105. *RNY,* Dec. 13, 1902, p. 837.

106. Amelia Shaw MacDonald, "Social Conditions among the Country Folk of New York State," Nov. 12, 1909, LHB, 21-2-541, box 4.

107. *Telephony,* Sept. 23, 1911, p. 386; USDA, Office of the Secretary, *Social and Labor Needs of Farm Women,* Report 103 (Washington, D.C.: GPO, 1915), p. 65; Atwood, "Telephony and Its Cultural Meanings," pp. 360–61.

108. *WF,* Jan. 7, 1910, p. 2.

109. MacDonald, "Social Conditions among the Country Folk," p. 14.

CHAPTER 2: TAMING THE DEVIL WAGON

1. *MW,* Nov. 6, 1902, p. 172; and June 16, 1904, p. 431.

2. Michael Berger, *The Devil Wagon in God's Country: The Automobile and Social Change in Rural America, 1893–1929* (Hamden, Conn.: Archon, 1979), chap. 1; Lowell J. Carr, "How the Devil-Wagon Came to Dexter: A Study of Diffusional Change in an American Community," *Social Forces,* 11 (1932): 64–70; "Bitter Fight All over the Country against Automobiles," *Chicago Tribune,* Aug. 17, 1902, p. 39; *Independent,* 61 (1906): 762; Clay McShane, *Down the Asphalt Path: The Automobile and the American City* (New York: Columbia Univ. Press, 1994), pp. 176–80.

3. See, e.g., *MW,* Feb. 9, 1905, pp. 968, 969; *RNY,* Aug. 13, 1904, p. 607; *MW,* May 8, 1902; *MA,* June 17, 1907, pp. 94–95. On the St. Louis group, see *MW,* Nov. 16, 1905, p. 383.

4. Carr, "How the Devil-Wagon Came to Dexter," p. 65.

5. Harold B. Chase, *Auto-biography: Recollections of a Pioneer Motorist, 1896–1911* (New York: Pageant, 1955), p. 135, quoted in Berger, *Devil Wagon,* p. 21.

6. *RNY,* July 23, 1904, p. 565; *Outlook,* 91 (1909): 832–33, on p. 833.

7. *RNY,* Aug. 13, 1904, p. 607; and Dec. 12, 1908, p. 958.

8. *WF,* July 30, 1909, p. 963.

9. *MA,* July 14, 1904, p. 14.

10. Eleanor Arnold, ed., *Buggies and Bad Times: Memories of Hoosier Homemakers* ([Indianapolis]: Indiana Extension Homemakers Association, [1984]), p. 35.

11. *MA,* Sept. 9, 1915, p. 15; Berger, *Devil Wagon,* pp. 13–14, 30, 88–90; James J. Flink, *The Automobile Age* (Cambridge: MIT Press, 1988), pp. 101, 169–70.

12. Quoted in *Telephony,* Mar. 1905, p. 277.

13. *RNY,* Aug. 4, 1906, p. 602; *MW,* Aug. 13, 1908, p. 683; *WF,* Sept. 17, 1909, p. 1177.

14. Berger, *Devil Wagon,* pp. 24–28; Reynold M. Wik, *Henry Ford and Grass-Roots America* (Ann Arbor: Univ. of Michigan Press, 1972), p. 17; *MW,* Dec. 24, 1903, p. 466; June 22, 1905, p. 569; Sept. 28, 1905, p. 38; and Oct. 24, 1907, p. 1842a; *MA,* July 7, 1904, p. 21.

15. James J. Flink, *America Adopts the Automobile, 1895–1910* (Cambridge: MIT Press, 1970), pp. 67–68; Wik, *Henry Ford,* p. 17; Joseph Interrante, "You Can't Go to Town in a Bathtub: Automobile Movement and the Reorganization of Rural American Space, 1900–1930," *Radical History Review,* 21 (1979): 151–68, quotation on p. 155, emphasis in the original; *MW,* Apr. 27, 1905, p. 211; and Nov. 30, 1905, p. 478; *MA,* July 30, 1908, p. 29; Carr, "How the Devil-Wagon Came to Dexter," p. 68; *Wilkes (N.C.) Chronicle,* July 14, 1909, reprinting an Indiana story; Howard L. Preston, *Dirt Roads to Dixie: Accessibility and Modernization in the South, 1885–1935* (Knoxville: Univ. of Tennessee Press, 1991), p. 48.

16. *MW,* Aug. 13, 1903, p. 753; June 1, 1905, p. 430; and May 10, 1906, p. 786.

17. Hal S. Barron, "And the Crooked Shall Be Made Straight: Public Road Administration and the Decline of Localism in the Rural North, 1870–1930," *Journal of Social History,* 26 (1992): 81–103; Ballard Campbell, "The Good Roads Movement in Wisconsin, 1890–1911," *Wisconsin Magazine of History,* 49 (1966): 273–93, quotation on p. 286; Paul S. Sutter, "Paved with Good Intentions: Good Roads, the Automobile, and the Rhetoric of Rural Improvement in the *Kansas Farmer,* 1890–1914," *Kansas History,* 18 (winter 1995–96): 284–99, quotation on p. 296; Preston, *Dirt Roads to Dixie,* pp. 21, 24, 37 (quotation), 64–65.

18. Wayne E. Fuller, *RFD: The Changing Face of Rural America* (Bloomington: Indiana Univ. Press, 1964), chap. 8.

19. *RNY,* Mar. 27, 1909, p. 338.

20. *LD,* 34 (1907): 753; *WF,* Nov. 22, 1907, p. 1366; Sept. 18, 1908, p. 1115; Oct. 30, 1908, p. 1331; May 21, 1909, p. 739 (quotation); Jan. 7, 1910, p. 10; and Jan. 6, 1911, p. 6; *PF,* Sept. 24, 1910, p. 745; Bruce E. Seely, *Building the American Highway System: Engineers as Policy Makers* (Philadelphia: Temple Univ. Press, 1987), p. 28; Barron, "And the Crooked Shall Be Made Straight," pp. 91–94.

21. *MW,* Feb. 9, 1905, p. 969; and Feb. 23, 1905, p. 431; *MA,* June 29, 1905, p. 6.

22. *NYT,* Nov. 15, 1905, sec. 3, p. 7; Mar. 2, 1906, p. 6; and Nov. 11, 1906, p. 2 (quotation).

23. *Grange Proc.,* (1908), p. 18; *WF,* Dec. 20, 1907, p. 1514; Jan. 3, 1908, p. 6; Feb. 14, 1908, pp. 215 (quotation), 245; and Sept. 25, 1908, p. 1171; *RNY,* Nov. 6, 1909, p. 961; Carr, "How the Devil-

Wagon Came to Dexter," pp. 68–69. The *Kansas Farmer* seems to have been more positive all along; see Sutter, "Paved with Good Intentions."

24. Berger, *Devil Wagon,* pp. 35–40, 47–51; Wik, *Henry Ford,* chaps. 3–4; Flink, *America Adopts the Automobile,* pp. 69–73, 82–85, 111; James Flink, *The Car Culture* (Cambridge: MIT Press, 1975), pp. 35, 53; F. Eugene Melder, "The 'Tin Lizzie's' Golden Anniversary," *American Quarterly,* 12 (1960): 466–81; Hal S. Barron, *Mixed Harvest: The Second Great Transformation in the Rural North, 1870–1930* (Chapel Hill: Univ. of North Carolina Press, 1997), pp. 195–97.

25. *MW,* Apr. 1, 1909, p. 71; *WF,* Feb. 24, 1911, p. 331; Peter J. Hugill, "Good Roads and the Automobile in the United States, 1880–1929," *Geographical Review,* 72 (1982): 327–49; Preston, *Dirt Roads to Dixie,* pp. 164–65 (quotation); Warren J. Belasco, *Americans on the Road: From Auto Camp to Motel, 1910–1945* (Cambridge: MIT Press, 1979), pp. 125–42; Fuller, *RFD,* p. 144; Wik, *Henry Ford,* pp. 19–20, 22–24, 26–29.

26. *WF,* Feb. 12, 1909, p. 251; and Aug. 19, 1910, p. 1091 (quotation); USDA, Office of the Secretary, *Social and Labor Needs of Farm Women,* Report 103 (Washington, D.C.: GPO, 1915), pp. 22– 25, quotation on p. 24. On the growing farm market, see *MW,* July 29, 1909, p. 734; *LD,* 41 (1910): 537–38, 1106–8; Wik, *Henry Ford,* pp. 21–33.

27. *MA,* Feb. 11, 1915, pp. 12, 18; Mar. 4, 1915, pp. 18–19; Apr. 1, 1915, p. 18; Apr. 8, 1915, pp. 20–21; and Apr. 22, 1915, p. 33; *WF,* June 4, 1915, p. 33; *NYT,* Dec. 23, 1917, sec. 5, p. 3.

28. *Grange Proc.,* (1904): 108–10; and (1905): 154; *MA,* June 17, 1907, pp. 94–95, quotation on p. 94; George W. Hilton and John F. Due, *The Electric Interurban in America* (Stanford: Stanford Univ. Press, 1960), chaps. 1–4.

29. Norman T. Moline, *Mobility and the Small Town, 1900–1930: Transportation Change in Oregon, Illinois,* Department of Geography Research Paper 132 (Chicago: Department of Geography, University of Chicago, 1971), pp. 40–49, quotation on p. 44.

30. U.S. Department of Commerce, Bureau of the Census, *Street and Electric Railways, 1902* (Washington, D.C.: GPO, 1905), p. 111 (quotation); Bureau of the Census, *Street and Electric Railways, 1907* (Washington, D.C.: GPO, 1910), p. 271.

31. Peter J. Hugill, "Technology Diffusion in the World Automobile Industry, 1885–1985," in *The Transfer and Transformation of Ideas and Material Culture,* ed. Peter J. Hugill and D. Bruce Dickson (College Station: Texas A&M Univ. Press, 1988), pp. 110–42. A further period of stabilization occurred in the mid-1920s with the advent of electric starters, closed bodies, and all-steel bodies; see Flink, *Automobile Age,* pp. 212–14.

32. *RNY,* Dec. 14, 1901, p. 836; and June 27, 1903, p. 467; *WF,* Jan. 8, 1909, p. 53. For other early examples, see *RNY,* Aug. 22, 1903, p. 595 (corn sheller); *MW,* Mar. 3, 1904, p. 1005 (sawing wood).

33. Ellis Parker Butler, "The Adventures of a Suburbanite, V: My Domesticated Automobile," *CL,* 17 (Feb. 1910): 417–19.

34. Roger B. Whitman, "The Automobile in New Roles," *CL,* 15 (Nov. 1908): 53; "Ford's Versatile Flivver," *Horseless Carriage Gazette,* 21 (Jan.–Feb. 1959): 8–19; Wik, *Henry Ford,* pp. 32–33; Berger, *Devil Wagon,* pp. 40–43; Flink, *America Adopts the Automobile,* p. 93.

35. *PF,* Aug. 7, 1920, p. 1340 (quotation); *MA,* June 3, 1915, p. 29.

36. Whitman, "The Automobile in New Roles," p. 53; *CL,* 15 (Apr. 1909): 636; *SA,* 102 (1910): 50–51; George E. Walsh, "Farming with Automobiles," *Review of Reviews,* 43 (Jan. 1911): 62–67; Charles M. Harger, "Automobiles for Country Use," *Independent,* 70 (1911): 1207–11; *RNY,* July 24, 1915, p. 935.

37. Interviews with Winfred Arnold, Nov. 28, 1994; Gerald Cornell, May 24, 1995; Jessie

Hamilton, Feb. 11, 1995; Leroy Harris, Apr. 4, 1995; Owen and Kathleen Howarth, Jan. 24, 1995; Stanley and Albina Konchar, Dec. 16, 1994; and Thena Whitehead, Feb. 11, 1995, SMI.

38. *PF,* Aug. 7, 1920, p. 1340.

39. The process, known technically as "interpretative flexibility," is a key element in the "social construction of technology" approach developed by Wiebe Bijker and Trevor Pinch. For its applicability here, see Ronald Kline and Trevor Pinch, "Users as Agents of Technological Change: The Social Construction of the Automobile in the Rural United States," *TC,* 37 (1996): 763–95, upon which much of this chapter is based.

40. Reynold M. Wik, *Steam Power on the American Farm* (Philadelphia: Univ. of Pennsylvania Press, 1953).

41. Cynthia Cockburn, *Machinery of Dominance: Women, Men, and Technical Know-how* (London: Pluto, 1985); Judy Wajcman, *Feminism Confronts Technology* (University Park: Pennsylvania State Univ. Press, 1991), pp. 38–40, 141–46; McShane, *Down the Asphalt Path,* p. 155.

42. McShane, *Down the Asphalt Path,* p. 163; Virginia Scharff, *Taking the Wheel: Women and the Coming of the Motor Age* (New York: Free Press, 1991), pp. 52–55.

43. Arnold, *Buggies and Bad Times,* pp. 30–44, quotations on pp. 40, 41; Ellis L. Kirkpatrick, *The Farmer's Standard of Living* (New York: Century, 1929), p. 159; Barron, *Mixed Harvest,* pp. 197–98; Jean Warren, "Use of Time in Its Relation to Home Management," Cornell University AES, *Bulletin,* no. 734, June 1940, p. 53; and interviews with Sylvia Schrumpf, Jan. 24, 1995; and Eva Watson, Feb. 21, 1995, SMI.

44. Harger, "Automobiles for Country Use," p. 1210.

45. *RNY,* Oct. 7, 1916, p. 1296. For other accounts, see Arnold, *Buggies and Bad Times,* p. 42; Barron, *Mixed Harvest,* p. 198. For the Missouri study, see Mary Neth, *Preserving the Family Farm: Women, Community, and the Foundations of Agribusiness in the Midwest, 1900–1940* (Baltimore: Johns Hopkins Univ. Press, 1995), p. 252.

46. *LD,* 51 (1915): 770; *RNY,* Apr. 13, 1912, p. 493, emphasis in the original.

47. See, e.g., *MA,* Mar. 18, 1915, p. 12; *Independent,* 73 (1912): 1091–92.

48. *FT,* July 15, 1908, p. 133 (quotation), FMCA, acc. 972. On the prevalence of this view, see *FT,* Feb. 7, 1914, p. 183, H. R. Harper, "The Automobile in the Farming Districts," *FT,* Dec. 1, 1908, pp. 6, 8; Harger, "Automobiles for Country Use," p. 1208; Wik, *Henry Ford,* chap. 4.

49. See, e.g., *WF,* Dec. 31, 1915, p. 1728; *RNY,* Jan. 11, 1919, p. 66; and Mar. 1, 1919, p. 396.

50. *RNY,* Sept. 21, 1918, p. 1089.

51. See, e.g., *MA,* July 2, 1908, p. 11; Rod Bantjes, "Improved Earth: Travel on the Canadian Prairies, 1920–1950," *Journal of Transport History,* 13 (1992): 115–40.

52. *RNY,* Dec. 14, 1901, p. 836; and Dec. 22, 1906, p. 945.

53. *RNY,* Jan. 19, 1907, p. 38; Feb. 8, 1913, p. 165; Mar. 8, 1913, p. 371; Mar. 29, 1913, p. 468; Aug. 3, 1913, p. 976; and June 13, 1914, p. 811 (quotation).

54. *RNY,* July 11, 1914, p. 910; Apr. 12, 1919, p. 647; and Dec. 6, 1919, p. 180.

55. *RNY,* June 22, 1907, p. 492; Feb. 21, 1920, p. 364; Jan. 10, 1925, p. 58; Mar. 21, 1925, p. 506; and Sept. 12, 1925, p. 1216; *WF,* Sept. 9, 1917, p. 1218.

56. *Horseless Age,* Nov. 4, 1903, p. 479; and photographs 188-20749, 188-4763, and 0338, FMCA, acc. 1660, box 9.

57. *FT,* July 1, 1908, p. 34; Harper, "The Automobile in Farming Districts," p. 6 (quotation). See, e.g., *FT,* Aug. 1, 1910, pp. 481–82 (grinding grain); Dec. 1910, pp. 112–13 (hauling produce); Nov. 1911, p. 34 (filling silo); Aug. 1, 1912, p. 357 (sawing wood); and Sept. 1913, pp. 509–11 (general); *Ford Sales Bulletin,* June 17, 1916, p. 195, FMCA, acc. 972, box 1913–1916 (plowing).

58. *FT,* Aug. 1912, p. 361; Melder, "The 'Tin Lizzie's' Golden Anniversary," p. 472. The Peoria poem originally had two more stanzas, including the one about driving the baby around the block; see *Peoria Transcript,* [ca. 1911], reel 1, vol. 1, p. 7, FMCA, acc. 7.

59. Two postcards from this twelve-card series are in FMCA, General Postcard Collection, box 2 (caricatures); two others are in the collection of the author. On the Ford company's attitude toward the jokes, see David L. Lewis, *The Public Image of Henry Ford: An American Folk Hero and His Company* (Detroit: Wayne State Univ. Press, 1976), pp. 121–26.

60. *FT,* Aug. 1, 1912, p. 357; *WF,* May 18, 1917, p. 817; *RNY,* July 21, 1917, p. 909.

61. *SA,* 117 (1917): 32; *WF,* Sept. 7, 1917, p. 1213; and Sept. 28, 1917, p. 1320; *MA,* Oct. 3, 1918, p. 104; and Nov. 6, 1919, p. 149; *RNY,* Jan. 11, 1919, p. 70; Dec. 6, 1919, p. 1802; and Nov. 20, 1948, p. 712.

62. *SA,* 89 (1903): 201; *Ford Sales Bulletin,* June 17, 1916, p. 195; *RNY,* Feb. 2, 1918, p. 150 (quotation); *MA,* Sept. 2, 1915, p. 46.

63. *WF,* Jan. 12, 1917, p. 49; Feb. 9, 1917, pp. 243, 270; Mar. 2, 1917, p. 401; Mar. 16, 1917, p. 510; and July 27, 1917, p. 1054; *MA,* May 17, 1917, p. 42; May 24, 1917, pp. 40–41; Nov. 22, 1917, pp. 71–74; Dec. 13, 1917, p. 46; Feb. 21, 1918, p. 9; and July 4, 1918, p. 42; *SA,* 116 (1917): 349; Victor W. Page, *The Model T Ford Car, Truck, and Tractor Conversion Kits* (New York: Henley, 1918), pp. 285–89; *Automotive Industries,* Mar. 6, 1919, pp. 518–19; *Tractor World,* Nov. 1919, p. 14; *RNY,* Sept. 27, 1919, p. 1409; and Dec. 12, 1931, p. 1208; *FJ,* Jan. 1923, p. 76; Paul C. Johnson, *Farm Power in the Making of America* (Des Moines, Iowa: Wallace-Homestead, 1978), pp. 126, 128; Floyd Clymer, *Henry's Wonderful Model T, 1908–1927* (New York: McGraw-Hill, 1955), pp. 164–65; Wik, *Henry Ford,* p. 33; Reynold M. Wik, "The Early Automobile and the American Farmer," in *The Automobile and American Culture,* ed. David L. Lewis and Laurence Goldstein (Ann Arbor: Univ. of Michigan Press, 1983), pp. 37–47, on p. 45; Berger, *Devil Wagon,* pp. 40–41.

64. According to Wik, *Henry Ford,* p. 33, and Robert C. Williams, *Fordson, Farmall, and Poppin' Johnny: A History of the Farm Tractor and Its Impact on America* (Urbana: Univ. of Illinois Press, 1987), p. 52, technical problems were key factors in the demise of the kits. The development of small tractors undoubtedly played a role, as well.

65. See ads from various manufacturers in *RNY,* from 1919 to 1950. The last ad we found for Pullford was Feb. 10, 1940.

66. Emily Schluenzen to Henry Ford, June 28, 1939, letter 25395; Leonard Dieler to Henry Ford, Aug. 3, 1939, letter 17015; Fred Desosivay to Henry Ford, June 3, 1940, letter 17021; and George Jallings to Henry Ford, Apr. 23, 1942, letter 36464, FMCA, acc. 380.

67. Conversation with Scott Crawford, Nov. 2, 1993 (1927 "Skeeter," in North Carolina); conversation with Raymond Kline, fall 1994 (1930s "Puddle Jumper," in Kansas); interview with Jessie Hamilton, Nov. 28, 1994 (Model T "Doodle-Bug," in New York), SMI.

68. John Matheson to Henry Ford, Nov. 28, 1908; F.L.K. to John Matheson, Dec. 1, 1908; W. W. Walker to Henry Ford, Nov. 26, 1908; and F.L.K. to W. W. Walker, Dec. 1, 1908, FMCA, acc. 2, box 28; J. M. Bullock to Henry Ford, Feb. 3, 1919; and G. S. Anderson to J. M. Bullock, Feb. 5, 1919, FMCA, acc. 62, box 78; *FT,* Aug. 1, 1912, p. 375.

69. "Altering Ford Cars," General Sales Letter 119, Feb. 28, 1916; "Truck Attachments and Special Bodies," General Sales Letter 242, Sept. 17, 1917; "Attachments to Ford Cars," General Sales Letter 267, Apr. 24, 1918, FMCA, acc. 78, box 1.

70. Wik, *Steam Power on the American Farm,* chap. 9; Williams, *Fordson, Farmall, and Poppin' Johnny;* R. Douglas Hurt, *Agricultural Technology in the Twentieth Century* (Manhattan, Kans.: Sunflower, 1991), chap. 1.

71. *MA*, Feb. 21, 1918, pp. 7–9; and Aug. 22, 1918, pp. 312–14; *Tractor World*, Aug. 1918, pp. 33–39; Mar. 1919, pp. 5–13; Nov. 1919, p. 14; *Automotive Industries*, Nov. 14, 1918, p. 849 (quotation); Williams, *Fordson, Farmall, and Poppin' Johnny*, p. 52. For examples of Pullford advertising its success at the Fremont trials, see *MA*, Oct. 11, 1917, p. 40; *RNY*, Feb. 1, 1919, p. 102; Johnson, *Farm Power*, p. 128. On these contests, see Reynold M. Wik, "Nebraska Tractor Shows and the Beginning of Power Farming," *Nebraska History*, 64 (1983): 193–208.

72. *Tractor World*, Nov. 1918, p. 10. On Ford's tractor experiments, see photograph number 833.63702, 1907, FMCA, acc. 1660; *MA*, June 3, 1915, p. 18; "Cost of Experimental Work on Tractor," [ca. Feb. 1916], FMCA, acc. 62, box 87; Wik, *Henry Ford*, pp. 84–86; Williams, *Fordson, Farmall, and Poppin' Johnny*, pp. 47–48.

73. See, e.g., "Fordson Fills Breach When Power Fails," *Ford News*, Oct. 1, 1923, p. 4, about a Fordson running a private electric plant in Kentucky when it was closed for repairs.

74. James R. Wren and Genevieve J. Wren, *Motor Trucks of America* (Ann Arbor: Univ. of Michigan Press, 1979), p. 69. On the Model TT chassis, see James Flink, "Unplanned Obsolescence," *American Heritage of Invention and Technology*, (summer 1996): 58–62.

75. Johnson, *Farm Power*, p. 119; Eleanor Arnold, ed., *Party Lines, Pumps, and Privies: Memories of Hoosier Homemakers* ([Indianapolis]: Indiana Extension Homemakers Association, [1984]), pp. 65–66; interviews with Goldie Jarvis, Nov. 28, 1994; John Nichols, May 22, 1995; and Sylvia Schrumpf, Jan. 24, 1995, SMI.

76. William L. Cavert, "Sources of Power on 538 Minnesota Farms" (Ph.D. diss., Cornell University, 1929); Jack Temple Kirby, *Rural Worlds Lost: The American South, 1920–1960* (Baton Rouge: Louisiana State Univ. Press, 1987), p. 342.

77. Sally Clarke, "New Deal Regulation and the Revolution in American Farm Productivity: A Case Study of the Diffusion of the Tractor in the Corn Belt, 1920–1940," *JEH*, 51 (1991): 101–23.

78. *NYT*, Sept. 16, 1917, sec. 7, p. 5.

79. See, e.g., Barron, *Mixed Harvest*, pp. 198–200.

80. *Technical World Magazine*, 18 (1912): 298–300; Moline, *Mobility and the Small Town*, chaps. 5–6; Berger, *Devil Wagon*, chaps. 5–6; Wik, *Henry Ford*, pp. 31–32; James H. Shideler, "Flappers and Philosophers and Farmers: Rural-Urban Tensions of the Twenties," *AH*, 47 (1973): 283–99, quotations on pp. 297, 298. Unlike Berger, Wik says the car actually increased church attendance.

81. Kirby, *Rural Worlds Lost*, pp. 55, 148, 185, 256–59 (quotation).

82. Malcom M. Willey and Stuart A. Rice, *Communication Agencies and Social Life* (New York: McGraw-Hill, 1933), pp. 54, 57 (quotation), 78.

83. Interrante, "You Can't Go to Town in a Bathtub," p. 157.

84. Bantjes, "Improved Earth," quotations on pp. 131–32.

85. Cavert, "Sources of Power on 538 Minnesota Farms," p. 7; Fuller, *RFD*, pp. 165–66.

86. See, e.g., Berger, *Devil Wagon*, chap. 2; Scharff, *Taking the Wheel*, pp. 142–45; Angela E. Davis, "'Valiant Servants': Women and Technology on the Canadian Prairies, 1910–1940," *MH*, 25 (spring 1993): 33–42, on p. 36.

87. *WF*, Sept. 9, 1910, p. 1179; and May 26, 1911, p. 871; *FT*, Jan. 1913, pp. 166–67, on p. 166; Florence E. Ward, "The Farm Woman's Problems," *JHE*, 12 (1920): 437–57, on p. 439. For the prevalence of this view, see Ina Z. Crawford, "The Use of Time by Farm Women," University of Idaho AES, *Bulletin*, no. 146, Jan. 1927, p. 8; *WF*, Jan. 7, 1910, p. 29 (ad); and May 12, 1911, p. 805 (ad); Katherine Jellison, *Entitled to Power: Farm Women and Technology, 1913–1963* (Chapel Hill: Univ. of North Carolina Press, 1993), pp. 51–53.

88. USDA, *Social and Labor Needs of Farm Women*, quotations on pp. 15, 70; Mary Meek Atkeson, *The Woman on the Farm* (New York: Century, 1924), p. 95; Leonard S. Reich, "From the Spirit of St. Louis to the SST: Charles Lindberg, Technology, and the Environment," *TC*, 36 (1995): 351–93, on p. 356; *LD*, 67 (Nov. 13, 1920): 52–53, quotation on p. 52; *RNY*, Mar. 14, 1925, p. 482. See also *PF*, Oct. 16, 1920, p. 1679.

89. See, e.g., Jellison, *Entitled to Power*, pp. 122–24.

90. Neth, *Preserving the Family Farm*, pp. 246–47; Claude S. Fischer, *America Calling: A Social History of the Telephone to 1940* (Berkeley: Univ. of California Press, 1992). On the auto and suburban women in the United States, see Ruth Schwartz Cowan, *More Work for Mother: The Ironies of Household Technology from the Open Hearth to the Microwave* (New York: Basic Books, 1983), pp. 82–85, 173–74.

91. *Tractor World*, Sept. 1918, p. 40; and Oct. 1918, p. 32; Jan. 1919, p. 16; and May 1919, pp. 33, 35; *LD*, 63 (Oct. 25, 1919): 68; Jellison, *Entitled to Power*, chap. 5.

92. On the time-use studies, see Ronald Kline, "Ideology and Social Surveys: Reinterpreting the Effects of 'Laborsaving' Technology on American Farm Women," *TC*, 38 (1997): 355–85.

93. Maud Wilson, "Use of Time by Oregon Farm Homemakers," Oregon State Agricultural College AES, *Bulletin*, no. 256, Nov. 1929, pp. 16, 23.

94. *RNY*, Dec. 6, 1919, p. 1804.

95. *WF*, Jan. 8, 1909, p. 53.

96. USDA, *Social and Labor Needs of Farm Women*, pp. 66–72, quotation on p. 65.

97. Belasco, *Americans on the Road*, pp. 74–75.

98. Fuller, *RFD*, pp. 168–70; John H. Kolb, "Rural Primary Groups: A Study of Agricultural Neighborhoods," University of Wisconsin AES, *Research Bulletin*, no. 51, Dec. 1921, on p. 28.

99. Kolb, "Rural Primary Groups," on p. 76.

100. Neth, *Preserving the Family Farm*, pp. 259–62.

CHAPTER 3: DEFINING MODERNITY IN THE HOME

1. *WF*, Jan. 31, 1910, p. 86.

2. Ellis L. Kirkpatrick, *The Farmer's Standard of Living* (New York: Century, 1929), pp. 133–34, 143.

3. *Report of the Commission on Country Life* (1909; reprint, New York: Sturgis and Walton, 1911), pp. 103, 104. On the Country Life movement and farm women, see Katherine Hempstead, "Agricultural Change and the Rural Problem: Farm Women and the Country Life Movement" (Ph.D. diss., University of Pennsylvania, 1992); Katherine Jellison, *Entitled to Power: Farm Women and Technology, 1913–1963* (Chapel Hill: Univ. of North Carolina Press, 1993), chap. 1.

4. "Is This the Trouble with the Farmer's Wife?" *Ladies Home Journal*, Feb. 1909, p. 5.

5. Martha Bensley Bruère and Robert Bruère, "The Revolt of the Farmer's Wife!," *HB*, Nov. 1912, pp. 539 (quotation), 550, 580; Dec. 1912, pp. 601–2, 621; Jan. 1913, pp. 15–16, 37; Feb. 1913, pp. 67–68, 92; Mar. 1913, pp. 115–16; and "After the Revolt," *HB*, May 1913, pp. 235, 248.

6. Edward B. Mitchell, "The American Farm Woman As She Sees Herself," in USDA, *Yearbook of Agriculture, 1914* (Washington, D.C.: GPO, 1915), pp. 311–18; USDA, Office of the Secretary, *Social and Labor Needs of Farm Women*, Report 103 (quotation on p. 11); USDA, Office of the Secretary, *Domestic Needs of Farm Women*, Report 104; USDA, Office of the Secretary, *Educational Needs of Farm Women*, Report 105; USDA, Office of the Secretary, *Economic Needs of Farm Women*, Report 106 (Washington, D.C.: GPO, 1915); *NYT*, May 30, 1915, sec. 5, pp. 14–15;

LD, 63 (Dec. 20, 1919): 74, 78; *Extension Service News* (of the New York State College of Agriculture), 2 (1919): 77–78, on p. 78.

7. USDA, *Social and Labor Needs of Farm Women*, p. 24.

8. Pitirim A. Sorokin and Carle C. Zimmerman, *Principles of Rural–Urban Sociology* (New York: Henry Holt, 1929), pp. 264–73; Pitirim A. Sorokin, Carle Zimmerman, and Charles Galpin, *A Systematic Source Book in Rural Sociology*, 3 vols. (Minneapolis: Univ. of Minnesota Press, 1930–32), 3:236–50.

9. Paul Betters, *The Bureau of Home Economics: Its History, Activities, and Organization* (Washington, D.C.: Brookings Institution, 1930); Carolyn Goldstein, "Mediating Consumption: Home Economics and American Consumers, 1900–1940" (Ph.D. diss., University of Delaware, 1994), chap. 2.

10. Florence Ward, "The Farm Woman's Problems," *JHE*, 12 (1920): 437–57, on pp. 449–50. On Ward, see Gladys L. Baker, "Women in the U.S. Department of Agriculture," *AH*, 50 (1976): 190–201.

11. *WF*, Aug. 7, 1925, p. 1009.

12. Hempstead, "Agricultural Change and the Rural Problem," pp. 101–21, on p. 116. On the urban base of the movement, see William L. Bowers, *The Country Life Movement in America, 1900–1920* (Port Washington, N.Y.: Kennikat, 1974).

13. Hempstead, "Agricultural Change and the Rural Problem," p. 201 (quotation); *Report of the Commission on Country Life*, p. 20. On the agrarianism of this statement, see Richard S. Kirkendall, "The Agricultural Colleges: Between Tradition and Modernization," *AH*, 60 (1986): 3–21, on p. 13.

14. See, e.g., *WF*, Sept. 17, 1909, p. 1164; and Jan. 7, 1910, p. 2; USDA, *Domestic Needs of Farm Women*, p. 66; Hempstead, "Agricultural Change and the Rural Problem," p. 201; Charles Galpin, *Rural Life* (New York: Century, 1918), chaps. 5–6; *LD*, 67 (Oct. 2, 1920): 56–57.

15. Mary Meek Atkeson, *The Woman on the Farm* (New York: Century, 1924), pp. 21–24, on p. 24.

16. Louise Stanley, "The Development of Better Farm Homes," *Agricultural Engineering*, 7 (1926): 129–30, on p. 130.

17. *WF*, Feb. 19, 1909, p. 262; *PF*, Apr. 17, 1915, p. 381; USDA, *Social and Labor Needs of Farm Women*, p. 5.

18. USDA, *Social and Labor Needs of Farm Women*, p. 47.

19. *WF*, Sept. 17, 1909, p. 1164.

20. *LD*, 67 (Nov. 13, 1920): 52, 55.

21. Atkeson, *Woman on the Farm*, p. 297; *RNY*, Mar. 14, 1929, p. 482.

22. Jellison, *Entitled to Power*, pp. 27–30. Mary Neth, *Preserving the Family Farm: Women, Community, and the Foundations of Agribusiness in the Midwest, 1900–1940* (Baltimore: Johns Hopkins Univ. Press, 1995), p. 237, argues that Hoag's portrayal of mutuality is more representative of the actual status of farm women than the crop-correspondent survey.

23. Roland Marchand, *Advertising the American Dream: Making Way for Modernity, 1920–1940* (Berkeley: Univ. of California Press, 1985); Jellison, *Entitled to Power*, chap. 2; Neth, *Preserving the Family Farm*, pp. 201–2; Hal S. Barron, *Mixed Harvest: The Second Great Transformation in the Rural North, 1870–1930* (Chapel Hill: Univ. of North Carolina Press, 1997), pp. 225–39.

24. *RNY*, Mar. 16, 1912, p. 37; and Mar. 30, 1912, p. 439; June 22, 1912, p. 720; Apr. 13, 1912, p. 503; June 8, 1912, p. 689; Aug. 3, 1912, p. 837; and Dec. 21, 1912, p. 1273.

25. Bruère and Bruère, "Revolt of the Farmer's Wife!," quotations on pp. 539, 550. On the

fireless cooker, see Madge J. Reese, "Farm Home Conveniences," *Farmers' Bulletin* (of the USDA), no. 927, Mar. 1918, pp. 4–7.

26. USDA, *Domestic Needs of Farm Women*, pp. 11, 23, 25; USDA, *Social and Labor Needs of Farm Women*, p. 63.

27. Angela E. Davis, "'Valiant Servants': Women and Technology on the Canadian Prairies, 1910–1940," *MH*, 25 (spring 1993): 33–42; *PF*, Apr. 17, 1915, p. 381.

28. *WF*, June 11, 1909, p. 818.

29. USDA, *Social and Labor Needs of Farm Women*, p. 21.

30. See Lowry Nelson, *Rural Sociology: Its Origins and Growth in the United States* (Minneapolis: Univ. of Minnesota Press, 1969); Harry C. McDean, "Professionalism in the Rural Social Sciences, 1896–1919," *AH*, 58 (1984): 373–92.

31. USDA, *Domestic Needs of Farm Women*, pp. 22–30, quotation on p. 22.

32. Donald B. Kraybill, *The Riddle of Amish Culture* (Baltimore: Johns Hopkins Univ. Press, 1989); "Results Striking in Farm Home Survey," *Extension Service News* (of the New York State College of Agriculture), 2, no. 6 (June 1919): 49; Joseph Interrante, "You Can't Go to Town in a Bathtub: Automobile Movement and the Reorganization of Rural American Space, 1900–1930," *Radical History Review*, 21 (1979): 151–68, quotation on p. 151.

33. USDA, *Social and Labor Needs of Farm Women*, p. 55.

34. Atkeson, *Woman on the Farm*, pp. 47, 170.

35. USDA, *Domestic Needs of Farm Women*, p. 25.

36. Kirkpatrick, *Farmer's Standard of Living*, pp. 136–37, 143, 159. On the acetylene systems, see *RNY*, Nov. 30, 1907, p. 870; and Sept. 28, 1912, p. 999 (ad); *PF*, Sept. 17, 1910, p. 728.

37. Maureen Ogle, *All the Modern Conveniences: American Household Plumbing, 1840–1890* (Baltimore: Johns Hopkins Univ. Press, 1996).

38. *PF*, July 30, 1910, pp. 604–5; Aug. 20, 1910, p. 659; and Jan. 16, 1915, p. 49.

39. *WF*, June 11, 1915, p. 863; Aug. 21, 1915, p. 771; and Nov. 11, 1915, p. 1469 (quotation); "Sewage and Sewerage of Farm Homes," *Farmers' Bulletin* (of the USDA), no. 1227, Jan. 1922.

40. David E. Nye, *Electrifying America: Social Meanings of a New Technology, 1880–1940* (Cambridge: MIT Press, 1990), p. 292; Kenneth E. Merrill, *Kansas Rural Electric Cooperatives: Twenty Years with the REA* (Lawrence: Univ. of Kansas Center for Research in Business, 1960), pp. 3–4; Michael K. Green,"A History of the Public Rural Electrification Movement in Washington to 1942" (Ph.D. diss., University of Idaho, 1967), pp. 73–74; James C. Williams, "Otherwise a Mere Clod: California Rural Electrification," *IEEE Technology and Society Magazine*, 7 (Dec. 1988): 13–27.

41. Raymond C. Miller, *Kilowatts at Work: A History of the Detroit Edison Company* (Detroit: Wayne State Univ. Press, 1957), p. 232.

42. *FJ*, Jan. 1925, pp. 11, 46; D. Clayton Brown, *Electricity for Rural America: The Fight for the REA* (Westport, Conn.: Greenwood, 1980), p. 5; NELA, *Progress in Rural and Farm Electrification for the Ten-Year Period, 1921–1931* (New York: NELA, 1932), p. 5.

43. Brown, *Electricity for Rural America*, pp. 13–15; E. J. Coil, "Analysis of the Dun and Bradstreet Study of Pre-REA Rural Electric Cooperatives," Apr. 21, 1937, JMC, box 97, quotations on pp. i, 2; *Co-operation*, Jan. 1931, p. 8; and Feb. 1932, pp. 25–26; D. Clayton Brown, "North Carolina Rural Electrification: Precedent of the REA," *NCHR*, 59 (1982): 109–24, on p. 114.

44. *REN*, Jan. 1937, pp. 5–7; *Co-operation*, Jan. 1926, p. 13; Merrill, *Kansas Rural Electric Cooperatives*, pp. 8–9.

45. *FJ*, Jan. 1926, pp. 11, 65; E. F. Chestnut, "Rural Electrification in Arkansas, 1935–1940: The Formative Years," *Arkansas Historical Quarterly*, 46 (1987): 215–60, on p. 221.

46. George Morse, "How Farmers Can Secure Electric Service by Cooperative Effort," *General Bulletin* (of the Pennsylvania Department of Agriculture), no. 412, Sept. 15, 1925, pp. 4–6; *FJ*, Jan. 1926, pp. 11, 65; Udo Rall, "A Study of Cooperative Consumer Associations for Rural Electrification," May 16, 1935, REAA, entry 11, box 29, pp. 5–6; "History of the Development of the Morrisons Cove Light and Power Company," n.d., MLC, box 203.

47. Quoted in *Co-operation*, Oct. 1930, p. 171.

48. Carol A. Lee, "Wired Help for the Farm: Individual Electric Generating Sets for Farms, 1880–1930" (Ph.D. diss., Pennsylvania State University, 1989), chaps. 3–4.

49. Robert W. Righter, *Wind Energy in America: A History* (Norman: Univ. of Oklahoma Press, 1996), chap. 4.

50. J. F. Forrest to K. J. T. Ekblaw, Feb. 27, 1922; and E. Wienecke to K. J. T. Ekblaw, Feb. 28 and Mar. 13, 1922, KJTE, box 1.

51. *WF*, Dec. 13, 1907, p. 1470.

52. Lee, "Wired Help," pp. 130–33, 165, 181; Stuart W. Leslie, *Boss Kettering: Wizard of General Motors* (New York: Columbia Univ. Press, 1983), pp. 58–60.

53. Lee, "Wired Help," pp. 133–46, 199–201; Ruth Schwartz Cowan, *More Work for Mother: The Ironies of Household Technology from the Open Hearth to the Microwave* (New York: Basic Books, 1983).

54. *RNY*, May, 27, 1916, p. 815; "Advertising Campaign for Use in Local Newspapers by Western Electric Power and Light Dealers," n.d., ATTA, location 96-01-623-06, p. 83; J. E. Bullard, "Advertising Farm and Lights Sets to the Farmer," *Electrical Contractor-Dealer*, July 1919, ATTA, location 96-01-623-06; Lee, "Wired Help," pp. 160–61, 174–79.

55. Lee, "Wired Help," pp. 202–5.

56. Joseph Dexter, field record, in R. F. Bucknam, "An Economic Study of Farm Electrification in New York State," 1927, microfilmed field records, CHE, reel 143.

57. See, e.g., the field records of Merritt Elwell, C. H. Blencoe, A. Dennis, Leland Kendler, George Prosser, CHE.

58. Lee, "Wired Help," pp. 166, 181; NELA, *Progress in Rural and Farm Electrification*.

59. Joann Vanek, "Keeping Busy: Time Spent in Housework, United States, 1920–1970" (Ph.D. diss., University of Michigan, 1973); Joann Vanek, "Time Spent in Housework," *SA*, 231 (Nov. 1974): 116–20; Joann Vanek, "Household Technology and Social Status: Rising Living Standards and Status and Residence Differences in Housework," *TC*, 19 (1978): 361–75; Joann Vanek, "Work, Leisure, and Family Roles: Farm Households in the United States, 1920–1955," *Journal of Family History*, 5 (1980): 422–31; Ruth Schwartz Cowan, "The 'Industrial Revolution' in the Home: Household Technology and Social Change in the Twentieth Century," *TC*, 17 (1976): 1–23; Cowan, *More Work for Mother*.

60. See *PF*, Feb. 12, 1910, pp. 130–31; USDA, *Social and Labor Needs of Farm Women*, p. 59; T. N. Carver, "The Organization of a Rural Community," in USDA, *Yearbook of Agriculture, 1914*, pp. 89–138, on pp. 135–36; Atkeson, *Woman on the Farm*, p. 304; Frank Sincebaugh, field record, in Bucknam, "Economic Study of Farm Electrification."

61. On the time studies, see Ronald Kline, "Ideology and Social Surveys: Reinterpreting the Effects of 'Laborsaving' Technology on American Farm Women," *TC*, 38 (1997): 355–85, which forms the basis for part of this chapter.

62. Goldstein, "Mediating Consumption," pp. 21–22, 110–11, 141; Stanley, "Development of Better Farm Homes," p. 130.

63. Maud Wilson, "Use of Time by Oregon Farm Homemakers," Oregon State Agricultural College AES, *Bulletin*, no. 256, Nov. 1929, pp. 37, 39, 46, emphasis in the original.

64. Inez F. Arnquist and Evelyn H. Roberts, "The Present Use of Work Time of Farm Homemakers," State College of Washington AES, *Bulletin*, no. 234, July 1929, pp. 26–27; Margaret Whittemore and Bernice Neil, "Time Factors in the Business of Homemaking in Rural Rhode Island," Rhode Island State College AES, *Bulletin*, no. 221, Sept. 1929, pp. 19–20.

65. Jessie E. Richardson, "The Use of Time by Rural Homemakers in Montana," Montana State College AES, *Bulletin*, no. 271, Feb. 1933, p. 23.

66. Jean Warren, "Use of Time in Its Relation to Home Management," Cornell University AES, *Bulletin*, no. 734, June 1940, pp. 42 (quotation), 49, 74, 77–79.

67. Grace E. Wasson, "The Use of Time by South Dakota Farm Homemakers," South Dakota State College AES, *Bulletin*, no. 247, Mar. 1930; Ina Z. Crawford, "The Use of Time by Farm Women," University of Idaho AES, *Bulletin*, no. 146, Jan. 1927; Blanche M. Kuschke, "Allocation of Time by Employed Married Women in Rhode Island," Rhode Island State College AES, *Bulletin*, no. 267, July 1938.

68. E. W. Lehmann and F. C. Kingsley, "Electric Power for the Farm," University of Illinois AES, *Bulletin*, no. 332, June 1929, pp. 375, 401–3. On Lehmann's employment by Illinois utility companies, see E. W. Lehmann to Rachel Mason, Aug. 21, 1928; and Max Hoagland to E. W. Lehmann, Dec. 11, 1928, EWL, box 5.

69. Hildegarde Kneeland, "Women on Farms Average Sixty-three Hours Work Weekly in Survey of Seven Hundred Homes," in USDA, *Yearbook of Agriculture, 1928* (Washington, D.C.: GPO, 1929), pp. 620–22, on p. 621.

70. Hildegarde Kneeland, "Is the Modern Housewife a Lady of Leisure?" *Survey*, 62 (1929): 301–2, 331, 333, 336, on p. 302.

71. Hildegarde Kneeland to Louise Stanley, July 17, 1930, BHE, entry 2, box 550. On Lehmann's project, see Eloise Davison to E. W. Lehmann, July 8, 1930; E. W. Lehmann to Vera Meacham, July 16, 1930; and E. W. Lehmann to E. A. White, Sept. 17, 1930, EWL, box 5.

72. USDA, Bureau of Nutrition and Home Economics, "The Time Costs of Homemakers: A Study of Fifteen Hundred Rural and Urban Households," 1944, NAL. Kneeland ignored the question of time saved by household technology in other articles she published on the time studies. See Hildegarde Kneeland, "Women's Economic Contribution in the Home," *Annals of the American Academy of Political and Social Science*, 143 (May 1929): 33–40; Hildegarde Kneeland, "Leisure of Home Makers Studied for Light on Standards of Living," in USDA, *Yearbook of Agriculture, 1932* (Washington, D.C.: GPO, 1932), pp. 562–64; Hildegarde Kneeland, "Home-making in This Modern Age," *Journal of the American Association of University Women*, 27 (1934): 75–79.

73. M. Ruth Clark and Greta Gray, "The Routine and Seasonal Work of Nebraska Farm Women," University of Nebraska AES, *Bulletin*, no. 238, Jan. 1930; Jellison, *Entitled to Power*, pp. 62–63. See also Neth, *Preserving the Family Farm*, pp. 239–40.

CHAPTER 4: TUNING IN THE COUNTRY

1. Susan J. Douglas, *Inventing American Broadcasting, 1899–1922* (Baltimore: Johns Hopkins Univ. Press, 1987). On the number of stations, see Malcom M. Willey and Stuart A. Rice, *Communication Agencies and Social Life* (New York: McGraw-Hill, 1933), p. 196.

2. Willey and Rice, in *Communication Agencies and Social Life*, pp. 192–93, discuss the so-called "metropolitan factor" on the basis of economics, cultural patterns, and programming for metropolitan tastes. On WLS, see Susan Smulyan, *Selling Radio: The Commercialization of American Broadcasting, 1920–1934* (Washington, D.C.: Smithsonian Institution Press, 1994), pp. 23, 25.

3. David Rutland, *Behind the Front Panel: The Design and Development of 1920s Radios* (Philomath, Oreg.: Wren, 1994), chap. 15; Michael B. Schiffer, *The Portable Radio in American Life* (Tucson: Univ. of Arizona Press, 1991), pp. 57, 59, 90–91.

4. *Radio Broadcast*, 16 (1930): 140–41, quotation on p. 141; Arthur Capper, "What Radio Can Do for the Farmer," in *Radio and Education* (Chicago: Univ. of Chicago Press, 1932), 223–40, on p. 236.

5. On electricity as a possible factor of the urban-rural difference in the ownership of radios, see Willey and Rice, *Communication Agencies and Social Life*, p. 194.

6. Morse Salisbury, "Radio and the Farmer," *Annals of the American Academy of Political and Social Science*, 177 (Jan. 1935): 141–46.

7. *WF*, Jan. 9, 1925, p. 42; Feb. 20, 1925, p. 277; and Sept. 11, 1925, p. 1167.

8. Electrical Merchandising and Radio Retailing, *Appliance Specifications and Directory, Including Refrigerators and Radio Sets, 1936* (New York: McGraw-Hill, 1936), pp. 122–35.

9. Ivan Bloch to W. E. Herring, May 4, 1936, FDR, office file 1570, box 1.

10. *WF*, Oct. 12, 1935, p. 22; Nov. 9, 1935, p. 13; Dec. 7, 1935, p. 21; Jan. 18, 1936, p. 46; Apr. 11, 1936, p. 29; Sept. 26, 1936, p. 22; and Nov. 7, 1936, p. 15; Robert W. Righter, *Wind Energy in America: A History* (Norman: Univ. of Oklahoma Press, 1996), pp. 100–101.

11. Interviews with Winifred Arnold, Nov. 28, 1994; and Gerald Cornell, May 24, 1995, SMI.

12. Erik Barnouw, *A Tower in Babel: A History of Broadcasting in the United States*, 3 vols. (New York: Oxford Univ. Press, 1966), 1:210.

13. Interview with Leroy Harris, Apr. 4, 1995, SMI.

14. See, e.g., *CL*, 41 (Feb. 1922): 63; *LD*, 73 (Apr. 15, 1922): 29; 73 (June 24, 1922): 23; and 74 (Sept. 23, 1922): 28; *WF*, Feb. 20, 1925, p. 262; Roland Marchand, *Advertising the American Dream: Making Way for Modernity, 1920–1940* (Berkeley: Univ. of California Press, 1985), pp. 89–94; Douglas, *Inventing American Broadcasting*, pp. 149, 306; and Smulyan, *Selling Radio*, pp. 20–21, who seems to accept the Country Life rhetoric at face value.

15. *LD*, 84 (Mar. 14, 1925): 23–24.

16. *WF*, Jan. 9, 1925, p. 42; *PF*, June 13, 1925, p. 680; *LD*, 100 (Feb. 23, 1929): 34; *RNY*, Sept. 20, 1924, pp. 1207–8.

17. Capper, "What Radio Can Do for the Farmer," quotations on pp. 238, 239, 240.

18. Edmund de S. Brunner, *Radio and the Farmer: A Symposium on the Relation of Radio to Rural Life* (New York: Radio Institute of the Audible Arts, 1935), quotations on pp. 5, 11.

19. See, e.g., *PF*, Feb. 1936, p. 33.

20. De S. Brunner, *Radio and the Farmer*, p. 20.

21. See, e.g., Robert S. Bader, *Hayseeds, Moralizers, and Methodists: The Twentieth-Century Image of Kansas* (Lawrence: Univ. Press of Kansas, 1988).

22. W. A. Wheeler, "Know Your Markets," in USDA, *Yearbook of Agriculture, 1920* (Washington, D.C.: GPO, 1921): 127–46, quotation on p. 135; *Current Opinion*, 72 (1922): 403–5; H. C. Wallace, "Report of the Secretary," in USDA, *Yearbook of Agriculture, 1922* (Washington, D.C.: GPO, 1923): 1–82, on pp. 22–23; W. A. Wheeler, "Down on the Farm in 1923," *Radio Broadcast*,

2 (1923): 212–14; Howard M. Gore, "Report of the Secretary," in USDA, *Yearbook of Agriculture, 1924* (Washington, D.C.: GPO, 1925): 1–96, on pp. 52–55, 63; *LD*, 99 (Nov. 3, 1928): 75; John C. Baker, *Farm Broadcasting: The First Sixty Years* (Ames: Iowa State Univ. Press, 1981), pp. 8–12, chaps. 3–4; Reynold M. Wik, "The USDA and the Development of Radio in America," *AH*, 62 (1988): 177–88.

23. Gore, "Report of the Secretary," pp. 54, 55; I. W. Dickerson, "Radio Helps Weather Forecasting," *WF*, Jan. 23, 1925, p. 119.

24. *WF*, Sept. 11, 1925, p. 1167.

25. W. M. Jardine, "Report of the Secretary," in USDA, *Yearbook of Agriculture, 1926* (Washington, D.C.: GPO, 1927), pp. 1–124, quotation on p. 56; Baker, *Farm Broadcasting*, pp. 10, 14, 26–29; Wik, "The USDA and the Development of Radio," p. 184.

26. Josephine F. Hemphill, "Broadcasting Home Economics from the U.S. Department of Agriculture," *JHE*, 19 (1927): 275–78; Smulyan, *Selling Radio*, p. 89.

27. *LD*, 77 (June 30, 1923): 25–26; *Ford News*, June 15, 1923, p. 8; *Radio Broadcast*, 5 (1924): 261–65; Lois P. Dowdle, "Radio and Extension Teaching," *JHE*, 19 (1927): 252–56; Salisbury, "Radio and the Farmer"; *WF*, Aug. 17, 1935, p. 469; Baker, *Farm Broadcasting*, pp. 29–36, 49–52.

28. Smulyan, *Selling Radio*.

29. *LD*, 73 (Apr. 22, 1922): 28; *WF*, Aug. 28, 1925, p. 1097 (quotation); *RNY*, Jan. 27, 1923, p. 109; Ashley C. Dixon, "What Radio Means to a Rocky Mountain Rancher," *Radio Broadcast*, 4 (1924): 192–95.

30. Katherine Jellison, *Entitled to Power: Farm Women and Technology, 1913–1963* (Chapel Hill: Univ. of North Carolina Press, 1993), pp. 59–60.

31. De S. Brunner, *Radio and the Farmer*, pp. 20, 26–27; *FJ*, Jan. 1926, p. 36; *WF*, Aug. 28, 1925, p. 1097; Reynold M. Wik, "Radio in the 1920s: A Social Force in South Dakota," *South Dakota History*, 11 (1980–81): 93–110, on p. 105.

32. Mrs. F. H. Unger, "What Radio Means to the Farmer," *RNY*, Aug. 1, 1925, pp. 1049–50, quotation on p. 1049; Dickerson, "Radio Helps Weather Forecasting"; and "What the Farmer Listens To," *Radio Broadcast*, 9 (1926): 316–17, quotation on p. 316. The entertainment function of the radio is emphasized by Mary Neth, *Preserving the Family Farm: Women, Community, and the Foundations of Agribusiness in the Midwest, 1900–1940* (Baltimore: Johns Hopkins Univ. Press, 1995), pp. 253–55, who notes the family aspects of listening to the radio and how it meshed well with rural traditions, work, and socialization.

33. "What the Farmer Listens To," p. 316.

34. C. W. Steffler, "Classifying the Invisible Audience," *Commerce and Finance*, 17 (1928): 2271.

35. Hooper-Holmes Bureau, *Columbia's RFD Audience: An Analysis of the Columbia Broadcasting System's Rural Audience* ([New York]: CBS, 1938), pp. 16–19.

36. Quoted in *LD*, 79 (Dec. 15, 1923): 25.

37. "What the Farmer Listens To;" and de S. Brunner, *Radio and the Farmer*, pp. 23–24.

38. *PF*, Aug. 8, 1925, p. 753; and Aug. 29, 1925, p. 823; Unger, "What Radio Means to the Farmer."

39. *WF*, Aug. 28, 1925, p. 1087; and Sept. 25, 1925, p. 1248 (ad).

40. "The Effect of Broadcasting upon Rural Life," *Educational Broadcasting, 1936*, ed. C. S. Marsh (Chicago: Univ. of Chicago Press, 1937), pp. 250–59, on pp. 252, 254–55. Morse Salisbury, "Radio and Home Economics Extension," *JHE*, 23 (1931): 847–49, also complained that these shows were too much like lectures and did not involve the "homemaker" enough. On Wherry's column, see Jellison, *Entitled to Power*, pp. 56–58.

41. Capper, "What Radio Can Do for the Farmer," p. 232.

42. Steffler, "Classifying the Invisible Audience."

43. Eleanor Arnold, ed., *Party Lines, Pumps, and Privies: Memories of Hoosier Homemakers* ([Indianapolis]: Indiana Extension Homemakers Association, [1984]), pp. 160 (quotation), 163; interviews with Gerald Cornell, May 24, 1995, and Dorothy Gracey, Jan. 21, 1995, SMI; Wik, "Radio in the 1920s," pp. 100–101.

44. E. L. Kirkpatrick, J. H. Kolb, Creagh Inge, and A. F. Wileden, "Rural Organizations and the Farm Family," University of Wisconsin AES, *Research Bulletin*, no. 96, Nov. 1929, pp. 24, 53; E. L. Kirkpatrick, P. E. McNall, and May L. Cowles, "Farm Family Living in Wisconsin," University of Wisconsin AES, *Research Bulletin*, no. 114, Jan. 1933, p. 23; E. L. Kirkpatrick, Rosalind Tough, and May L. Cowles, "How Farm Families Meet the Emergency," University of Wisconsin AES, *Research Bulletin*, no. 126, Jan. 1935, pp. 15 (quotation), 39; Bruce L. Melvin, "The Sociology of a Village and the Surrounding Territory," Cornell University AES, *Bulletin*, no. 523, May 1931, pp. 15, 33.

45. Maud Wilson, "Use of Time by Oregon Farm Homemakers," Oregon State Agricultural College AES, *Bulletin*, no. 256, Nov. 1929, p. 21; Jessie E. Richardson, "The Use of Time by Rural Homemakers in Montana," Montana State College AES, *Bulletin*, no. 271, Feb. 1933, p. 26; Blanche M. Kuschke, "Allocation of Time by Employed Married Women in Rhode Island," Rhode Island State College AES, *Bulletin*, no. 267, July 1938, p. 15.

46. Joint Committee on Radio Research, *The Joint Committee Study of Rural Radio Ownership and Use in the United States* ([New York]: CBS and NBC, 1939), pp. 4, 6.

47. Smulyan, *Selling Radio*, pp. 23–26. See also Hal S. Barron, *Mixed Harvest: The Second Great Transformation in the Rural North, 1870–1930* (Chapel Hill: Univ. of North Carolina Press, 1997), pp. 215–25.

48. Capper, "What Radio Can Do for the Farmer," p. 229; Salisbury, "Radio and the Farmer." For Fischer's excellent criticism of the impact-imprint model, see Claude S. Fischer, *America Calling: A Social History of the Telephone to 1940* (Berkeley: Univ. of California Press, 1992), chap. 1.

49. W. F. Ogburn and S. C. Gilfillan, "The Influence of Invention and Discovery," in *Recent Social Trends in the United States: Report of the President's Research Committee on Social Trends*, 2 vols. (New York: McGraw-Hill, 1933): 1:122–66, on pp. 153–54. The authors do say that many of these effects are speculative.

CHAPTER 5: CREATING THE REA

1. *LD*, 79 (Oct. 27, 1923): 30 (quotation), citing *NYT*, Oct. 7, 1923, p. 14.

2. U.S. Department of Commerce, Bureau of the Census, *The Statistical History of the United States from Colonial Times to the Present* (New York: Basic Books, 1976), p. 827; Michael Adas, *Machines as the Measure of Men: Science, Technology, and Ideologies of Western Dominance* (Ithaca: Cornell Univ. Press, 1989).

3. Thomas P. Hughes, *Networks of Power: Electrification in Western Society, 1880–1930* (Baltimore: Johns Hopkins Univ. Press, 1983); David E. Nye, *Electrifying America: Social Meanings of a New Technology, 1880–1940* (Cambridge: MIT Press, 1990); Ronald C. Tobey, *Technology as Freedom: The New Deal and the Electrical Modernization of the American Home* (Berkeley: Univ. of California Press, 1996).

4. David Schap, *Municipal Ownership in the Electric Utility Industry: A Centennial View* (New York: Praeger, 1986), pp. 9, 98–99.

5. Keith R. Fleming, *Power at Cost: Ontario Hydro and Rural Electrification, 1911–1958* (Montreal: McGill–Queen's Univ. Press, 1992), chaps. 2–5; Bayla Singer, "Power Politics," *IEEE Technology and Society Magazine,* 7 (Dec. 1988): 20–27; Leonard DeGraaf, "Corporate Liberalism and Electric Power System Planning in the 1920s," *Business History Review,* 64 (1990): 1–31.

6. Preston J. Hubbard, *Origins of the TVA: The Muscle Shoals Controversy, 1920–1932* (Nashville: Vanderbilt Univ. Press, 1961); Philip J. Funigiello, *Toward a National Power Policy: The New Deal and the Electric Utility Industry, 1933–1941* (Pittsburgh: Univ. of Pittsburgh Press, 1973), chap. 1.

7. D. Clayton Brown, *Electricity for Rural America: The Fight for the REA* (Westport, Conn.: Greenwood, 1980), chaps. 1, 3; Fleming, *Power at Cost,* pp. 16, 48–49, 63–72; *FJ,* Aug. 1923, pp. 13, 15.

8. *Report of the Giant Power Survey Board* (Harrisburg, Penn: Telegraph Printing, [1924]), pp. vii, 6–7, 163–64; Jean Christie, *Morris Llewellyn Cooke: Progressive Engineer* (New York: Garland, 1983), chaps. 2–4; Thomas P. Hughes, "Technology and Public Policy: The Failure of Giant Power," *IEEE Proceedings,* 64 (1976): 1361–71; Morris Cooke to A. G. Connally, June 23, 1924, MLC, box 202 (quotation).

9. *Report of the Giant Power Survey Board,* pp. 117–40, 243–307, 411–80; "The Farm Electrified," *General Bulletin* (of the Pennsylvania Department of Agriculture), no. 407, July 1925, pp. 5–6.

10. Christie, *Cooke,* p. 58; Tobey, *Technology as Freedom,* p. 51; Martha Bruère to George Morse, Mar. 19, 1924; George Morse to Martha Bruère, Mar. 24, 1924; and Morris Cooke to Martha Bruère, July 3, 1924, MLC, box 186; Martha Bruère, "Following the Hydro," *Survey,* 51 (1924), 591–94; Cooke, "Light and Power: Planning the Electrical Future," *Survey,* 67 (1932): 607–11.

11. *FJ,* Apr. 1926, p. 28; Brown, *Electricity for Rural America,* pp. 19, 31; Michael K. Green, "A History of the Public Rural Electrification Movement in Washington to 1942" (Ph.D. diss., University of Idaho, 1967), chap. 3; C. A. Sorenson, "Rural Electrification: A Story of Social Pioneering," *Nebraska History,* 25 (1944): 257–70.

12. Lee Prickett to Emil Lehmann, May 13, 1930, EWL, box 5 (letterhead); Brown, *Electricity for Rural America,* chap. 1; John Carmody to Judson King, July 26, 1944, JMC, box 83 (quotation).

13. NELA, *Progress in Rural and Farm Electrification for the Ten-Year Period, 1921–1931* (New York: NELA, 1932), p. 11; George Kable, "Progress Report Number 1," Dec. 1, 1925, BHE, entry 6, box 604; Eloise Davison, "Electricity and the Farm Home," *JHE,* 18 (1926): 215–16; *CREA Bulletin,* May 12, 1926, p. 3; June 15, 1927, p. 7; and Oct. 10, 1928, pp. 9–10.

14. Eloise Davison, "Standards for the Selection of Household Equipment," *JHE,* 20 (1928): 879–81; "Roundtable on Research Regarding the House," *JHE,* 27 (1935): 529–30; Iona R. Logie, "Home Economist to a Nation," in *Careers in the Making,* 2d ed. (New York: Harpers, 1942), pp. 145–51; Nye, *Electrifying America,* p. 276; D. Clayton Brown, "North Carolina Rural Electrification: Precedent of the REA," *NCHR,* 59 (1982): 109–24, p. 119, n. 22; P. C. Welch to Emil Lehmann, May 28, 1930, EWL, box 5 (CREA letterhead). On the Electric Home and Farm Authority, see Gregory B. Field, "'Electricity for All': The Electric Home and Farm Authority and the Politics of Mass Consumption, 1932–1935," *Business History Review,* 64 (1990): 32–60. On the CREA's part in the National Rural Electrification Project, see NELA, *Progress in Rural and Farm Electrification,* p. 11.

15. "Project Leaders Conference at University of Minnesota . . . ," June 21–22, 1927, CNT, box 1; P. C. Welch to Emil Lehmann, May 28, 1930; Eloise Davison to Emil Lehmann, July 8, 1930; Emil Lehmann to Vera Meacham, July 16, 1930; and Emil Lehmann to E. A. White, Sept. 17, 1930, EWL, box 5.

16. Fred Shepperd to Emil Lehmann, Aug. 14, 1930 (quotation); J. M. Birks to Emil Lehmann, Sept. 27, 1932; Emil Lehmann to J. M. Birks, Sept. 29, 1932; Emil Lehmann to George Kable, Oct. 2, 1937; and W. H. Colwell to Emil Lehmann, Sept. 5, 1944, EWL, box 5.

17. *FJ,* Apr. 1926, p. 28 (quotation); NELA, *Progress in Rural and Farm Electrification.*

18. J. P. Schaenzer, "Rural Electrification in the United States," conference paper, June 19, 1935, p. 4, NAL; Raymond C. Miller, *Kilowatts at Work: A History of the Detroit Edison Company* (Detroit: Wayne State Univ. Press, 1957), p. 233.

19. *LD,* 107 (Nov. 29, 1930): 5–7; 114 (Oct. 15, 1932): 12–13; and 115 (Nov. 29, 1933): 8; William M. Emmons III, "Franklin D. Roosevelt, Electric Utilities, and the Power of Competition," *JEH,* 54 (1993): 880–907; Forrest McDonald, *Insull* (Chicago: Univ. of Chicago Press, 1962), chaps. 11–12; Thomas K. McGraw, *TVA and the Power Fight, 1933–1939* (Philadelphia: J. B. Lippincott, 1971). On the dissolution of the NELA, see *LD,* 115 (Feb. 25, 1933): 36; and Christie, *Cooke,* p. 126.

20. My account of the creation of the REA, from February 1934 to May 1936, is crafted from Funigiello, *Toward a National Power Policy,* chaps. 5–6; Brown, *Electricity for Rural America,* chaps. 4–6; and Christie, *Cooke,* chap. 9, who have used the same archives and agree on the basic details of the story. I give references only for quotations, critical documents, new material, and interpretations that differ from theirs. I explore more fully than they the decisions to use co-ops and to rely on state farm bureaus in the beginning of the program.

21. Morris Cooke, "National Plan for the Advancement of Rural Electrification under Federal Leadership and Control with State and Local Cooperation and as a Wholly Public Enterprise," [Feb. 1934], MLC, box 203, quotations on pp. 5, 10, 11.

22. Cooke, "National Plan," pp. 1, 4.

23. Morris Cooke to Harold Ickes, Feb. 13, 1934, quoted in Christie, *Cooke,* pp. 167–68; REA, "Conference of Cooperative Representatives and Rural Electrification Administration," June 6, 1935, NAL, quotation on p. 10. On promoting the REA to ameliorate the lot of farm women, see Katherine Jellison, *Entitled to Power: Farm Women and Technology, 1913–1963* (Chapel Hill: Univ. of North Carolina Press, 1993), pp. 98–103.

24. Morris Cooke to Secretary [Harold] Ickes, Oct. 22, 1934, REAA, entry 11, box 11, emphasis in the original.

25. *LD,* 119 (May 18, 1935): 8.

26. *PF,* May 1935, p. 5; *FJ,* June 1935, p. 12; *WF,* June 8, 1935, p. 16. On Poe, see Brown, "North Carolina Rural Electrification," pp. 115–16.

27. "Rural Electrification," Apr. 8, 1935, REAA, entry 11, box 31; Cooke, "Electrify America Now," radio address, May 18, 1935, REAA, entry 13, box 9, on p. 5. All historians of the REA report that Cooke favored private utilities at this time. Yet when the REA wanted to publish something to this effect in an annual report, Cooke declared, "When I say there is absolutely no basis in fact for this statement, I am expressing the truth mildly." See Morris Cooke to John Carmody, Apr. 13, 1939, MLC, box 147.

28. Udo Rall, "A Study of Cooperative Consumer Associations for Rural Electrification," May 16, 1935, REAA, entry 11, box 29, p. 11, quotation on p. 5.

29. REA, "Conference of Cooperative Representatives," quotations on pp. 8, 18.

30. H. S. Person, "The Rural Electrification Administration in Perspective," *AH*, 24 (1950): 70–89, on p. 76; Joe Jenness, "Cooperator through Necessity: Morris Llewellyn Cooke," in *Great American Cooperators,* ed. Joseph Knapp (Washington, D.C.: American Institute of Cooperation, 1967), pp. 130–37, on p. 133.

31. Morris Cooke to C. W. Warburton, May 29, 1935, REAA, entry 11, box 35; Morris Cooke to C. W. Warburton, June 10, 1935, REAA, entry 11, box 7.

32. Boyd Fisher to Morris Cooke, June 20, 1935, and July 10, 1935, REAA, entry 11, box 14; "REA, Light and Power for the Farm: What It Means and How to Get It," sent out on July 23, 1935, REAA, entry 11, box 29. On Fisher, see Christie, *Cooke*, p. 42.

33. Rural Electrification Committee of Privately Owned Utilities to Morris Cooke, July 24, 1935; Morris Cooke to W. W. Freeman, July 31, 1935; and REA news release, July 31, 1935, REAA, entry 11, box 30.

34. George Norris to Morris Cooke, Aug. 1, 1935, REAA, entry 11, box 26; Morris Cooke to Franklin Roosevelt, Aug. 2, 1935 (quotation), REAA, entry 11, box 30; "Address Delivered by Rural Electrification Administrator Morris L. Cooke . . . ," Aug. 9, 1935, REAA, entry 13, box 9; and "REA, Light and Power for the Farm."

35. *WF,* Aug. 17, 1935, p. 5; and Sept. 28, 1935, p. 12.

36. Brown, *Electricity for Rural America,* p. 52.

37. *REN,* Nov. 1935, p. 6; REA news releases, Nov. 4 and 6, 1935, REAA, entry 11, box 28.

38. Morris Cooke, talk reported in "Transcript of the REA General Staff Conference," Feb. 1–5, 1937, JMC, box 109, vol. 1, pp. 8–9.

39. During the Giant Power project, Cooke learned about successful co-ops, like the one in Morrison's Cove, Pennsylvania, and George Morse, a Giant Power engineer, wrote a study of these and other co-ops as a feasible rural electric alternative. See Morris Cooke to Harold Ickes, Oct. 22, 1934, REAA, entry 11, box 11, appendix B, pp. 3–4; and George Morse, "How Farmers Can Secure Electric Service by Cooperative Effort," *General Bulletin* (of the Pennsylvania Department of Agriculture), no. 412, Sept. 15, 1925, on pp. 4–6.

40. *FJ,* June 1935, p. 12; *LD,* 121 (June 13, 1936): 7–8, on p. 8.

41. Murray D. Lincoln with David Kapp, *Vice President in Charge of Revolution* (New York: McGraw-Hill, 1960), p. 132.

42. Joseph G. Knapp, *The Advance of American Cooperative Enterprise: 1920–1945* (Danville, Ill.: Interstate, 1973), p. 358. For Hull's attendance at the meeting, see REA, "Conference of Cooperative Representatives," p. 39.

43. Knapp, *Advance of American Cooperative Enterprise,* pp. 350–51; George M. Foster Jr., "A Study of the Rural Electrification Project in Boone County, Indiana" (B.S. thesis, Purdue University, June 1939), copy in REAA, entry 88, box 4, p. 2.

44. Boyd Fisher to John Carmody, Mar. 12, 1937, REAA, entry 13, box 13.

45. *REN,* June 1936, p. 25; and Sept. 1936, p. 29; Boyd Fisher to Morris Cooke, Sept. 23, 1935, REAA, entry 11, box 14.

46. Boyd Fisher to Morris Cooke, July 25, 1935, Sept. 23, 1935, and Nov. 9, 1935, REAA, entry 11, box 14; Morris Cooke to Edward O'Neal, Nov. 22, 1935, REAA, entry 11, box 11; "Interview between Mr. Joseph Marion and Mr. Boyd Fisher," Dec. 13, 1935, attached to Boyd Fisher to Morris Cooke, Dec. 17, 1935, REAA, entry 11, box 14; *REN,* Dec. 1935, p. 18; and Apr. 1936, p. 25 (quotation); Morris Cooke to Sam Rayburn, Jan. 10, 1936, REAA, entry 11, box 30; Morris

Cooke to Clarence Poe, Mar. 10, 1936, REAA, entry 11, box 28; Morris Cooke to Murray Lincoln, Apr. 8, 1936 (telegram), REAA entry 11, box 20; Boyd Fisher to Morris Cooke, Nov. 24, 1936, REAA, entry 13, box 14.

47. Person, "Rural Electrification Administration," p. 78.

48. Morris Cooke to Clarence Poe, Mar. 10, 1936, REAA, entry 11, box 28; *REN,* Jan. 1937, p. 11.

49. *REN,* Sept. 1935, p. 21; Oct. 1935, pp. 16–18; Nov. 1935, p. 12; and Jan.–Feb. 1936, pp. 11–12.

50. Boyd Fisher to Morris Cooke, Apr. 10, 1936, REAA, entry 11, box 14; Boyd Fisher, "Cooperatives," REA in-service lecture, attached to Jack Levin to John Carmody, Apr. 24, 1937, REAA, entry 13, box 19, p. 25. On the beginnings of the Development Division, see "Interview between Mr. Joseph Marion and Mr. Boyd Fisher," p. 1; and Boyd Fisher to Mr. Sears, Jan. 7, 1936, REAA, entry 11, box 14.

51. Boyd Fisher to Morris Cooke, [Jan. 1936], REAA, entry 11, box 14; REA Organizational Chart, Sept. 5, 1936, JMC, box 83 (quotation). On Liter, see Boyd Fisher to Morris Cooke, Dec. 17, 1936, REAA, entry 11, box 14; Oneta Liter to John W. Asher Jr., Nov. 16, 1950, REAA, entry 45, box 3.

52. *REN,* Nov. 1936, pp. 11–12, on p. 11; Brown, "North Carolina Rural Electrification," pp. 121–22.

53. E. G. Cort to Franklin D. Roosevelt, telegram, Oct. 1, 1936; and Elmer Benson to Franklin D. Roosevelt, Oct. 1, 1936, REAA, entry 11, box 30.

54. *REN,* Jan. 1937, pp. 11–12; "Transcript of the REA General Staff Conference," vol. 1, p. 42.

55. Morris Cooke to Rudolph Forster, July 15, 1936, FDR, office file 1570, box 1; John Carmody to Morris Cooke, May 29, 1936, JMC, box 61 (résumé).

56. Fisher, "Cooperatives," p. 16.

57. John Carmody to H. H. Fagan, Apr. 15, 1937; John Carmody to O. J. Grau, Apr. 24, 1937 (quotation); John Carmody to Carl Thompson, Apr. 29, 1937; John Carmody to Claude Johnson, May 3, 1937 (quotation); and John Carmody to Earl Dean, May 13, 1937, JMC, box 86.

58. Boyd Fisher to John Carmody, Mar. 1, 1938, REAA, entry 13, box 14.

59. John Carmody to Nelson Mandernach, Dec. 13, 1937; and John Carmody to Arthur Greenwood, Mar. 17, 1938, JMC, box 87; *Report of Rural Electrification Administration, 1938* (Washington, D.C.: GPO, 1939), pp. 144–45; Mark C. Stauter, "The Rural Electrification Administration: A New Deal Case Study" (Ph.D. diss., Duke University, 1973), pp. 95–98.

60. Morris Cooke to Murray Lincoln, Aug. 13, 1936, and Sept. 3, 1936, REAA, entry 11, box 20.

61. Murray Lincoln to John Carmody, Aug. 25, 1937, JMC, box 82; *Business Week,* Nov. 6, 1937, p.47 ff.; John Carmody to John Robertson, Dec. 13, 1937; John Carmody to A. H. Howalt, Dec. 16, 1937, JMC, box 87; John Carmody to Delaware Rural Electric Cooperative, OH-24, July 20, 1938, JMC, box 88; Lincoln, *Vice President in Charge of Revolution,* p. 145; Knapp, *Advance of American Cooperative Enterprise,* p. 351.

62. *REA Progress Bulletin,* no. 14, Oct. 12, 1937, MLC, box 147; *Report of Rural Electrification Administration, 1937* (Washington, D.C.: GPO, 1938), pp. 8–10, 91.

63. Person, "Rural Electrification Administration," p. 77.

64. Morris Cooke to C. W. Warburton, June 10, 1935, REAA, entry 11, box 7, p. 3; Morris Cooke to Murray Lincoln, Apr. 23, 1936, REAA, entry 11, box 20.

65. Emily KneuBuhl to Morris Cooke, Oct. 9, 1935; Emily KneuBuhl to M. L. Ramsay,

Jan. 9, 1936; Emily KneuBuhl, "Report of Utilization Section," Feb. 17, 1936; and Emily KneuBuhl, "Report of the Utilization Section," May 19, 1936, REAA, entry 11, box 19; *REN,* Aug. 1936, pp. 3–4; REA Organizational Chart, Sept. 6, 1936, JMC, box 83; Emily KneuBuhl to Eleanor Roosevelt, telegram, Dec. 15, 1936, ER, box 1388.

66. Morris Cooke to John Carmody, Boyd Fisher, W. Herring, M. L. Ramsay, and Emily KneuBuhl, Oct. 19, 1936, REAA, entry 11, box 5.

67. Emily KneuBuhl to John Carmody, Apr. 1, 1937; John Carmody to [Utilization Staff], Apr. 8, 1937, REAA, entry 13, box 23; REA news release, Apr. 12, 1937, REAA, entry 13, box 29; "Report of Observation Trip to Tennessee Valley," attached to Mary Taylor to Emily KneuBuhl, Apr. 7, 1936, REAA, entry 19, box 2, p. 11; Emily KneuBuhl to Perry Taylor, May 22, 1937, REAA, entry 13, box 23; John Carmody to Morris Cooke, June 28, 1937, MLC, box 147; John Carmody to Emily KneuBuhl, July 1,1937, REAA, entry 13, box 23; Emily KneuBuhl to Eleanor Roosevelt, [ca. Dec. 1938], ER, box 1509.

68. John Carmody to W. W. Clark, May 25, 1937, JMC, box 86; Louise Stanley to Clara Nale, June 5, 1937, BHE, entry 3, box 579; *REN,* Feb. 1938, pp. 5–6; M. L. Wilson to All Extension Directors, July 1, 1940, REAA, entry 17, box 4; Clara Nale, "Home Economists in the Field of Rural Electrification," *JHE,* 30 (1938): 223–25, biographical information on p. 288; John Carmody to Rayborn Sullivan, Mar. 25, 1937, JMC, box 86; Ernest Collins to Boyd Fisher, Dec. 18, 1936, REAA, entry 11, box 7; Oneta Liter to John Asher Jr., Nov. 16, 1950, REAA, entry 45, box 3; Elva Bohannan to Oscar Meier, Feb. 29, 1942, REAA, entry 76, box 1; Lenore Sater, address, REA Annual Administrative Conference of Staff and Field Personnel, typescript, Jan. 9–13, 1939, JMC, box 91, pp. 570–82, on p. 581.

69. Lee Prickett to Emil Lehmann, May 13, 1930, EWL, box 5; Lee Prickett, REA field report, PA-15, Dec. 4, 1937, JMC, box 83.

70. Funigiello, *Toward a National Power Policy,* pp. 165–67; John Carmody to Morris Cooke, July 1, 1939, MLC, box 147; John Carmody to Judson King, July 25, 1944, JMC, box 83; John Carmody résumé, [post-1945], JMC, box 61.

71. Lyndon Johnson to Franklin Roosevelt, July 29, 1939; and Franklin Roosevelt to Lyndon Johnson, Aug. 2, 1939, FDR, office files, 1570a, box 3; see also Robert A. Caro, *The Years of Lyndon Johnson,* vol. 1 *The Path to Power* (New York: Random House, 1982), chap. 28; and Jordan A. Schwartz, *The New Dealers: Power Politics in the Age of Roosevelt* (New York: Knopf, 1993), chap. 12. Neither Caro nor Schwartz mentions the REA offer. For one of the many congratulatory letters on Johnson's decision to stay in Congress, see Earl White to Lyndon Johnson, Aug. 10, 1939, LBJ, House of Representatives files, box 174.

72. *REN,* Oct.–Nov. 1949, p. 9; Funigiello, *Toward a National Power Policy,* pp. 167–68; Morris Cooke to Harry Slattery, Sept. 14, 1939, HS, box 11 (quotation).

CHAPTER 6: STRUGGLING FOR LOCAL AUTONOMY

1. Joseph Knapp, *The Advance of American Cooperative Enterprise: 1920–1945* (Danville, Ill.: Interstate, 1973).

2. Vincent Nicholson to Paul Appleby, July 1, 1939, HS, box 11.

3. *REN,* June 1937, pp. 13–14; REA Annual Administrative Conference of Staff and Field Personnel, typescript, Jan. 9–13, 1939, JMC, box 91, especially pp. 12–21.

4. Dora Haines and Udo Rall to Harry Slattery, Oct. 4, 1939, REAA, entry 76, box 1.

5. For an argument that the New Deal created this framework for the entire country, see

Ronald C. Tobey, *Technology as Freedom: The New Deal and the Electrical Modernization of the American Home* (Berkeley: Univ. of California Press, 1996).

6. Mercer Johnston to Boyd Fisher, Dec. 16, 1936, REAA, entry 11, box 7; Transcript of the REA General Staff Conference, Feb. 1–5, 1937, JMC, box 109, vol. 1: 63, 67.

7. C. O. Falkenwald to Boyd Fisher, Apr. 16, 1936; Boyd Fisher to Morris Cooke, Apr. 21, 1936 (quotation), and Nov. 9, 1936, REAA, entry 11, box 14.

8. F. J. Lund to Saul Gamer, Mar. 27, 1936, REAA, entry 11, box 14; Boyd Fisher to Morris Cooke, Apr. 8, 1936, REAA, entry 11, box 7 (quotation); John Carmody to F. B. Farrell, Feb. 28, 1939, JMC, box 103; C. O. Falkenwald to Boyd Fisher, Apr. 16, 1936, REAA, entry 11, box 14; Eugene Gilson to Boyd Fisher, Dec. 17, 1936; and J. H. Roger to Morris Cooke, Nov. 13, 1935, REAA, entry 11, box 7.

9. Morris Cooke to M. L. Ramsay, Sept. 11, 1935, REAA, entry 11, box 29; Morris Cooke to C. W. Warburton, Dec. 27, 1935, REAA, entry 11, box 35; Memorandum of Understanding between REA and U.S. Extension Service, Jan. 10, 1936, REAA, entry 11, box 34; C. O. Falkenwald to Boyd Fisher, Apr. 16, 1936, REAA, entry 11, box 14; Morris Cooke to David Weaver, Nov. 6, 1936, REAA, entry 11, box 35; David Weaver, "Cooperation in the Extension of Electric Service to the Farm," *Agricultural Engineering,* 17 (1936): 507–8; Mercer Johnston to Boyd Fisher, Dec. 16, 1936, REAA, entry 11, box 7; Boyd Fisher to Harry Slattery, Nov. 6, 1939, HS, box 11. On Weaver, see D. Clayton Brown, "North Carolina Rural Electrification: Precedent of the REA," *NCHR,* 59 (1982): 109–24, who does not mention this episode.

10. Jack Levin to John Carmody, May 5, 1937, REAA, entry 13, box 10.

11. Eugene Gilson to Boyd Fisher, Dec. 17, 1936; Walter Wolff to Boyd Fisher, Dec. 17, 1936; J. Warner Pyles to Boyd Fisher, Dec. 17, 1936; and Ernest Collins to Boyd Fisher, Dec. 18, 1936, REAA, entry 11, box 7.

12. Oscar Meier to George Munger, July 31, 1937, JMC, box 84.

13. Boyd Fisher to Jonathan Daniels, Oct. 17, 1938; and Boyd Fisher to John Carmody, Nov. 28, 1938 (quotation), REAA, entry 13, box 14.

14. Minutes of CREA meeting, Nov. 24, 1924, CNT, box 1; *NYT,* May 29, 1928, pp. 1, 10; J. P. Schaenzer, "Coordinated Rural Electrification Activities in the State of New York," *CREA Bulletin,* June 1938, pp. 4–11, on p. 10; D. E. Blandy to Carl Ladd, July 26, 1935, CEL, box 10; Carl Ladd to C. H. Chapin, Apr. 3, 1936, CEL, box 9.

15. Schaenzer, "Coordinated Rural Electrification Activities"; *REN,* July 1936, pp. 16–18; Lincoln D. Kelsey to Specialists and Others in the Colleges of Agriculture and Home Economics, Apr. 3, 1936, CHE, box 35 (quotation); Lincoln D. Kelsey to L. R. Simons, July 27, 1936, CEL, box 11; Gould P. Colman, *Education and Agriculture: A History of the New York State College of Agriculture at Cornell* (Ithaca: Cornell Univ. Press, 1963), pp. 414–28.

16. Lincoln Kelsey to C. E. Ladd, July 31, 1936, CEL, box 11.

17. Boyd Fisher to Morris Cooke, July 30, 1936, REAA, entry 11, box 14; Harry Slattery to David Smith, Aug. 8, 1940, REAA, entry 18, box 25; *REN,* Sept. 1943, p. 14.

18. *REN,* May 1936, pp. 25–26; C. O. Falkenwald to Boyd Fisher, Apr. 16–17 (quotation), 1936, REAA, entry 11, box 14; T. F. Carmickle to John Carmody, Mar. 16, 1939, KS-7; and Gilbert Jones to John Carmody, Mar. 10, 1939 (quotation), KS-21, REAA, entry 13, box 21.

19. Harold Severson, *Corn Belt: A Pioneer in Cooperative Power Production* (Humboldt, Iowa: Corn Belt Power Cooperative, 1972), p. 114; Oscar Meier to P. T. Cooper, Sept. 6, 1941, REAA, entry 76, box 1; Robert A. Caro, *The Years of Lyndon Johnson,* vol. 1, *The Path to Power* (New York: Random House, 1982), p. 524; D. Clayton Brown, *Electricity for Rural America: The*

Fight for the REA (Westport, Conn.: Greenwood, 1980), pp. 71–72; Boyd Fisher to John Carmody, Oct. 17, 1938, REAA, entry 13, box 14.

20. E. F. Chestnut, "Rural Electrification in Arkansas, 1935–1940: The Formative Years," *Arkansas Historical Quarterly*, 46 (1987): 215–60; *REN*, Apr. 1936, p. 22; Walter L. Wolff to C. O. Falkenwald, June 24, 1938, GA-20, REAA, entry 89, box 1; *REN*, Mar. 1936, p. 18; Severson, *Corn Belt*, especially pp. 13, 18; W. G. Morrison to C. O. Falkenwald, Oct. 25, 1938, LBJ, House of Representatives file, box 172; David Mitchell, "The Origins of the Robertson Electric Cooperative," *East Texas Historical Journal*, 25, no. 2 (1987): 71–79.

21. John Carmody to C. E. Jakway, July 7, 1944, JMC, box 85; Kenneth E. Merrill, *Kansas Rural Electric Cooperatives: Twenty Years with the REA* (Lawrence: Univ. of Kansas Center for Research in Business, 1960), especially p. 32; John Carmody to John Houston, Apr. 26, 1939, JMC, box 103.

22. C. O. Falkenwald to Boyd Fisher, Apr. 16, Apr. 17, Apr. 18, 1936; and Boyd Fisher to Morris Cooke, Nov. 28, 1936, REAA, entry 11, box 14.

23. O. J. Long to Boyd Fisher, Mar. 8, 1936, IA-46, REAA, entry 3, reel 28 (quotation); John Carmody to C. W. Warburton, July 16, 1937, JMC, box 86; John Carmody, "REA Generating Plants," June 6, 1958, JMC, box 82; Lyndon Johnson to R. J. Buchanan, Aug. 6, 1938, LBJ, House of Representatives file, box 175.

24. Harry Slattery to D. Z. McCormick, Oct. 24, 1939; and Harry Slattery to Boyd Fisher, Oct. 25, 1939, HS, box 11; M. L. Wilson to All Extension Directors, July 1, 1940; and Oscar Meier to Harry Slattery, Sept. 19, 1940, and Nov. 7, 1940, REAA, entry 17, box 4; Harold H. Beaty, "Summary of Letters from Extension Service Directors," Nov. 4, 1940, REAA, entry 76, box 3; *REN*, Jan.–Feb. 1941, pp. 12–13; Oscar Meier to Hugh Milton, June 27, 1941, REAA, entry 76, box 1.

25. Gary A. Donaldson, "A History of Louisiana's Rural Electric Cooperatives, 1937–1983" (Ph.D. diss., Louisiana State University, 1983), quotation on p. vii; D. Jerome Tweton, *The New Deal at the Grass Roots: Programs for the People in Otter Tail County, Minnesota* (St. Paul: Minnesota Historical Society Press, 1988), chap. 9; Allen H. Chessher, *Let There Be Light: A History of Guadalupe Valley Electric Cooperative* (San Antonio, Tex.: Naylor, 1964), pp. 11–12, 21–23.

26. *REN*, Oct. 1935, pp. 16–18, 25 (Indiana quotation); Mrs. John B. Merritt to Morris Cooke, June 19, 1935, REAA, entry 3, reel 1; Elwood P. Hain to Boyd Fisher, [Apr. 1936], REAA, entry 3, reel 23 (Iowa quotation).

27. Chestnut, "Rural Electrification in Arkansas," p. 219; Harold Severson, *Architects of Rural Progress: A Dynamic Story of the Electric Cooperatives as Service Organizations in Illinois* (n.p.: Association of Illinois Electric Cooperatives, [1966?]), pp. 71–71, 84, 188, 220, 248; Felix Rhea to F. B. Bastion, Sept. 1, 1936, WA-14, REAA, entry 4, box 56; Clay Seward to Lyndon Johnson, Apr. 21, 1939, LBJ, House of Representatives file, box 172; Harold Severson, *Determination Turned on the Power: A History of the Eastern Iowa Light and Power Cooperative* (n.p., [1964?]), pp. 17, 24, 26–27, 32; Margaret Anderson, field report, Jan. 28, 1941, KS-32, REAA, entry 86, box 8. For further instances, see Severson, *Corn Belt*, pp. 39, 85, 106, 114, 119, 130, 138; Harold Severson, *The Night They Turned on the Lights: The Story of the Electric Power Revolution in the North Star State* (n.p.: Midwest Historical Features, 1962), pp. 134, 147; Harold Severson, *Out of the Dark Ages: A History of the Agralite Cooperative, Benson, Minnesota* (Benson, Minn., 1965), p. 9; Tweton, *New Deal at the Grass Roots*, chap. 9; Caro, *Path to Power*, chap. 28; Chessher, *Let There Be Light*, pp. 25–26; Lemont K. Richardson, *Wisconsin REA: The Struggle to Extend Electricity to Rural Wisconsin, 1935–1955* (Madison: Univ. of Wisconsin AES, Apr. 1961), p. 44.

28. William Nivison to John Carmody, Nov. 5, 1937, JMC, box 84; Severson, *Determination Turned on the Power*, p. 17.

29. Severson, *Architects of Rural Progress*, p. 143; Severson, *Corn Belt*, pp. 106, 114, 130; Severson, *The Night They Turned on the Lights*, p. 31; Tweton, *New Deal at the Grass Roots*, pp. 142–43; C. A. Sorenson, "Rural Electrification: A Story of Social Pioneering," *Nebraska History*, 25 (1944): 257–70, on p. 268; Chessher, *Let There Be Light*, p. 25; Caro, *Path to Power*, p. 525; Udo Rall to Thumb Electric Cooperative, June 7, 1939, MI-37, REAA, entry 89, box 2.

30. *REN*, Apr. 1939, passim; Oct. 1939, pp. 20–21; and Apr. 1941, p. 10; *Live Wire* newsletter, Mar. 1941, TX-89, REAA entry 89, box 3.1, emphasis in the original; *REN*, Nov. 1943, pp. 18–19, quotation on p. 19; Frank J. Busch, "History of Montana Rural Electric Cooperatives" (Ph.D. diss., University of Montana, 1975), pp. 193–94.

31. John Carmody to Benton Rural Electric Association, WA-83, Aug. 12, 1938, JMC, box 88.

32. *Report of Rural Electrification Administration, 1938* (Washington, D.C.: GPO, 1939), p. 75; Severson, *The Night They Turned on the Lights*, pp. 84, 250; Severson, *Determination Turned on the Power*, p. 25 (quotation); Chessher, *Let There Be Light*, p. 27. See also Jane Adams, *The Transformation of Rural Life: Southern Illinois, 1890–1990* (Chapel Hill: Univ. of North Carolina Press, 1994), pp. 152–53.

33. Severson, *The Night They Turned on the Lights*, p. 143; Severson, *Corn Belt*, p. 136; Tweton, *New Deal at the Grass Roots*, p. 143; Richardson, *Wisconsin REA*, pp. 46–47.

34. J. W. Pyles to John Carmody, June 26, 1937, JMC, box 82; Caro, *Path to Power*, p. 528. For further instances, see Severson, *Determination Turned on the Power*, pp. 27–28; Tweton, *New Deal at the Grass Roots*, pp. 140–43; and Busch, "History of Montana Electric Cooperatives."

35. John Carmody to Governor Henry Horner, Feb. 26, 1937, JMC, box 86; *REN*, Apr. 1938, p. 19; John Carmody to A. D. Taylor, Apr. 15, 1938, JMC, box 88.

36. John Carmody to Leslie Potter, July 15, 1938, JMC, box 88; George Munger, remarks, REA Annual Administrative Conference (1939), pp. 116–21; J. W. Pyles, remarks, REA Annual Administrative Conference (1939), pp. 371–80, quotation on p. 372; D. W. Teare to John Carmody, Nov. 6, 1937 (quotation), JMC, box 84.

37. Thelma Wilson to John Carmody, Nov. 7, 1937, JMC, box 84.

38. George Munger to Wallace Theobald, Feb. 11, 1939, IN-14, REAA, entry 86, box 5; "Utilization Analysis," July 10, 1939, IA-9, REAA, entry 86, box 6; Loren Jenks to Carl Schlich, Aug. 5, 1939, AL-20, REAA, entry 86, box 1; George Munger to Utilization Representatives, June 29, 1937, [GA-7, ID-4], REAA, entry 13, box 29; A. R. Tucker to George Munger, June 14, 1938, TX-21, REAA, entry 86, box 21.

39. *Steele-Waseca Sparks* newsletter, Oct. 1939, MN-53, REAA, entry 86, box 11.

40. Mary Taylor to John Carmody, Nov. 8, 1937, JMC, box 84.

41. Walter Moulton, field report, July 5, 1938, TX-50, REAA, entry 86, box 21; G. P. Easley to George Munger, Aug. 14, 1940, NM-8, REAA, entry 86, box 15; J. W. Pyles to Walter Moulton, Apr. 11, 1938, TX-21, REAA, entry 86, box 21; L. P. Zimmerman to Loren Jenks, Nov. 1, 1939, MN-53, REAA, entry 86, box 11; Richard Dell to John Carmody, Nov. 4, 1937 (quotation), JMC, box 84.

42. Frank Soukup to Boyd Fisher, May 12, 1938, IA-15, REAA, entry 86, box 6; Richard Dell to Fisher, July 13, 1938, CA-6, REAA, entry 86, box 1; T. S. Jackson to George Munger, May 27, 1940 (quotation), GA-31, REAA, entry 86, box 2.

43. William Nivison to John Carmody, Nov. 5, 1937, JMC, box 84; C. A. Winder to Board

of Directors, May 23, 1940, IA-74, REAA, entry 89, box 1; Udo Rall, "A List of Words and Their Meaning for REA Cooperators," Nov. 30, 1938, REAA, entry 13, box 39.

44. *REN*, July 1937, p. 23; C. D. Blair to R. J. Beamish Jr., Dec. 8, 1937, AL-9, REAA, entry 3, reel 1.

45. *Report of Rural Electrification Administration, 1937* (Washington, D.C.: GPO, 1938), pp. 48–49; *Report of REA, 1938*, pp. 104–5; *Report of Rural Electrification Administration, 1939* (Washington, D.C.: GPO, 1940), pp. 143–45; *Report of the Administrator of the Rural Electrification Administration, 1940* (Washington, D.C.: GPO, 1940), pp. 25–26; *Report of the Administrator of the Rural Electrification Administration, 1941* (Washington, D.C.: GPO, 1941), p. 18; John Carmody to F. I. Kiley, Mar. 14, 1939, JMC, box 103; Carl Cox to Ben Creim, Dec. 5, 1938; and Harry Lamberton to V. D. Nicholson, Dec. 22, 1938 (quotation), NM-9, REAA, entry 8, box 109; Roswell Garts to Paul Appleby, May 21, 1939, IA-5, REAA, entry 15, box 11.

46. Don F. Hadwiger and Clay Cochran, "Rural Telephones in the United States," *AH*, 58 (1984): 221–38, on p. 226. Claude Fischer, "Technology's Retreat: The Decline of Rural Telephony in the United States," *Social Science History*, 11 (1987): 295–327, argues that factors like increasing costs for modernizing lines were not a sufficient cause for the decline.

47. [D. B. Lancaster], field report, Oct. 1939, TX-40, REAA, entry 86, box 21; C. D. Dunlap to Harry Slattery, June 16, 1940, OH-41, REAA, entry 13, box 39; C. A. Winder, Operations Memorandum 49, Feb. 1, 1940, Operations Memorandum 60, July 17, 1940, and Operations Memorandum 71, Dec. 15, 1940, REAA, entry 13, box 33.

48. Mercer G. Johnston for Boyd Fisher, "Division of Operations Supervision," Aug. 26, 1938, REAA, entry 13, box 14, quotations on pp. 1, 4, emphasis in the original.

49. Boyd Fisher to John Carmody, Jan. 21, 1938, REAA, entry 13, box 14; Boyd Fisher, address, REA Annual Administrative Conference (1939), p. 8.

50. Boyd Fisher to John Carmody, Nov. 24, 1937, REAA, entry 13, box 14; REA Annual Administrative Conference of Staff and Field Personnel, typescript, Jan. 5–14, 1938, REAA, entry 96, box 5, on Jan. 13, pp. 3, 4, 5e, 7.

51. John Carmody to B. Fisher, M. Ramsay, and A. Walters, July 12, 1938, REAA, entry 13, box 14.

52. Boyd Fisher to Dora Haines, Nov. 10, 1937, REAA, entry 13, box 14; Dora Haines and Udo Rall to Harry Slattery, Oct. 4, 1939, REAA, entry 76, box 1, p. 2; Udo Rall to Mercer Johnston, Aug. 2, 1940, REAA, entry 88, box 1.

53. Elva Bohannan, field report, Jan. 25–28, 1941, PA-4, REAA, entry 89, box 2, emphasis in the original.

54. See, e.g., Clara Nale to Victoria Harris, Jan. 29, 1941, KY-18, REAA, entry 89, box 1; Elva Bohannan, field report, Feb. 1, 1941, PA-22, REAA, entry 89, box 3.1.

55. John Carmody to Boyd Fisher, Aug. 14, 1937, TX-49, REAA, entry 13, box 13; Dora Haines and Udo Rall to C. A. Winder, Oct. 4, 1939, REAA, entry 76, box 1, p. 2; Boyd Fisher to John Carmody, Feb. 21, 1940 (quotation), GA-51, REAA, entry 18, box 16.

56. Lyndon Johnson to Fritz Engelhard, May 18, 1938; and Homer Wade to Lyndon Johnson, May 24, 1939, LBJ, House of Representatives file, box 175; "Short History of the Development of Pedernales Electric Cooperative," attached to Lee McWilliams to Robert Conrod, Aug. 26, 1939, LBJ, House of Representatives file, box, 173; Caro, *Path to Power,* chap. 28. For examples of progress reports, see Lee McWilliams to Lyndon Johnson, May 4, 1940, June 30, 1940, and Aug. 17, 1940, LBJ, House of Representatives file, box 182.

57. Udo Rall to Helen Moss, June 23, 1938, TX-76, REAA, entry 89, box 3.1; Mason Har-

rell to Lyndon Johnson, May 15, 1939, LBJ, House of Representatives file, box 171; Udo Rall to Allyn Walters, Oct. 2, 1943, TX-76, REAA, entry 89, box 3.1.

58. John Carmody to E. H. Williams, Sept. 22, 1937, WV-10, JMC, box 87; Udo Rall to Frank Robinson, Aug. 14, 1940, REAA, entry 88, box 1; W. E. Herring to Board of Directors, Sept. 23, 1942, VA-11, REAA, entry 89, box 3.1; Boyd Fisher to John Carmody, July 26, 1938, NE-44, REAA, entry 13, box 14; C. A. Winder to R. L. Maynard, Aug. 2, 1940, GA-45, REAA, entry 88, box 1; Evelyn Bloome, field report, Oct. 20, 1939, IA-19, REAA, entry 86, box 6. On Winder, see *REN*, Nov. 1938, p. 5.

59. Boyd Fisher to Walter Wolff, Apr. 24, 1938, GA-51, REAA, entry 13, box 14.

60. Dora Haines and Udo Rall to C. A. Winder, Oct. 4, 1939, REAA, entry 76, box 1, p. 2; transcript of telephone conversation between George Munger and James Bevis, Dec. 5, 1939, AL-28, REAA, entry 86, box 1; Boyd Fisher to John Carmody, July 21, 1938, telegram, NE-?, and Aug. 9, 1938, telegram, MO-?, REAA, entry 13, box 14; W. O. Zervas, field report, Oct. 28, 1937, MN-56, REAA, entry 13, box 29.

61. J. C. Mortensen to John Carmody, June 14, 1938, TX-48, REAA, entry 86, box 21; John Carmody to Fruitbelt Electric Cooperative, Aug. 24, 1938, MI-38, JMC, box 88. For reports of directors' selling appliances in Ohio, Mississippi, and Tennessee, see Boyd Fisher to John Carmody, June 28, 1937, OH-50, REAA, entry 13, box 14; Special Membership Reports for MS-17, TN-20, and TN-38, 1940, REAA, entry 88, box 2B.

62. Boyd Fisher to John Carmody, Sept. 30, 1938, REAA, entry 19, box 2; Udo Rall to W. E. Herring, Dec. 21, 1940, TN-1, Jan. 4, 1941, AL-18; Special Membership Reports for OK-18 and TN-20, 1940, REAA, entry 88, box 2B.

63. Richard Dell to Boyd Fisher, June 26, 29, 1938, CA-6, REAA, entry 86, box 1; Elva Bohannan to Oscar Meier, Jan. 29, 1941, PA-00, REAA, entry 89, box 2; Lee Lloyd to George Munger, Nov. 22, 1940, IL-2, REAA, entry 86, box 4.

64. John Carmody to W. W. Weatherwax, Oct. 16, 1937; John Carmody to H. J. Strong, Dec. 21, 1937; John Carmody to William Jacobsen, Jan. 17, 1938; and John Carmody to Guy Gillette, Mar. 11, 1938, IA-9, JMC, box 87; Boyd Fisher to John Carmody, Mar. 18, 1938 (quotation), AR-11, REAA, entry 13, box 14; Boyd Fisher to John Carmody, Feb. 21, 1940, GA-51, REAA, entry 18, box 16; Boyd Fisher to John Carmody, telegram, July 23, 1938, NE-44, REAA, entry 13, box 14.

65. Dora Haines and Udo Rall to C. A. Winder, Oct. 4, 1939, REAA, entry 76, box 1.

66. Victoria Harris, remarks, in REA Annual Administrative Conference (1938), on Jan. 14, p. 2; John Becker to Harry Slattery, Oct. 4, 1939, HS, box 11; E. M. Faught to George Munger, July 16, 1940, TX-49, REAA, entry 86, box 21.

67. Special Membership Reports for MS-1, MS-17, and MS-49, Dec. 1940, REAA, entry 88, box 2B.

68. Boyd Fisher, remarks, in REA Annual Administrative Conference (1938), on Jan. 14, p. 4.

69. Udo Rall to Boyd Fisher, Jan. 10, 1938, REAA, entry 19, box 2; Boyd Fisher to John Carmody, June 9, 1938, REAA, entry 13, box 14. See, e.g., "By-Laws of the Red River Valley Co-Operative Power Association," Apr. 8, 1940, MN-74, REAA, entry 89, box 2.

70. C. A. Winder, "REA Cooperative Letters," no. 1, Nov. 29, 1938, no. 2, Dec. 16, 1938, and no. 3, Dec. 28, 1938, REAA, entry 13, box 39; "REA Cooperative Letter," no. 9, Mar. 8, 1940, REAA, entry 88, box 10; Harry Slattery, "Revised REA Cooperative Letter," no. 6, Sept. 13, 1940, REAA, entry 13, box 33; Rall, "A List of Words and Their Meaning for REA Cooperators."

71. *REN*, June 1938, p. 19; and Sept. 1938, p. 23; John Carmody to McDonald County

Women's Extension Club Council, June 7, 1939, MO-48, JMC, box 103; Winder, "REA Cooperative Letter," no. 9.

72. "Pre-Allotment Procedure for New Projects," Washington, D.C., revised Oct. 10, 1940, FDR, Department of Agriculture files, box 1, p. 2.

73. Boyd Fisher to Morris Cooke, Jan. 15, 1936, REAA, entry 11, box 14; Morris Cooke to Perry Taylor and Russell Cook, Feb. 21, 1936, REAA, entry 11, box 33.

74. Udo Rall to Frank Robinson, Oct. 11, 1940; and Frank Robinson to Board of Directors, Oct. 30, 1940, TX-45, REAA, entry 89, box 3.1

75. Alva B. Davis to W. R. Moulton, July 15, 1940; George Munger to Alva B. Davis, Aug. 1 (quotation), Aug. 27, Oct. 18, 1940; Alva B. Davis to George Munger, Aug. 22, 1940 (quotation); Margaret Anderson, field reports, Aug. 26, 1940, to May 12–17, 1941; and George Munger to Margaret Anderson, Nov. 18, 1940, KS-32, REAA, entry 86, box 8; *REN*, May 1942, pp. 6–7.

76. Frank Wilson to Franklin Roosevelt, May 11, 1938, FDR, office file 1570, box 3; John Carmody to Frank Wilson, June 20, 1938, MI-37, JMC, box 88; Merrill, *Kansas Rural Electric Cooperatives,* pp. 144, 188.

77. *REN,* Oct. 1939, p. 19; Donaldson, "Louisiana's Rural Electric Cooperatives, 1937–1983," p. 128; Severson, *Architects of Rural Progress,* pp. 247, 254; M. Rempel to Dora Haines, letterhead, Nov. 28, 1940, AR-18, REAA, entry 89, box 1; Merrill, *Kansas Rural Electric Cooperatives,* pp. 144, 147, 170, 188; Mrs. O. E. Bolon to John Carmody, Mar. 16, 1939, MT-13, ER, box 1489; D. B. Lancaster to George Munger, letterhead, Sept. 10, 1940, TX-40, REAA, entry 86, box 21; Frank Wilson to Franklin Roosevelt, letterhead, May 11, 1939, FDR, office file 1570, box 3; "Rural Lines Incorporate," *Steuben (N.Y.) Advocate,* June 13, 1941.

78. *REN,* Oct. 1939, p. 19; *RL,* Sept. 1956, p. 8; Severson, *Architects of Rural Progress,* pp. 246–55, on p. 247.

79. David Long, "'We're Not Isolated Now!': Anna Boe Dahl and the REA," *Montana,* 39 (spring 1989): 18–23. On the Farmers' Union and the REA, see Mary Neth, *Preserving the Family Farm: Women, Community, and the Foundations of Agribusiness in the Midwest, 1900–1940* (Baltimore: Johns Hopkins Univ. Press, 1995), pp. 211–12.

80. Mrs. O. E. Bolon to John Carmody, Mar. 16, 1939; John Carmody to Eleanor Roosevelt, Mar. 29, 1939 (quotation); and Eleanor Roosevelt to John Carmody, Apr. 8, 1939, MT-13, ER, box 1489; John Carmody to Mrs. O. E. Bolon, Apr. 18, 1939 (quotation), JMC, box 103.

CHAPTER 7: LIGHTS IN THE COUNTRY

1. REA, "Conference of Cooperative Representatives and Rural Electrification Administration," June 6, 1935, NAL, quotation on p. 9.

2. S. P. Lyle, "Factors Indicating Specific Needs for Electrification Research in the U.S. Department of Agriculture," July 20, 1937, JMC, box 84; George Munger to Wallace Theobald, Feb. 11, 1939, IN-14, REAA, entry 86, box 5.

3. U.S. Department of Commerce, Bureau of the Census, *The Statistical History of the United States from Colonial Times to the Present* (New York: Basic Books, 1976), p. 827; *Electrical World,* Nov. 4, 1939, p. 101. For the range of state studies, see Wendell E. Keeper, "Consumption, Costs, and Uses of Electricity on New York State Farms" (Ph.D. diss., Cornell University, 1938); E. B. Lewis, "Use of Electricity on Nebraska Farms, 1920–1934," University of Nebraska AES, *Bulletin,* no. 289, June 1934.

4. George Munger to Utilization Representatives, June 29, 1937, REAA, entry 13, box 29, table.

5. See, e.g., Dora Haines and Udo Rall to Administrator [Harry Slattery], Oct. 4, 1939, p. 1, REAA, entry 76, box 1; *Current News,* Aug. 1940, MN-35, REAA, entry 86, box 11, p. 5; *Report of the Administrator of the Rural Electrification Administration, 1941* (Washington, D.C.: GPO, 1941), p. 1; Harold Severson, *Architects of Rural Progress: A Dynamic Story of the Electric Cooperatives as Service Organizations in Illinois* (n.p.: Association of Illinois Electric Cooperatives, [1966?]), p. 220.

6. George Munger to Utilization Representatives, June 29, 1937, REAA, entry 13, box 29, table.

7. "Utilization Division Field Staff," Nov. 1, 1939, REAA, entry 17, box 4. On kitchen parties, see *REN,* Dec. 1939, pp. 3–5.

8. For examples of these activities, see Enola Guthrie, field report, Dec. 17, 1937, TX-00, JMC, box 84; Elva Bohannan, field report, Dec. 30, 1937, PA-15, JMC, box 83; Mary Alice Willis, field report, Feb. 8, 1938, AL-9, REAA, entry 86, box 1; Victoria Harris, field report, summarized in George Munger to George L. Morris, May 27, 1938, MN-48, REAA, entry 86, box 11; *REN,* July 1938, pp. 17–19; Louisan Mamer, Sept. 27–29, 1938, IA-9, REAA, entry, 86, box 6; Oneta Liter to George Munger, June 6, 1939, TX-30, REAA, entry 86, box 21; Lee Prickett, field report, Dec. 4, 1937, PA-15, JMC, box 83; James Cobb, field report, Feb. 8, 1938, AL-9, REAA, entry 86, box 1; Walter Moulton, field report, May 11, 1938, extracted in Moulton to F. H. Robinson, Feb. 9, 1940, TX-21, REAA, entry 86, box 21; Richard Dell to Boyd Fisher, June 26, 1938, CA-6, REAA, entry 86, box 1.

9. Oneta Liter, field report, extracted in George Munger to Boyd Fisher, June 3, 1938, TX-49, REAA, entry 86, box 21.

10. See, e.g., Edwin Wallace to George Munger, Dec. 22, 1939, AL-22, REAA, entry 86, box 1; Lee McWilliams to Lyndon Johnson, Mar. 26, 1940, LBJ, House of Representatives file, box 182.

11. *Report of Rural Electrification Administration, 1939* (Washington, D.C.: GPO, 1940), pp. 76–80; *REN,* Dec. 1938, pp. 3–6, quotation on p. 5; Oct. 1939, pp. 6–8; Jan. 1940, p. 16; and June 1940, pp. 7–9; D. W. Teare to Thomas Connor, July 22, 1940, IA-5, REAA, entry 86, box 6; "1939 Farm Equipment Tour Itinerary: Revised," Nov. 21, 1939; D. W. Teare to Oscar Meier, Mar. 13, 1941; and D. W. Teare to C. A. Winder, Aug. 25, 1941, REAA, entry 76, box 2. For sample programs, see "REA Electric Farm Equipment Tour: The Systems Part," booklet, [ca. Feb. 1941], REAA, entry 16, box 7. On the showing of the Lorentz films, see Robert Craig to Paul Appleby, July 29, 1939, REAA, entry 13, box 12.

12. George Munger to John Carmody, June 15, 1938, NE-4; July 1, 1938, KY-33; and Oct. 13, 1938, AR-18, REAA, entry 13, box 29; *REN,* Nov. 1938, pp. 3–15, 18–19, 29; James Cobb, field report, Dec. 10, 1938, TX-11, REAA, entry 86, box 21; Thelma Wilson, field reports, June 26–30, 1939, GA-39, REAA, entry 86, box 2; *Marion County (Tex.) Courier,* Nov. 21, 1939, TX-30, REAA, entry 86, box 21.

13. *Memphis Commercial Appeal,* July 29, 1938 (quotation); and *Arkansas Gazette,* July 29, 1938, JMC, box 104. On the ironies of the New Deal regarding sharecropping, see Pete Daniel, *Breaking the Land: The Transformation of Cotton, Tobacco, and Rice Cultures since 1880* (Urbana: Univ. of Illinois Press, 1985).

14. *Louisville Courier Journal,* June 16, 1938, KY-40, JMC, box 104; *REN,* Nov. 1938, pp. 10–11; Deward Clayton Brown, "Rural Electrification in the South" (Ph.D. diss., University of California at Los Angeles, 1970), p. 274.

15. *REN*, July 1938, p. 27; *Report of REA, 1939*, p. 63.

16. Thelma Wilson, field report, extracted in George Munger to John Carmody, July 1, 1938, KY-33; and Loren Jenks, field report, extracted in George Munger to John Carmody, Oct. 13, 1938, REAA, entry 13, box 29.

17. Thelma Wilson, field report, June 28, 1939, GA-39, REAA, entry 86, box 2.

18. *REN,* Feb. 1938, p. 22 (quotation); Feb. 1939, pp. 24–25; and Apr. 1944, p. 21.

19. See, e.g., *REN,* Aug. 1939, cover.

20. *REN,* July 1937, pp. 7–8; Katherine Jellison, "'Let Your Corn Stalks Buy a Maytag': Prescriptive Literature and Domestic Consumption in Rural Iowa," *Palimpsest,* 69 (1988): 132–39, on p. 133.

21. See the back pages of *REN,* Oct. 1938 and Jan., Feb., May, and July 1939.

22. "Report of Observation Trip to Tennessee Valley," attached to Mary Taylor to Miss [Emily] KneuBuhl, Apr. 7, 1936, REAA, entry 19, box 2, p. 13; *Report of Rural Electrification Administration, 1938* (Washington, D.C.: GPO, 1939), p. 16; Harry Slattery to George Munger, Oct. 11, 1939, HS, box 11.

23. *REN,* Nov. 1936, pp. 17, 30; Dec. 1936, pp. 22–23; Feb. 1939, p. 25; May 1939, p. 27; Apr. 1940, p. 5; and May 1946, pp. 12–13, 22; Florence L. Hall, "Report of Home Demonstration Work, 1937," *Extension Service Circular* (of the USDA), no. 294 (Nov. 1938): 2; C. W. Warburton, "Aims and Objectives of Home Demonstration Work," (USDA) *Extension Service Circular,* no. 265 (June 1937); C. W. Warburton, address, REA Annual Administrative Conference of Staff and Field Personnel, typescript, Jan. 9–13, 1939, JMC, box 91, pp. 547–59; Gladys Gallup and Florence L. Hall, "Progress in Home Demonstration Work," (USDA) *Extension Service Circular,* no. 319 (Feb. 1940); Gladys Gallup, "The Effectiveness of the Home Demonstration Program of the Cooperative Extension Service of the United States Department of Agriculture in Reaching Rural People and in Meeting Their Needs" (Ph.D. diss., George Washington University, 1943), p. 133.

24. Oscar Meier to Clara Nale et al., Sept. 20, 1940; and M. L. Wilson to All Extension Directors, Apr. 16, 1940, and July 1, 1940, REAA, entry 17, box 4; Clara Nale to C. A. Winder, Dec. 16, 1940, REAA, entry 76, box 1 (quotation).

25. Harold H. Beaty, "Summary of Letters from Extension Service Directors," Nov. 4, 1940, REAA, entry 76, box 3; Udo Rall, "The REA Program and the Extension Service," June 3, 1941, REAA, entry 76, box 1, quotation on p. 4.

26. *REN,* Oct. 1936, p. 10; [Lo Petree?] to George Munger, AL-18, Mar. 22, 1939, REAA, entry 86, box 1; James Cobb, field report, Dec. 10, 1938, REAA, entry 86, box 21.

27. George Munger to Administrator, Jan. 20, 1940 (quotation), and Mar. 21, 1940, REAA, entry 13, box 29. For an example of Munger's urging co-ops to hire a specialist, see George Munger to Wallace Theobald, Feb. 11, 1939, IN-14, REAA, entry 86, box 5.

28. Margaret Anderson, field report (sample form), Aug. 26–31, 1940, KS-32, REAA, entry 86, box 8; Cecil Beyler, "Purposes of Utilization Department of Marshall County Rural Electric Membership Corporation," [1939], IN-9, REAA, entry 86, box 5.

29. Cecil Beyler, "Marshall County Utilization Program, June 1–Dec. 31, 1940," IN-9, REAA, entry 86, box 6.

30. Margaret Anderson, field reports, Aug. 26, 1940, to Jan. 18, 1941, KS-32, REAA, entry 86, box 8, quotations in Aug. 26–31, 1940, and Dec. 2–7, 1940.

31. *REN,* July 1941, p. 19.

32. See, e.g., *Radiant Eye Announcer,* July 1940, AL-30; *The Nodak Neighbor,* Apr. 1960,

ND-?, CW, box 69; *REA News: The Monthly Messenger of Craighead Electric Co-op*, July 10, 1940, AR-9, REAA, entry 89, box 1.

33. See, e.g., *Line Up*, Apr. 1953, GA-74, CW, box 18; *Kilowatt Ours*, Mar. 1950, IN-?, CW, box 69; *Current News*, Aug. 1940; *Lighter Way*, Nov. 1940, TX-30, REAA, entry 86, box 21; *Live Wire*, Mar. 1941, TX-89, REAA, entry 89, box 3.1.

34. *Radiant Eye Announcer*, Nov. 1940, p. 6, AL-30, REAA, entry 89, box 1; *Current News*, Aug. 1940, p. 5.

35. Mary Willis to John Carmody, Nov. 4, 1937, JMC, box 84; Robert Tisinger to Loren Jenks, Oct. 5, 1938, GA-34, REAA, entry 86, box 2; Warren Hallum to J. W. Pyles, Apr. 7, 1938, MN-5, REAA, entry 86, box 11; Margaret Anderson, field report, Sept. 23–28, 1940, KS-32, REAA, entry 86, box 8.

36. See, e.g., H. A. Rosseth to George Munger, Mar. 29, 1939, MN-35, REAA, entry 86, box 11; Eugene Mulder to George Munger, Aug. 18, 1939, IA-2, REAA, entry 86, box 6; Utilization Summary, Apr. 1, 1940, TX-49, REAA, entry 86, box 21.

37. Oscar Meier, remarks, REA Annual Administrative Conference (1939), p. 381; and D. W. Teare to C. A. Winder, [1941], REAA, entry 76, box 2.

38. George Munger to E. V. Ellis, Mar. 2, 1938, AL-21, REAA, entry 86, box 1; Margaret Anderson, field report, Apr. 28 to May 3, 1941, KS-32, REAA, entry 86, box 8; W. P. Coppinger, field report, Aug. 5, 1938, IN-9, REAA, entry 86, box 5.

39. Emily KneuBuhl to Morris Cooke, Mar. 5, 1936, REAA, entry 11, box 19; Boyd Fisher, "Cooperatives," typescript, Apr. 1, 1937, p. 14, attached to Jack Levin to Administrator, Apr. 24, 1937, REAA, entry 13, box 19; Loren Jenks to E. V. Ellis, Jan. 28, 1938, AL-21; Loren Jenks to A. W. Redd, June 10, 1939, AL-25, REAA, entry 86, box 1; John Carmody to S. O. Bland, June 3, 1938, VA-28, JMC, box 88; John Carmody to Sandy Stewart, Apr. 10, 1939, IL-18, JMC, box 103; *Electrical Merchandising*, Dec. 1937, pp. 6–7, 26, 79; George Munger to D. M. Pollock, June 8, 1939 (quotation), GA-35, REAA, entry 86, box 2.

40. Ann Schoenfeld, "'A Great Advance over the General Level of Taste Displayed by Our Government in Official Design': Posters for the REA," paper presented at the annual meeting of the Society for the History of Technology, Oct. 18, 1997, Pasadena, Calif.

41. Robert L. Snyder, *Pare Lorentz and the Documentary Film* (Norman: Univ. of Oklahoma Press, 1968), p. 121; Robert Craig to Harry Lamberton, Apr. 15, 1939, REAA, entry 16, box 7.

42. Edith Hertz to Harry Lamberton, Apr. 20, 1939, REAA, entry 16, box 7, emphasis in the original.

43. John Carmody to Lowell Mellett, Apr. 26, 28, 1939 (quotation); and Lowell Mellett to John Carmody, Apr. 27, 1939, REAA, entry 16, box 7.

44. The script is attached to Arch Mercey to Marion Ramsay, Aug. 7, 1939, REAA, entry 16, box 7.

45. Harlow Person, "Suggestive Comments on Script for REA Film," Aug. 21, 1939, REAA, entry 16, box 7.

46. Udo Rall to Administrator, Aug. 24, 1939, REAA, entry 76, box 1. On Steinbeck's suggestion for a fire scene, see Marion Ramsay to Arch Mercey, Aug. 9, 1939, REAA, entry 16, box 7. Steinbeck is mentioned as a possible narrator in the Aug. 7 script.

47. Marion Ramsay to Arch Mercey, Aug. 9, 1939, REAA, entry 16, box 7.

48. Joris Ivens, "Collaboration in Documentary," *Films*, 2 (spring 1940): 30–42; Ivens, *The*

Camera and I (New York: International Publishers, 1969), quotations on pp. 137, 209; William Alexander, *Films on the Left: American Documentary Film from 1931 to 1942* (Princeton: Princeton Univ. Press, 1981), pp. 113–25.

49. Snyder, *Pare Lorentz*, pp. 126–30; Marion Ramsay to [William] Phillips, Jan. 29, 1940; Harry Slattery to Pare Lorentz, n.d. [responding to May 3, 1940, letter], REAA, entry 16, box 7; Harry Slattery to Marguerite LeHand, Aug. 19, 1940, and Franklin Roosevelt to Missy [Le-Hand], Aug. 24, 1940, FDR, office file 1570, box 2; *REN*, Oct. 1940, pp. 8–9; Dec. 1940, p. 19; and Jan.–Feb. 1941, p. 21.

50. Ivens, *The Camera and I*, pp. 189, 195, 197; Snyder, *Pare Lorentz*, pp. 123–24. Ivens wanted to focus on the public-private battle but could not do so for political reasons. He also said later that there was not enough about cooperatives in the film.

51. Harry Slattery to Staff, Oct. 9, 1940; and "Contract between REA and RKO on Power and the Land," [1940], REAA, entry 16, box 7; *REN*, Jan.–Feb. 1941, p. 21.

52. *Variety Film Reviews, 1938–1942*, vol. 6 (New York: Garland, 1983), entry at Oct. 2, 1940; *Time*, Oct. 14, 1940, pp. 114–15; *REN*, Nov. 1940, pp. 13–14; Dec. 1940, p. 19; and Jan.–Feb. 1941, p. 21; P. T. Hartung, "I Remember, I Remember," *Commonweal*, Dec. 20, 1940, pp. 232–33; *Magazine of Art*, Jan. 1941, p. 43; Snyder, *Pare Lorentz*, p. 192.

53. For an account of this phenomenon in Illinois, see Jane Adams, "Resistance to 'Modernity': Southern Illinois Farm Women and the Cult of Domesticity," *American Ethnologist*, 20 (1993): 89–113.

54. Thelma Wilson to John Carmody, Nov. 7, 1937; Elbert Karns to John Carmody, Nov. 6, 1937; and Ray Hugus to John Carmody, Nov. 7, 1937, JMC, box 84; James Cobb, field report, Nov. 22, 1938; and Walter Moulton to F. H. Robinson, Feb. 9, 1940 (extracts of field reports), TX-21, REAA, entry 86, box 21; Oneta Liter to Bernice Smith, Aug. 3, 1942, IN-1, REAA, entry 89, box 1.

55. Walter Zervas, field report, Aug. 9, 1938, WI-9, REAA, entry 86, box 25.

56. J. W. Pyles to Walter Moulton, Apr. 11, 1938, TX-21, REAA, entry 86, box 21; Willard Luft, remarks, REA Regional Conference, Dallas, Tex., May 23–26, 1939, REAA, entry 96, box 6, p. 147

57. Wallace Theobald to George Munger, Jan. 18, 1939, IN-14, REAA, entry 86, box 5; A. R. Tucker to George Munger, May 8, 1939, TX-21, REAA, entry 86, box 21; H. E. Antle to Loren Jenks, July 12, 1939, and Feb. 5, 1940, IN-6, REAA, entry 86, box 5; T. S. Jackson to Loren Jenks, Nov. 20, 1939, GA-31, REAA, entry 86, box 2; James Cobb to C. O. Faulkenwald, June 7, 1940, TX-49, REAA, entry 86, box 21; L. A. Killough to P. W. Smith, Aug. 10, 1940, AL-22, REAA, entry 86, box 1.

58. George Munger to John Carmody, Dec. 9, 1938, REAA, entry 13, box 29; George Munger to Wallace Theobald, Feb. 11, 1939, IN-14, REAA, entry 86, box 5.

59. Margaret Anderson, field report, Sept. 23–28, 1940, KS-32, REAA, entry 86, box 8; *Current News*, Aug. 1940, p. 2; Harriet Harris, field report, Feb. 4, 1941, PA-15, REAA, entry 89, box 2; *Report of REA, 1938*, pp. 61–62.

60. *Report of REA, 1938*, p. 62; George Halonen Jr. to George Munger, Sept. 27, 1939, WI-16; George Munger to Wallace Theobald, Feb. 11, 1939; and W. L. Rahiser, utilization reports, June 1939 to Nov. 1940, TX-49, REAA, entry 86, box 21; "Monthly Report: Operating Statistics," June 1940, MO-30, REAA, entry 16, box 10; Utilization Division memo to MN-4, [early 1938], REAA, entry 86, box 11.

61. For examples of pre-REA appliance surveys, see *Electricity on the Farm*, 3 (Feb. 1930): S6; and Jean Warren, "Use of Time in Its Relation to Home Management," Cornell University AES, *Bulletin*, no. 734, June 1940, p. 11.

62. William Nivison to John Carmody, Nov. 5, 1937, JMC, box 84 (quotation).

63. For the urban case, see Joel A. Tarr and Gabriel DuPuy, eds., *Technology and the Rise of the Networked City in Europe and America* (Philadelphia: Temple Univ. Press, 1988); Jane Busch, "Cooking Competition: Technology on the Domestic Market in the 1930s," *TC*, 24 (1983): 222–45; Mark Rose, "Urban Environments and Technological Innovation: Energy Choices in Denver and Kansas City, 1900–1940," *TC*, 25 (1984): 503–39, on p. 536; David E. Nye, *Electrifying America: Social Meanings of a New Technology, 1880–1940* (Cambridge: MIT Press, 1990), p. 268.

64. *REN*, July 1938, pp. 4–10; Jan. 1940, pp. 6–8; and Oct. 1940, pp. 10–11; J. Stewart Wilson to Robert Craig et al., Aug. 8, 1941, REAA, entry 17, box 4.

65. J. G. McClean to Morris Cooke, Aug. 7, 1935, REAA, entry 11, box 11, p. 4.

66. V. L. Heid, quoted in C. A. Winder to All System Managers and Superintendents, Apr. 22, 1940, REAA, entry 17, box 5; G. Gordon Downey to George Munger, July 2, 1938, IN-8, REAA, entry 86, box 5; George Halonen Jr. to George Munger, June 28, 1940, MN-34; and Matt Wilson to George Munger, May 21, 1940, MN-10, REAA, entry 86, box 11. On the high cost of wiring and electric appliances as a factor, see Katherine Jellison, *Entitled to Power: Farm Women and Technology, 1913–1963* (Chapel Hill: Univ. of North Carolina Press, 1993), pp. 103, 111–12. On resistance to debt as an economic strategy, see Mary Neth, *Preserving the Family Farm: Women, Community, and the Foundations of Agribusiness in the Midwest, 1900–1940* (Baltimore: Johns Hopkins Univ. Press, 1995), pp. 193–94.

67. Eleanor Arnold, ed., *Party Lines, Pumps, and Privies: Memories of Hoosier Homemakers* ([Indianapolis]: Indiana Extension Homemakers Association, [1984]), p. 89.

68. *Report of REA, 1938*, p. 118.

69. Arnold, *Party Lines, Pumps, and Privies*, quotation on p. 67; Jellison, "Let Your Corn Stalks Buy a Maytag," ad on p. 139; *Electrical Merchandising*, Sept. 1938, pp. 32–33; and Sept. 1939, pp. 14–15, 49.

70. Miles Nelson to George Munger, June 13, 1939, IA-27, REAA, entry 86, box 6.

71. *Appliance Specifications and Directory, 1936* (New York: McGraw-Hill, 1936), pp. 9–10; *Electrical Merchandising*, Jan. 1940, p. 14; Nye, *Electrifying America*, pp. 275–72.

72. Blanche M. Kuschke and Margaret Whittemore, "Home Refrigeration in Rural Rhode Island," Rhode Island State College AES, *Bulletin*, no. 239, Sept. 1933; *Appliance Specifications, 1936*, pp. 92–97. Electric refrigerators were advertised for as low as one hundred dollars in a Muncie newspaper in 1934; see Nye, *Electrifying America*, p. 356.

73. Carolyn Goldstein, "Mediating Consumption: Home Economics and American Consumers, 1900–1940" (Ph.D., diss., University of Delaware, 1994), pp. 142–76; Susan Strasser, *Never Done: A History of American Housework* (New York: Pantheon, 1982), chaps. 14–16; Ruth Schwartz Cowan, *More Work for Mother: The Ironies of Household Technology from the Open Hearth to the Microwave* (New York: Basic Books, 1983), chaps. 6–7.

74. Miriam Rapp, "Fuels Used for Cooking Purposes in Indiana Rural Homes," Purdue University AES, *Bulletin*, no. 339, May 1930.

75. *Electrical Merchandising*, Aug. 1940, p. 23; Harold F. Williamson, Ralph L. Andreano, Arnold R. Daum, and Gilbert C. Klose, *The American Petroleum Industry*, 2 vols. (Evanston: Northwestern Univ. Press, 1963), 2:421–22.

76. Loren Jenks to H. E. Antle, Apr. 12, 1939, IN-6, REAA, entry 86, box 5. On the bot-

tled-gas competition, see Miles Nelson to George Munger, June 13, 1939, IA-27, REAA, entry 86, box 6; Glen Hammerbeck to REA Information Section, June 3, 1940, WI-16, REAA, entry 86, box 25; and Harold Severson, *Determination Turned on the Power: A History of the Eastern Iowa Light and Power Cooperative* (n.p. [1964?]), p. 39.

77. Taylor, "Report of Observation Trip to Tennessee Valley," on pp. 3, 12.

78. Rapp, "Fuels Used for Cooking," pp. 8–9; Lewis, "Use of Electricity on Nebraska Farms," p. 29.

79. Interviews with Eva Watson, Feb. 21, 1995; and Dorothy Gracey, Jan. 21, 1995, SMI.

80. Rapp, "Fuels Used for Cooking," pp. 12–13; Arnold, *Party Lines, Pumps, and Privies,* p. 18; interviews with George Woods, Mar. 18, 1995; Gerald Cornell, May 24, 1995; and Thena Whitehead, Feb. 11, 1995, SMI.

81. Elva Bohannan, field report, Feb. 1, 1941, PA-22, REAA, entry 89, box 3.1.

82. Eleanor Arnold, ed., *Voices of American Homemakers* (Bloomington: Indiana Univ. Press, 1985), pp. 179, 182, 180. See also Arnold, *Party Lines, Pumps, and Privies,* pp. 11–22.

83. Morris Cooke to John Carmody, Jan. 28, 1938, MLC, box 147.

84. *REN,* Aug. 1937, pp. 13–15, on p. 14.

85. See, e.g., John Carmody to Michael F. Garrett, Apr. 15, 1938; and John Carmody to Harry Hopkins, May 19, 1938, JMC, box 88; John Carmody, remarks, REA Annual Administrative Conference (1939), pp. 384–87; Willard Luft, remarks, REA Regional Conference, Dallas, Tex., May 23–26, 1939, REAA, entry 96, box 6, pp. 145–50; Willard Luft, "Plumbing," attached to George Munger to Harry Slattery, Jan. 3, 1940, REAA, entry 13, box 29.

86. On Delco owners with plumbing, see, e.g., George Weissert to George Munger, July 7, 1938, IN-9, REAA, entry 86, box 5.

87. Homer Ummel to George Munger, July 8, 1938, IN-7, REAA, entry 86, box 5 (quotation); Luft, "Plumbing," p. 1.

88. Willard Luft, remarks, REA Regional Conference, Dallas, Tex., May 23–26, 1939, REAA, entry 96, box 6, pp. 145–50, on p. 150.

89. Arnold, *Party Lines, Pumps, and Privies,* p. 47.

90. E. W. Lehmann and F. C. Kingsley, "Electric Power for the Farm," University of Illinois AES, *Bulletin,* no. 332, June 1929, p. 421; A. R. Tucker to C. A. Winder, Jan. 11, 1939, TX-21, REAA, entry 86, box 21.

91. George Kable, remarks, REA Annual Administrative Conference (1939), p. 628; and A. R. Tucker to C. A. Winder, Nov. 11, 1939, TX-21, REAA, entry 86, box 21.

92. *REN,* May 1941, pp. 28–29.

93. *Report of REA, 1939,* p. 18 (quotation); *Report of the Administrator of the Rural Electrification Administration, 1940* (Washington, D.C.: GPO, 1940), pp. 3–4.

94. George Munger, "Activities of the Utilization Division," in REA Annual Administrative Conference (1939), pp. 111–31, on p. 128; Miles Nelsen to George Munger, June 13, 1939 (quotation), IA-27, REAA, entry 86, box 6; D. B. Lancaster to Oscar Meier, Mar. 31, 1939; T. R. Qualls to Lee Prickett, Sept. 5, 1939; and Lee Prickett to T. R. Qualls, Oct. 13, 1939, TX-49, REAA, entry 86, box 21.

95. Margaret Anderson, field report, Feb. 3–8, 1941, REAA, entry 86, box 8.

96. Sears, Roebuck, and Co., "Rural Electrification Comparative Cost Data," 1938, UIA, on p. 2.

97. Clara Nale, "Interesting Farm Women in Rural Electrification," conference paper, June 22, 1939, NAL, p. 2.

98. Clara O. Nale, "Home Economics in the Field of Rural Electrification," *JHE*, 30 (1938): 223–25, on p. 225, emphasis added.

99. *REN*, July 1938, pp. 17–19.

100. D. W. Teare to Oscar Meier, June 15, 1941, REAA, entry 76, box 2.

101. Edwin T. Layton Jr., "Science, Business, and American Engineering," in *The Engineers and the Social System*, ed. Robert E. Perrucci and Joel E. Gerstl (New York: Wiley, 1969), pp. 51–72; Goldstein, "Mediating Consumption," chaps. 4–6.

102. Claire E. Gilbert, "The Development of Selected Aspects of Home Demonstration Work in the United States" (Ph.D. diss., Cornell University, 1958), p. 131.

103. *PF*, June 1935, p. 24; and Dec. 1936, p. 48 (quotation); Louise J. Peet and Ruth Pratt Koch, "Lighting the Home with Liquid Fuels," *JHE*, 36 (1944): 354–58, quotation on p. 354.

104. Lincoln Kelsey to Extension Staff Members, Apr. 15, 1936, CHE, box 35; Florence E. Wright, "Rural Home Lighting Problems," conference paper, Oct. 15–17, 1936, copy in *Annual Report of the New York State College of Home Economics at Cornell University, 1936* (Ithaca, N.Y.: New York State College of Home Economics, 1936), pt. 2, quotation on p. 2; *Electrical Merchandising*, Dec. 1939, pp. 14–15, 44.

105. Taylor, "Report of Observation Trip to Tennessee Valley," pp. 2, 14.

106. Dora Haines and Udo Rall, "The Place and Function of an REA Education Program," Oct. 4, 1939, REAA, entry, 76, box 1, p. 5; Rall,"REA Program and the Extension Service," p. 3.

107. "Summary of Letters from Extension Service Directors," Nov. 4, 1940, REAA, entry 76, box 3.

108. "REA Load-Building Program Rebuked by House Group," *Electrical World*, Mar. 24, 1941, copy in REAA, entry 76, box 1 (quotation); *REN*, Jan.–Feb. 1941, pp. 7–8, 20.

109. Oscar Meier to Robert Craig, June 5, 1940; Oscar Meier to George Munger, July 27, 1940; Oscar Meier to M. L. Ramsay, Aug. 14 and Sept. 19, 1940; Oscar Meier, "Social Aims of the Electrification Program, National Standpoint," Nov. 16, 1940; and Oscar Meier to C. A. Winder, Feb. 18, 1941, REAA, entry 17, box 4; Oscar Meier to C. A. Winder, Dec. 6, 1940, and Jan. 18, 1941, REAA, entry 76, box 3.

CHAPTER 8: COMPLETING THE JOB

1. Gilbert C. Fite, *American Farmers: The New Minority* (Bloomington: Indiana Univ. Press, 1981), p. 118.

2. Ibid., chaps. 5–8; John L. Shover, *First Majority, Last Minority: The Transforming of Rural Life in America* (De Kalb: Northern Illinois Univ. Press, 1976).

3. *REN*, Apr. 1942, p. 2; *Report of the Administrator of the Rural Electrification Administration, 1942* (Washington, D.C.: GPO, 1942), pp. 4–5 (quotation); Mark C. Stauter, "The Rural Electrification Administration: A New Deal Case Study" (Ph.D. diss., Duke University, 1973), p. 201.

4. *Report of the REA, 1942*, pp. 6–7; Oscar Meier to M. M. Samuels, Jan. 13, 1942; and Oscar Meier to C. A. Winder, Feb. 2, 1942, REAA, entry 76, box 1; Oscar Meier to Harry Slattery, Feb. 24, 1942, REAA, entry 17, box 4; "REA and WPB Experience: Wartime Work," attached to Louisan Mamer to John Asher, Nov. 13, 1950, REAA, entry 45, box 3.

5. *REN*, Nov. 1942, p. 9; Sept. 1942, p. 7; Jan. 1943, p. 9; and Mar. 1942, p. 17.

6. *Report of the REA, 1942*, pp. 6, 11–12, quotation on p. 11; *Report of the Administrator of the Rural Electrification Administration, 1943* (Washington, D.C.: GPO, 1943), pp. 2–5; *Report of the*

Administrator of the Rural Electrification Administration, 1944 (Washington, D.C.: GPO, 1944), pp. 3–7; Stauter, "Rural Electrification Administration," pp. 205, 215; D. Clayton Brown, *Electricity for Rural America: The Fight for the REA* (Westport, Conn.: Greenwood, 1980), pp. 82–87.

7. Harry Slattery to Charles Poletti, Nov. 6, 1941; C. O. Falkenwald to Administrator, Aug. 4, 1942; and C. O. Falkenwald to C. A. Winder, Aug. 12, 1942, REAA, Slattery files, box 25; "REA Approves Loans for State Power Units," *NYT,* Nov. 15, 1941; "Plan to Open Rural Electric Line Wednesday," *Jamestown (N.Y.) Post-Journal,* Feb. 14, 1944; "Portion of Farmer-Owned Electric System Energized," *Jamestown (N.Y.) Post-Journal,* Feb. 17, 1944; *REN,* July 1942, p. 16; Sept. 1943, p. 14; and Aug. 1944, p. 3.

8. *Report of the REA, 1944,* p. 9; *Report of the Administrator of the Rural Electrification Administration, 1945* (Washington, D.C.: GPO, 1945), pp. 6–7; *Report of the Administrator of the Rural Electrification Administration, 1946* (Washington, D.C.: GPO, 1946), pp. 33–34; Dean Albertson, *Roosevelt's Farmer: Claude R. Wickard in the New Deal* (New York: Columbia Univ. Press, 1961), pp. 384–99; Stauter, "Rural Electrification Administration," chap. 8; Brown, *Electricity for Rural America,* chap. 8.

9. Pete Daniel, "Going among Strangers: Southern Reactions to World War II," *Journal of American History,* 77 (1990): 886–911.

10. See *Report of the Administrator of the Rural Electrification Administration, 1951* (Washington, D.C.: GPO, 1951), p. 36.

11. Calculated from Jerry L. Anderson, ed., *Rural Electric Fact Book* (Washington, D.C.: NRECA, [1960]), pp. 63–97.

12. *Report of the REA, 1946,* p. 9; *Report of the Administrator of the Rural Electrification Administration, 1947* (Washington, D.C.: GPO, 1947), pp. 19–20, 23 (quotation), 25–28; *Report of the Administrator of the Rural Electrification Administration, 1948* (Washington, D.C.: GPO, 1948), pp. 10–15; *Report of the Administrator of the Rural Electrification Administration, 1949* (Washington, D.C.: GPO, 1950), pp. 7–9; *Report of the Administrator of the Rural Electrification Administration, 1950* (Washington, D.C.: GPO, 1950), pp. 12–13, 20–21; *Report of the REA, 1951,* pp. 17–20; *Report of the Administrator of the Rural Electrification Administration, 1952* (Washington, D.C.: GPO, 1952), p. 27; Claude Wickard to John Staumbaugh, Feb. 20, 1952, REAA, entry 27, box 1 (quotation).

13. *Report of the REA, 1946,* p. 24; *REN,* Aug.–Sept. 1948, pp. 14–15, 18; Frank Peebles to Craig-Botetourt Electric Cooperative, Apr. 23, 1940, REAA, entry 89, box 3.1; *REN,* June–July 1949, pp. 8–9.

14. *Report of the REA, 1947,* p. 27 (quotation); *Report of the REA, 1950,* pp. 20–21. A copy of the fake dollar bill is attached to Kermit Overby to Administrator, June 1, 1949, REAA, entry 25, box 10.

15. Joe Belden and Associates, "Texas Public Opinion on Rural Electric Co-ops," Jan. 1949, REAA, entry 45, box 20; "New Co-Op Leaders Voice Their Thoughts," Mar. 20–25, 1950, REAA, entry 25, box 18, p. 8 (Wild quotation); "Group Discussion on Co-Op Principles and Practices," n.d., attached to ibid., p. 4 (quotation); and "What Electric Co-Op Members Should Know," n.d., attached to ibid.

16. See, e.g., *Report of the REA, 1947,* p. 27; Board of Directors Minutes, Nov. 8, 1947, AL-18, REAA, entry 77, box 1; Charles Brannan, secretary of agriculture, memorandum 1307, Mar. 24, 1952, REAA, entry 25, box 7, p. 1.

17. Neil MacNeil and Harold W. Metz, *The Hoover Report, 1953–1955: What It Means to You as a Citizen and Taxpayer* (New York: Macmillan, 1956), pp. 143–46, 307.

18. Claude Wickard, "Statistical Notes," [May 1948], CW, box 18, p. 8.

19. Porter Hardy Jr. to Claude Wickard, Sept. 15, 1949; Claude Wickard to Porter Hardy Jr., Sept. 23, 1949; and Porter Hardy Jr. to Co-ops, Nov. 15, 1949, REAA, entry 25, box 9; and Committee on Expenditures in the Executive Departments, *A Study of Certain Operations of the Rural Electrification Administration,* House Report 2908, Washington, D.C., 1950, copy in REAA, entry 25, box 9.

20. Kermit Overby to Administrator, Jan. 28, 1952, REAA, entry 25, box 10.

21. *Report of the REA, 1946,* pp. 1, 4; *Report of the REA, 1947,* pp. 4, 19–20; Board of Directors Minutes, July 9, 1947, NY-20, REAA, entry 77, box 94; Board of Directors Minutes, Nov. 13, 1946, AL-9, REAA, entry 77, box 1; Mrs. Larry Hamilton to [Director of Agriculture Price Statistics], Aug. 1, 1950, REAA, entry 99, box 7.

22. Kermit Overby to Administrator, June 1, 1949, REAA, entry 25, box 10; Sheridan Maitland, "A Study of the REA Program of Cooperative Member Education," Feb. 28, 1950, REAA, entry 45, box 1; William Wise to Area Directors, June 3, 1952, REAA, entry 27, box 1.

23. Marion Abrams, field report, Aug. 15–20, 1949, FL-29, REAA, entry 45, box 19; Allyn Waters, field report, Dec. 8–9, 1952, NC-00, REAA, entry 87, box 12.

24. Claude Wickard, "Administrative Bulletin 55-R2," n.d., attached to Charles Samenow to William Wise et al., Mar. 3, 1952, REAA, entry 27, box 1; Board of Directors Minutes, Feb. 9, 1946, AL-18, REAA, entry 77, box 1.

25. Maitland, "Study of the REA Program," p. 10. For examples of such reports, see Joel Babb, field report, July 6–7, 15, 1949, GA-67, REAA, entry 45, box 19; Allyn Waters, field report, Dec. 8–9, 1952, NC-00, REAA, entry 87, box 12. For an example of a co-op that implemented area coverage, see Board of Directors Minutes, Aug. 12, 1947, NY-24, REAA, entry 77, box 94.

26. *Report of the REA, 1946,* pp. 22–23; *Report of the REA, 1947,* pp. 9–11; *Report of the REA, 1948,* pp. 24–25; *Report of the REA, 1949,* p. 8 (quotation).

27. *Report of the REA, 1950,* p. 8.

28. "Questions on the Telephone Program," [June 1950], REAA, entry 25, box 9, on pp. 5, 7.

29. *Report of the REA, 1946,* pp. 32–33, *Report of the REA, 1947,* pp. 12–13, *Report of the REA, 1948,* p. 10, *Report of the REA, 1949,* p. 15, *Report of the REA, 1951,* p. 29, *Report of the REA, 1952,* p. 34; *REN,* Oct.–Nov. 1952, pp. 19–21.

30. *REN,* Apr.–May 1950, p. 17.

31. *Directory of Texas Electric Cooperatives,* 1952; *Directory [of] Electric Cooperatives in Illinois,* 1952; and Iowa Rural Electric Cooperative Association, Women's Division, "Program, Third Annual Meeting," Sept. 21–22, 1949, REAA, entry 45, box 20; David Long, "'We're Not Isolated Now!': Anna Boe Dahl and the REA," *Montana,* 39 (spring 1989): 18–23. On Stevenson, see chap. 6 above and Louisan Mamer, field report, Mar. 26, 1948, NRECA Conference, REAA, entry 45, box 3.

32. "Directory of Rural Electric Cooperatives," *Co-Op Power,* Dec. 1949, pp. 28–50, on pp. 48, 50, copy in CW, box 68; Phillip Voltz, field report, Apr. 20–21, 1951, TX-30, REAA, entry 45, box 1; Bernard Krug, field report, Apr. 30–May 3, 1951, VT-10, REAA, entry 25, box 12. Scales managed from at least 1948 to 1950, Snow from at least 1949 to 1951, Robertson in 1949. Samuel Scales was manager of the co-op when it was energized in 1938 and at least to March 1940. See Samuel Scales to George Munger, June 11, 1938, and Mar. 13, 1940, TX-30, REAA, entry 86, box 21.

33. *Prairie Farmer,* May 7, 1949, copy in REAA, entry 27, box 3.

34. Claude Fischer, "Technology's Retreat: The Decline of Rural Telephony in the United States," *Social Science History*, 11 (1987): 295–327.

35. Florence E. Parker, "Cooperative Telephone Associations, 1936," *Monthly Labor Review*, 46 (1938): 392–413; Udo Rall to Kermit Overby, Oct. 7 and 11, 1949, REAA, entry 27, box 3; Richard G. Schmitt Jr., "Farmers' Mutual Telephone Companies," *Journal of Farm Economics*, 33 (1951): 134–39.

36. Lynn Robertson and Keith Amstutz, "Telephone Problems in Rural Indiana," Purdue University AES, *Bulletin*, no. 548, Sept. 1949, on p. 6.

37. Ibid, p. 7.

38. Chief Statistician's Division, AT&T, memorandum, Nov. 10, 1941, ATTA, location 38-02-02-06, including a copy of an editorial from *WF*.

39. See Don F. Hadwiger and Clay Cochran, "Rural Telephones in the United States," *AH*, 58 (1984): 221–38, on pp. 226, 229.

40. "Rural Telephone Service Using Carrier on Power Lines," *BLR*, 21 (1943): 145–48, quotation on p. 148. On the development of the system, see Western Electric Company, *Power Line Carrier Telephone Equipment*, booklet, 1924, ATTA, location 45-04-01-05; and R. D. Gibson, laboratory notebook, 1927, ATTA, location 123-10-01.

41. "Purpose and Functions of Joint Committee on Rural Telephone Service, United States Independent Telephone Association and Bell System," n.d., attached to Keith McHugh to [Bell operating companies], Nov. 22, 1944, ATTA, location 129-05-02-04.

42. Hadwiger and Cochran, "Rural Telephones in the United States," pp. 226–27. For an example of a co-op supporting Hill's bill, see Board of Directors Minutes, Jan. 4, 1945, AL-18, REAA, entry 77, box 1.

43. Keith McHugh to [Bell operating companies], Dec. 13, 1944, ATTA, location 129-05-02-04; *195 Bulletin*, Apr. 1945, pp. 3, 11, 15; "First Call Made over New Rural Power Line Carrier System . . . ," press release, Dec. 11, 1945; *BLR*, 24 (1946): 72–73.

44. "Report of Progress on Bell System Rural Telephone Development Activities," [1946], press release (quotation); Keith McHugh to W. C. Henry, Feb. 21, 1946; press release, Feb. 27, 1946; and "Radiotelephone Service for Farmers to be Tested by Bell System . . . ," press release, May 24, 1946, ATTA, location 129-05-04-02.

45. Hadwiger and Cochran, "Rural Telephones in the United States," p. 227.

46. *195 Bulletin*, Apr. 1946, p. 5; Nov. 1946, p. 6; and Aug. 1947, p. 4; *Report of the REA, 1947*, pp. 15–16; *Report of the REA, 1948*, p. 18; REA, *Joint Use of Facilities by REA Borrowers and Telephone Companies*, booklet, [1948]; and George W. Haggard to All REA Borrowers, Sept. 1, 1950, REAA, entry 63, box 1. On the negotiations about sharing poles, see Claude Wickard to Secretary of Agriculture, Jan. 4, 1946, REAA, entry 25, box 15. On the negotiations about the powerline carrier contract, see William J. Neal to Claude Wickard, Jan. 4, 1946, REAA, entry 25, box 15; W. H. Harrison to All Operating Vice Presidents, July 5, 1946; and J. J. Hanselman to General Commercial Managers, Aug. 30, 1946, June 26, 1947, and June 27, 1947, ATTA, location 556-01-03-06.

47. J. W. Emling, "Transmission Features of the M1 Carrier System," *BLR*, 26 (1948): 101–5, quotations on p. 101. For a summary of the system's development and earlier optimistic assessments of its technical capabilities, see J. M. Barstow, "Carrier Telephones for Farms," *BLR*, 25 (1947): 363–66; Lester Hochgraf, "The Subscriber Terminal for Rural Power-Line Carrier," *BLR*, 25 (1947); 413–17; J. M. Dunham, "Power Line Treatment for the M1 Carrier System," *BLR*, 26 (1948): 2–5.

48. REA Program Analyst, "Background Statement on Farm Telephones (Including 1945 Census Data)," Nov. 29, 1948, REAA entry 27, box 3; E. C. Weitzel to Claude Wickard, Jan. 27, 1949, REAA, entry 25, box 15, emphasis in the original.

49. William Neal to Grant Bell, Jan. 27, 1949, REAA, entry 25, box 12; Claude Wickard diary, Feb. 20–Mar. 2, 1949, Apr. 13–18, 1949, CW, box 21, quotations on Mar. 2, Feb. 22–23.

50. "Telephone Conversation between Clyde Bailey . . . and Claude Wickard . . . ," Oct. 19, 1949, REAA, entry 27, box 3; Hadwiger and Cochran, "Rural Telephones in the United States," pp. 228–30.

51. Kermit Overby to Claude Wickard, Nov. 17, 1949, REAA, entry 98, box 2; REA, "The Rural Telephone Program," draft of REA bulletin, Dec. 1949, REAA, entry 27, box 1, quotation on p. 5; *REN,* Feb.–Mar. 1950, pp. 3–4; Wickard, "Administrative Bulletin T-10," May 30, 1950; and Wickard, "Administrative Bulletin T-12," June 8, 1950, REAA, entry 98, box 2.

52. REA, "The Rural Telephone Program," pp. 8–9.

53. The bulletins, dating from May 30, 1950, to Nov. 15, 1951, are in REAA, entry 98, box 2.

54. *REN,* Apr.–May 1950, pp. 12–13; June–July 1950, pp. 20–21; Oct.–Nov. 1950, p. 19; Dec. 1949–Jan. 1950, pp. 12–13; and Oct.–Nov. 1951, p. 12; Arthur W. Gerth, minutes of A&L Division meeting, June 12–16, 1950, REAA, entry 25, box 9.

55. *REN,* Aug.–Sept. 1950, p. 22; and Oct.–Nov. 1951, p. 12; *Reports of the REA, 1950–1952.*

56. *REN,* Oct.–Nov. 1950, pp. 18–19, on p. 19.

57. "Joint Statement of Senator Lister Hill and Gordon Persons . . . ," Dec. 8, 1944, attached to Keith McHugh to [Bell operating companies], Dec. 13, 1944, ATTA, location 129-05-02-04.

58. *Report of the REA, 1950,* p. 33, emphasis added.

59. *REN,* Apr.–May 1950, pp. 12–13. For a regional survey of joint-use of facilities, see H. B. Lee to All Field Engineers, Region V, July 21, 1950, REAA, entry 63, box 1.

60. T. E. Orman, field report, Dec. 13–14, 1950, IA-506, REAA, entry 47, box 2. For the previous efforts to organize the co-op, see J. Ward Wray, field report, Oct. 16, 1950, ibid.; and T. E. Orman, field report, Oct. 16, 1950, ibid.

61. T. E. Orman, field report, Apr. 25–26, 1951, IA-506, REAA, entry 47, box 2.

62. See field reports from May 15–18, 1951, to Dec. 6, 1951, IA-506, REAA, entry 47, box 2.

63. J. Ward Wray, field report, Jan. 9, 1951; and T. E. Orman, field report, Mar. 27, 1951, IA-506, REAA, entry 47, box 2.

64. W. B. Bridgforth, field report, Aug. 22, 1951, IA-506, REAA, entry 47, box 2; "Phony Business," *Reader's Digest,* Oct. 1951, p. 71. For REA's rebuttal of the *Reader's Digest* charges, see Claude Wickard to Dewitt Wallace, Oct. 10, 1951, REAA, entry 25, box 10.

65. W. B. Bridgforth, field report, Dec. 6, 1951, IA-506, REAA, entry 47, box 2.

66. R. F. Nance, field report, Aug. 3, 1951, AL-500, REAA, entry 47, box 1; Owen Lynch to Winfield Denton, Mar. 7, 1950; and Claude Wickard to Winfield Denton, Apr. 12, 1950, IN-500, REAA, entry 25, box 15.

67. Daniel Corman to Claude Wickard, July 30, 1951, REAA, entry 25, box 15.

68. *REN,* Oct.–Nov. 1952, p. 15 (quotation); and Dec. 1952–Jan. 1953, pp. 18–19.

69. *REN,* Oct.–Nov. 1952, pp. 12–13; and Dec. 1952–Jan. 1953, pp. 20–21, 23, quotation on p. 21.

70. Daniel Corman to Claude Wickard, Feb. 12, 1952, MN-500, REAA, entry 63, box 2. On the movie, see *REN,* Dec. 1951–Jan. 1952, pp. 11–13.

71. J. P. Voegtli, field report, July 3, 1951, NY-500; and W. P. Bauer, field report, Mar. 9–12, 1951, NY-?, REAA, entry 47, box 5.

72. See, e.g., C. U. Samenow to Claude Wickard, Jan. 11, 1950, ND-500, REAA, entry 98, box 2; *REN,* Feb.–Mar. 1950, pp. 3–4; and Claude Wickard, "Administrative Bulletin T-31," June 9, 1950, REAA, entry 98, box 2.

73. *REN,* June–July 1951, pp. 20–21; and Aug.–Sept. 1951, pp. 18–19.

74. W. P. Bauer, field report, Mar. 9–12, 1951, NY-?, REAA, entry 47, box 5.

75. Claude Wickard, "Statement on REA Reorganization," Apr. 28, 1952, REAA, entry 87, box 7; *Report of the REA, 1952,* pp. 4–9.

76. Albertson, *Roosevelt's Farmer,* p. 400; Udo Rall to Claude Wickard, Mar. 18, 1953, CW, box 18; *REN,* June–July 1953, p. 3; Kansas Electric Cooperatives, news report, Sept. 9, 1953, REAA, entry 45, box 21.

77. Bud Wells, *Four Score for Dave Hamil: A Biography* (Wolfe City, Tex.: Henington, 1988). Yet Hadwiger and Cochran, "Rural Telephones in the United States," p. 236, states, on the basis of an interview with Hamil, that Hamil had drafted legislation to "remove REA funding authorization from the federal budget" but "could find no member of Congress to introduce it."

Chapter 9: (Re)forming Rural Life

1. Pete Daniel, "Going among Strangers: Southern Reactions to World War II," *Journal of American History,* 77 (1990): 886–911, on p. 897. Katherine Jellison, *Entitled to Power: Farm Women and Technology, 1913–1963* (Chapel Hill: Univ. of North Carolina Press, 1993), pp. 149–53, argues that this demand extended only to farm families who were profiting from the postwar boom.

2. *REN,* Jan.–Feb. 1941, pp. 7–8, 20; and June 1946, p. 20; *Report of the Administrator of the Rural Electrification Administration, 1941* (Washington, D.C.: GPO, 1941), pp. 9–10; *Report of the Administrator of the Rural Electrification Administration, 1946* (Washington, D.C.: GPO, 1946), p. 34.

3. Claude Wickard, "Administrative Memorandum BMPU-M1," Oct. 22, 1946, attached as fig. 2c in Glenn D. Wagner, "REA's Power Use and Member Education Program," Mar. 20, 1950, REAA, entry 45, box 24.

4. Wagner, "REA's Power Use and Member Education Program," pp. 4, 7, and fig. 3, p. 2.

5. Ibid., pp. 5, 14–19. On the continued employment of regional home economists who had worked for the REA before the war, see Oneta Liter, field report, Feb. 11, 1947, REAA, entry 45, box 3; Mary Alice Willis to A. W. Gerth, Mar. 12, 1951, REAA, entry 45, box 13; and C. Agnes Wilson to Dick Dell and Allen Arness, Oct. 30, 1951, REAA, entry 45, box 20.

6. On the number of prewar specialists, see George Munger to Administrator, Mar. 21, 1940, REAA, entry 13, box 29.

7. Clifton Bradley to Field Representatives, Region V, May 5, 1947, REAA, entry 45, box 20; [George Dillon?], "A Cooperative Education Program," speech given at an NRECA conference, Feb. 1–3, 1949, REAA, entry 25, box 18, p. 3; Wagner, "REA's Power Use and Member Education Program," figs. 47 and 48; *Report of the Administrator of the Rural Electrification Administration, 1950* (Washington, D.C.: GPO, 1950), pp. 39–40; *Report of the Administrator of the Rural Electrification Administration, 1951* (Washington, D.C.: GPO, 1951), pp. 41–42; George Dillon to George Haggard, Oct. 30, 1950; and George Dillon, "Comparison Chart," June 1951, REAA, entry 45, box 10; George Dillon to E. E. Karns, Aug. 5, 1952, REAA, entry 25, box 18.

8. John L. Shover, *First Majority, Last Minority: The Transforming of Rural Life in America* (De Kalb: Northern Illinois Univ. Press, 1976), pp. 143–47.

9. Oneta Liter, field report, Feb. 11, 1947, REAA, entry 45, box 3, p. 2; "New Co-Op Leaders Voice Their Thoughts," Mar. 20–25, 1950, REAA, entry 25, box 18, p. 4.

10. Dora Haines to Kermit Overby, Dec. 6, 1948, REAA, entry 88, box 12; Marion Abrams, field report, Aug. 15–20, 1949, FL-29, REAA, entry 45, box 19; M. E. Cadwallader, field report, July 12, 1949, KS-24, REAA, entry 45, box 21. The REA expected both women and men advisers to start at two hundred dollars a month, plus expenses; see William Callaway to Isaac Thacker, Apr. 14, 1949, KY-57, REAA, entry 45, box 19.

11. W. P. Nixon to H. D. Underhill, Apr. 2, 1951, KS-36, REAA, entry 45, box 21; Louisan Mamer, field report, Mar. 12–29, 1953, NE-00, REAA, entry 45, box 8; *RL,* June 1955, pp.7–9, quotation on p. 7.

12. George Dillon to E. E. Karns, Aug. 5, 1952, REAA, entry 25, box 18.

13. Chester Bennett, field report, Mar. 4, 1952, SC-35; L. W. Lynch, field report excerpt, Nov. 5–25, 1952, IA-00; and E. C. Collier, field report excerpt, Apr. 28, 1953, IL-21, REAA, entry 45, box 19.

14. Dillon, "Comparison Chart"; George Dillon to E. E. Karns, Aug. 5, 1952; and Marion Abrams, field report, Dec. 18–20, 1950, FL-15, REAA, entry 45, box 19; Reed Hutchinson, field report, Feb. 1–2, 1951, VA-29, REAA, entry 45, box 2; C. Agnes Wilson, field report excerpt, Jan. 12–16, 1953, OK-00, REAA, entry 45, box 21.

15. See, e.g., William Spivey, field report, Nov. 6–12, 1952, NM-00, REAA, entry 87, box 12; and G. E. Dillon to All Field Representatives, June 9, 1950, REAA, entry 45, box 18.

16. W. P. Nixon to H. D. Underhill, Apr. 2, 1951, KS-32, REAA, entry 45, box 21.

17. Sheridan Maitland, "A Study of the REA Program of Cooperative Member Education," Feb. 28, 1950, REAA, entry 45, box 1, p. 9.

18. George Dillon, comments on Reed Hutchinson, field report, Feb. 1–2, 1951, REAA, entry 45, box 2; Phillip Voltz, field report, Mar. 21, 1952, SC-35; and C. H. Wright, field report, Jan. 24–26, 1950, MI-40, REAA, entry 45, box 19.

19. Betty Williams, *Now and Then,* Nov. 1951, newsletter, REAA, entry 45, box 20.

20. Eleanor Delany, field report, May 16–17, 1951, GA-00; Elizabeth O'Kelley, field report, May 23–24, 1951, VA-41, REAA, entry 45, box 19; Earl Arnold, field report, Nov. 6–9, 1950, GA-00, REAA, entry 45, box 4.

21. Oneta Liter to John Asher Jr., Nov. 15, 1950, REAA, entry 45, box 3, p. 2.

22. The job description for regional home economists, which Liter probably helped write, also does not speak in terms of load building or sales. See Wagner, "REA's Power Use and Member Education Program," fig. 35.

23. *Report of the REA, 1950,* pp. 39–41, on p. 39. For a list of training schools, see George Dillon, "Participation in Training Schools by the Power Use Specialists: A&L Div.," June 1951, REAA, entry 45, box 10. On the potential of TV as a load builder, see, e.g., Phillip Voltz, field report, Mar. 19–20, 1952, SC-38, REAA, entry 45, box 1.

24. Board of Directors Minutes, July 17, 1947, NY-21, REAA, entry 77, box 94; William Dean, field report, Apr. 13–15, 1953, REAA, entry 45, box 19.

25. *REN,* Feb.–Mar. 1948, pp. 6–7; Allyn Walters to Kermit Overby, May 25, 1950; and O. E. Mabrey to Walters, Mar. 31, 1950, KS-27, REAA, entry 87, box 4.

26. *Kilowatt Ours,* Mar. 1950, IN-24, CW, box 19; *Vigilante,* Apr. 1950, MT-10, REAA, entry 45, box 21; *Current News,* Dec. 1953, IA-9, REAA, entry 45, box 20; *Line Up,* Apr. 1953, GA-74, CW, box 18; *Newsboy,* Aug. [1952], CO-14, REAA, entry 45, box 21.

27. Udo Rall to George Dillon, Mar. 18, 1953, NM-20, REAA, entry 45, box 21; *Power-Grams*, Sept. 1953, MS-29, REAA, entry 45, box 19.

28. On the selection of Willie Wirehand as the symbol, see *Newsboy*, Aug. 1952, p. 1.

29. Reed Hutchinson, field report, Oct. 10–11, 1951, NC-35, REAA, entry 45, box 19.

30. See, e.g., Oneta Liter, field report, Oct. 21–23, 1948, AR-12, REAA, entry 45, box 3; Elwood F. Olver to Manager and Electrification Adviser, Sept. 7, 1951, IA-00, REAA, entry 45, box 20, p. 3; Ray Cannon, field report, Oct. 20–22, 1952, KS-00, REAA, entry 45, box 21; Louisan Mamer, field report, Oct. 15–21, KY-00, REAA, entry 45, box 3; Katherine Bailey, field report, Apr. 28–30, 1953, WI-59, REAA, entry 45, box 19; *RL*, Feb. 1960, p. 18.

31. "Methods and Techniques Useful in Small Appliance Activity," [ca. 1953], REAA, entry 45, box 2.

32. *RL*, Nov. 1956, pp. 3–12.

33. C. H. Wright, field report, Jan. 24–26, 1950, MI-40, REAA, entry 45, box 19, p. 3.

34. Williams, *Now and Then*; Bernard Krug, field report, Apr. 30–May 3, 1951, VT-10, REAA, entry 25, box 12.

35. J. R. Cobb to A&L Field Representatives, Aug. 23, 1946, REAA, entry 45, box 3; *REN*, Mar. 1946, pp. 4–5, 15; and Oct.–Nov. 1949, pp. 18–19; *RL*, Nov. 1959, pp. 10–12.

36. *REN*, July 1944, pp. 6–7; Aug. 1944, pp. 10,15; Feb. 1945, p. 19; July 1945, p. 22; and Mar. 1947, pp. 12–13, 22. For examples of the Information Division's work in radio, see Phillip Voltz, field report, Feb. 12, 1952, IL-41, REAA, entry 45, box 1; O. E. Mabrey to Allyn Walters, Mar. 31, 1950, KS-27, REAA, entry 87, box 4; and James Ford, field report, Dec. 13, 1951, MS-22, REAA, entry 45, box 19.

37. See, e.g., *RL*, Apr. 1955, p. 15; and Mar. 1957, p. 15.

38. Elizabeth O'Kelley, field report, May 23–24, 1951, VA-41, REAA, entry 45, box 19; A. S. Arness, field report, Aug. 5–7 [1952 or 1953], REAA, entry 45, box 19.

39. See, e.g., Clate Cox, field report, June 10–12, 1952, AR-33, REAA, entry 45, box 21; Max Colbert, field report, Apr. 8–10, 1952, TX-7, REAA, entry 45, box 10; George Dillon to Assistant Regional Heads, Mar. 12, 1951, SD-23, REAA, entry 45, box 14; Elizabeth O'Kelley, field report, July 18, 1951, VT-8, REAA, entry 45, box 19.

40. *RL*, June 1955, p. 8. On the trend of determining women's needs in home economics, see Gladys Gallup, "The Effectiveness of the Home Demonstration Program of the Cooperative Extension Service of the United States Department of Agriculture in Reaching Rural People and in Meeting Their Needs" (Ph.D. diss., George Washington University, 1943).

41. *REN*, Dec. 1949–Jan. 1950, pp. 20–21; Clifton J. Bradley, field report, Jan. 3, 1951, MS-00; "Home and Farm Electrical Exposition," schedule, Apr. 20, 1950, REAA, entry 45, box 19; *Kilowatt News*, newsletter, Sept. 1951, TN-19, REAA, entry 45, box 1; "Two Hundred Thousand Will Attend Kentucky Rural Electric Cooperatives' Annual Meetings and Electric Farm Shows This Summer," KY-00, [1950]; and C. J. Ross to Directors of Kentucky Rural Electric Co-Op Corp., Mar. 5, 1952, REAA, entry 45, box 19; "Preview of the 1952 Annual Meeting and Electrical Farm Show," n.d., attached to J. W. Smith to George Dillon, June 17, 1952, REAA, entry 45, box 1; "Electric Fair," booklet, [1951], REAA, entry 45, box 19.

42. Katherine Bailey, field report excerpt, Nov. 24–25, 28, 1952, MN-70; and Elmer Smith, field report, Apr. 13–17, 1953, REAA, entry 45, box 19.

43. Gregory B. Field, "'Electricity for All': The Electric Home and Farm Authority and the Politics of Mass Consumption, 1932–1935," *Business History Review*, 64 (1990): 32–60, on p. 55.

44. Board of Directors Minutes, Sept. 7, 1946, AL-18, REAA, entry 77, box 1; *REN*, Apr.–May 1948, p. 10; and Feb.–Mar. 1950, p. 10; Ronald C. Tobey, *Technology as Freedom: The New Deal and the Electrical Modernization of the American Home* (Berkeley: Univ. of California Press, 1996).

45. See, e.g., George Dillon to Fred Hamilton, Feb. 27, 1953, NE-98, REAA, entry 45, box 21.

46. Vincent Nicholson to Wallace Veatch, Apr. 20, 1945, MO-23, REAA, entry 45, box 21; Claude Wickard, "Merchandising of Electrical Appliances of Doubtful Benefit to Cooperatives," *Illinois REA News*, Sept. 1946, excerpt in REAA, entry 45, box 20; Jennings Ray, field report, Nov. 9–15, 1949, IL-37, REAA, entry 45, box 20 (quotation); Phillip Voltz, field report, Feb. 4–8, 1952, IL-37, REAA, entry 45, box 1 (quotation); Clifton Bradley, field report, Feb. 27, 1952, KY-55, REAA, entry 45, box 19; Allyn Walters, field report, Dec. 8–9, 1952, NC-00, REAA, entry 87, box 12 (quotation); Phillip Voltz, field report, Feb. 28, 1952, IL-43, REAA, entry 45, box 1.

47. George Dillon to file, Feb. 28, 1952, REAA, entry 45, box 11.

48. E. C. Collier, field report excerpt, Apr. 14, 1953, IL-18, REAA, entry 45, box 19; Elwood Olver, *Electric Power Use Guide*, [1951], REAA, entry 45, box 20, on p. 6.

49. Robert Morris, field report excerpt, Sept. 15, 1952, WI-21; and Elizabeth O'Kelley, field report, July 18, 1951, VT-8, REAA, entry 45, box 19.

50. Elwood Olver, "Iowa's Power Use Program," June 1952, IA-00, REAA, entry 45, box 21, on p. 27; James Ford, field report, Dec. 13, 1951, MS-22, REAA, entry 45, box 19.

51. Reed Hutchinson and Asher Young, field report, Apr. 9–27, 1951, ME-16, REAA, entry 45, box 19.

52. Jane Stern and Michael Stern, "Neighboring," *New Yorker*, Apr. 15, 1991, pp. 78–93; Claire Gilbert, "The Development of Selected Aspects of Home Demonstration Work in the United States" (Ph.D. diss., Cornell University, 1958), p. 161.

53. Louisan Mamer, field report, July 1–18, 1953, REAA, entry 45, box 2.

54. *Rural Lines*, which replaced *Rural Electrification News* with the change in administration in 1954, did, however, use the term *load building* extensively before 1960.

55. Daniel Teare, "Wiring the Farmstead and Organizing the Farm for Electric Power," Mar. 7, 1946, NAL.

56. Joe F. Davis, "Economic Studies of Farm Electrification," *Agricultural Engineering*, 31 (1950): 565–68, quotations on pp. 565, 567; *REN*, Dec. 1952–Jan. 1953, pp. 9–11. The Iowa part of the bureau's study is summarized in *REN*, Feb.–Mar. 1951, pp. 16–17. For the full reports, see Oscar Steanson and Joe F. Davis, "Electricity on Farms in the Upper Piedmont of Georgia," University of Georgia AES, *Bulletin*, no. 263, June 1950; Howard J. Bonser and Joe F. Davis, "Electricity on Farms and in Rural Homes in the East Tennessee Valley," University of Tennessee AES, *Bulletin*, no. 221, Apr. 1951.

57. Davis, "Economic Studies of Farm Electrification," pp. 567–68.

58. For an example of these instructions, see Oneta Liter, "Cold Facts," Apr. 1952, REAA, entry 45, box 3, p. 3.

59. Maud M. Wilson, "Use of Electricity in Oregon Rural Homes," *JHE*, 39 (1947): 337–40; Davis, "Economic Studies of Farm Electrification"; H. H. London and Robert W. Adams, "Uses of Electricity: An Educational Problem," *JHE*, 50 (1948): 247–48.

60. Wilson, "Use of Electricity in Oregon Rural Homes," p. 339.

61. Lucile W. Reynolds and Emma G. Holmes, "House Improvements of Forty Ohio Farm Families," *JHE*, 39 (1947): 341–44, on p. 341; *REN*, Feb.–Mar. 1951, p. 17; Dorothy Dickins, "The Farm Home Improves Its Equipment," *JHE*, 40 (1948): 567–70, quotation on pp. 567–68.

On the generational factor in adopting new household technologies, see Mary Neth, *Preserving the Family Farm: Women, Community, and the Foundations of Agribusiness in the Midwest, 1900–1940* (Baltimore: Johns Hopkins Univ. Press, 1995), pp. 195–96.

62. Jane Adams, *The Transformation of Rural Life: Southern Illinois, 1890–1990* (Chapel Hill: Univ. of North Carolina Press, 1994), p. 210.

63. *REN*, Feb.–Mar. 1951, pp. 3–5, on p. 4; Louisan Mamer, "Electricity Pays Its Way in the Rural Home," conference paper, Mar. 11, 1952, REAA, entry 45, box 3, pp. 16–17.

64. *RL*, Jan. 1956, pp. 5–7, quotation on p. 5; Sept. 1956, pp. 4–5; and Mar. 1959, pp. 8–10.

65. Wilson, "Use of Electricity in Oregon Rural Homes," pp. 339–40; Dickins, "Farm Home Improves Its Equipment," p. 569.

66. Carroll Willis, "Who's Afraid of the REA?" *Butane/Propane News*, Apr. 1944, copy in REAA, entry 45, box 16.

67. Cited in "Liquid Petroleum Gas Usage and Its Relation to Future Consumption of Electricity on Farms," draft report, Mar. 13, 1952, REAA, entry 45, box 2, p. 5.

68. E. C. Weitzell, "New Mexico 21 Lincoln: Special Field Study," Nov. 21, 1951, REAA, entry 45, box 20, pp. 3 (quotation), 5.

69. Harold Darst to Gentlemen, July 15, 1950, OH-1, REAA, entry 45, box 16.

70. "New Co-op Leaders Voice Their Thoughts," Mar. 1950, REAA, entry 25, box 18, p. 6.

71. Kermit Overby to Carl Hamilton, Nov. 17, 1947, REAA, entry 45, box 16; Elwood Olver, "Range Promotion Program," Sept. 7, 1951, IA-00, REAA, entry 45, box 20, p. 1.

72. Olver, "Range Promotion Program," p. 1; Phillip W. Voltz, field report, Feb. 4–8, 1952, IL-37, REAA, entry 45, box 1; Louisan Mamer, field report, Oct. 15–21, 1949, KY-00, REAA, entry 45, box 3.

73. See, e.g., Clate Cox, field report, June 10–12, 1952, AR-33, REAA, entry 45, box 21.

74. *RL*, June 1958, pp. 10–11.

75. *RL*, Oct. 1957, pp. 21–22.

76. "Partial Results of Bell Company Survey, Iowa, 1945," attached to E. C. Weitzel to Claude Wickard et. al., Jan. 23, 1951, REAA, entry 25, box 15.

77. *REN*, Apr.–May 1953, pp. 18–19.

78. Compare table A.4 with USDA, Bureau of Agricultural Economics, "Preliminary Report on Survey of Radios, Telephones, and Electricity on Farms, 1947," Dec. 1947, NAL.

79. Dickins, "Farm Home Improves Its Equipment," p. 569.

80. *REN*, Aug.–Sept. 1950, p. 20; and Aug.–Sept. 1951, pp. 20–21.

81. USDA, Bureau of Agricultural Economics, *Attitudes of Rural People toward Radio Service: A Nation-Wide Survey of Farm and Small-Town People* (Washington, D.C., 1946), pp. 1, 12, 16–17.

82. U.S. Department of Commerce, Bureau of the Census, *United States Census of Agriculture, 1954: General Report*, vol. 2, *Statistics by Subjects* (Washington, D.C.: GPO, 1956–57), p. 201; *REN*, Feb.–Mar. 1952, p. 21; *Wisconsin REA News*, May 1953, p. 1, CW, box 69.

83. *REN*, June–July 1949, pp. 10–11.

84. Mamer, "Electricity Pays Its Way," pp. 12–14, 25. On the popularity of urban TV shows on the farm, see Jellison, *Entitled to Power*, pp. 161–62.

85. John Carmody, remarks, in REA Annual Administrative Conference of Staff and Field Personnel, typescript, Jan. 5–14, 1938, REAA, entry 96, box 5, p. 30.

86. *Report of Rural Electrification Administration, 1937* (Washington, D.C.: GPO, 1938), pp. vi, 33.

87. Mamer, "Electricity Pays Its Way," p. 7.

88. See, e.g., Deward Clayton Brown, "Rural Electrification in the South" (Ph.D. diss., University of California at Los Angeles, 1970), chap. 11.

89. Elizabeth O'Kelley, field report, May 23–24, 1951, VA-41; and Walter Rich, field report, June 4–6, 1951, SC-40, REAA, entry 45, box 19. On blacks in the New Deal, see Roger Biles, *A New Deal for the American People* (DeKalb: Northern Illinois Univ. Press, 1991), chap. 9; and Pete Daniel, *Breaking the Land: The Transformation of Cotton, Tobacco, and Rice Cultures since 1880* (Urbana: Univ. of Illinois Press, 1985), chaps. 4–6.

90. James Ford, field report, Dec. 11, 1951, MS-22, REAA, entry 45, box 19.

91. Eleanor Delaney, field report, Oct. 1–3, 1951, SC-32, REAA, entry 45, box 19.

92. *Report of REA, 1937*, p. 74; *Report of Rural Electrification Administration, 1938* (Washington, D.C.: GPO, 1939), p. 46; Oscar Meier to C. A. Winder, Sept. 3, 1941, REAA, entry 76, box 1.

93. *REN*, Aug.–Sept. 1948, pp. 12–13; Oct.–Nov. 1948, pp. 12–13, 20; Oct.–Nov. 1949, pp. 18–19; Apr.–May 1950, p. 19; June–July 1950, pp. 3–4, 16–17; and Aug.–Sept. 1950, pp. 3–6.

94. Deborah Fink and Dorothy Schwieder, "Iowa Farm Women in the 1930s," *Annals of Iowa*, 49 (1989): 570–90, on p. 585.

95. Gertrude Dieken, "What Electricity Should Mean to the Farm Wife," conference paper, Nov. 7, 1946, pp. 1, 4; and Louisan Mamer, field report on NRECA meeting, Mar. 10, 1948, REAA, entry 45, box 3 (quoting Zikmund).

96. Ruth Schwartz Cowan, *More Work for Mother: The Ironies of Household Technology from the Open Hearth to the Microwave* (New York: Basic Books, 1983); *REN*, Dec. 1937, pp. 9–10, quotation on p. 9; *Report of REA, 1938*, p. 118; Mamer, "Electricity Pays Its Way," quotation on p. 2.

97. Eleanor Arnold, *Party Lines, Pumps, and Privies: Memories of Hoosier Homemakers* ([Indianapolis]: Indiana Extension Homemakers Association, [1984]), p. 49, ellipsis in the original.

98. See Ronald Kline, "Ideology and Social Surveys: Reinterpreting the Effects of 'Laborsaving' Technology on American Farm Women," *TC*, 38 (1997): 355–85. For an argument that adjustments of two sets of time study data for family size, education, and farm residence indicate that the time spent on household work for married women dropped about one hour per day over forty years, see W. Keith Bryant, "A Comparison of the Household Work of Married Females: The Mid-1920s and the Late 1960s," *Family and Consumer Sciences Research Journal*, 24 (1996): 358–84. For critiques of Bryant's methodology and a response, see Jane Kolodinsky, "Issues in the Estimation of Women's Household Time Use: Some Thoughts from the Field," ibid., 385–92; Cathleen D. Zick, "Assessing the Past and Future of Research on Household Work: A Comment on Bryant's Article," ibid., 393–400; and W. Keith Bryant, "Trends in the Household Work of Married Females: Response to Commentaries," ibid., 401–5.

99. Marianne Muse, "Time Expenditures on Homemaking Activities in 183 Vermont Farm Homes," University of Vermont AES, *Bulletin*, no. 530, June 1946, pp. 43–44, 49, 52–53, 59–60.

100. Elizabeth Wiegand, "Use of Time by Full-time and Part-time Homemakers in Relation to Home Management," Cornell University AES, *Memoir*, no. 330 (July 1954): 39 (quotation), 42–43.

101. May L. Cowles and Ruth P. Dietz, "Time Spent in Homemaking Activity by a Selected Group of Wisconsin Farm Homemakers," *JHE*, 48 (1956): 29–35.

102. Sarah L. Manning, "Time Use in Household Tasks by Indiana Families," Purdue University AES, *Research Bulletin*, no. 837, Jan. 1968.

103. Mamer, "Electricity Pays Its Way," quotations on pp. 7, 9.

104. Paulena Nickell and Jean M. Dorsey, *Management in Family Living* (New York: Wiley, 1942); 2d ed., 1950, pp. 111–24; 3d ed., 1963, pp. 103–12; 4th ed., 1967, pp. 127–32. On Dorsey's earlier statements, see *Household Management and Kitchens*, ed. John M. Gries and James Ford, *President's Conference on Home Building and Home Ownership*, vol. 9 (Washington, D.C.: President's Conference on Home Building and Home Ownership, 1932), pp. 26–32, on p. 30.

105. Joann Vanek, "Keeping Busy: Time Spent in Housework, United States, 1920–1970" (Ph.D. diss., University of Michigan, 1973); Joann Vanek, "Time Spent in Housework," *SA*, 231 (Nov. 1974): 116–20.

106. Kathryn E. Walker and Margaret E. Woods, *Time Use: A Measure of Household Production of Family Goods and Services* (Washington, D.C.: Center for the Family of the American Home Economics Association, 1976), p. 32.

107. Cleo Fitzsimmons and Nellie L. Perkins, "Fifty Farm Kitchens," *JHE*, 37 (1945): 567–70.

108. Jellison, *Entitled to Power*, chap. 6. See also Sara Elbert, "Amber Waves of Grain: Women's Work in New York Farm Families," in *"To Toil the Livelong Day": America's Women at Work, 1780–1980*, ed. Carol Groneman and Mary Beth Norton (Ithaca: Cornell Univ. Press, 1987), pp. 250–68; Sandra Schackel, "Ranch and Farm Women in the Contemporary West," in *The Rural West since World War II*, ed. R. Douglas Hurt (Lawrence: Univ. Press of Kansas, 1998), pp. 99–118. Jellison cites Cowan's thesis approvingly for the prewar period (on pp. 53–54) but speculates that Wilson's 1929 study masked the fact that women could choose to spend the time saved by technology to do farmwork (pp. 62–63).

109. London and Adams, "Uses of Electricity," p. 248; Prudence S. Connor, Helen M. Seyse, and Lila Thurston, "Extent of Information about Electrical Equipment," *JHE*, 38 (1946): 356–58, on p. 357.

110. Lynn Robertson and Keith Amstutz, "Telephone Problems in Rural Indiana," Purdue University AES, *Bulletin*, no. 548, Sept. 1949, quotation on p. 18.

111. Lana F. Rakow, *Gender on the Line: Women, the Telephone, and Community Life* (Urbana: Univ. of Illinois Press, 1992), on pp. 9, 58.

Conclusion

1. The foreground of this image, showing just the two men and the map, heads the article by J. B. Hutson, "Acreage Allotments, Market Quotas, and Commodity Loans," in USDA, *Yearbook of Agriculture, 1940* (Washington, D.C.: GPO, 1940), pp. 551–65. The original photograph was later published in USDA, *Yearbook of Agriculture, 1962* (Washington, D.C.: GPO, 1962), p. 561.

2. See, e.g., Richard Hofstadter, *The Age of Reform: From Bryan to FDR* (New York: Vintage, 1955), pp. 39–40, 101.

3. Katherine Jellison, *Entitled to Power: Farm Women and Technology, 1913–1963* (Chapel Hill: Univ. of North Carolina Press, 1993), pp. 171–72, notes that such images, which equated farm life with suburban living, were common in ads on the women's pages of farm publications of this period.

4 Robert G. Yeck, "House Plans," in USDA, *Yearbook of Agriculture, 1965* (Washington, D.C.: GPO, 1965), pp. 13–16, caption on p. 14; Lyle M. Mamer, "Refrigeration," ibid., pp. 134–38, on p. 137; Lloyd E. Partain, "Country Vacations," ibid., pp. 299–301, on p. 299; John W. Rockey, "Disposal of Wastes," ibid., pp. 38–42, on p. 41.

5. Kathleen Babbitt, "Legitimizing Nutrition Education: The Impact of the Great Depression," in *Rethinking Home Economics: Women and the History of a Profession,* ed. Sarah Stage and Virginia B. Vincenti (Ithaca: Cornell Univ. Press, 1997), pp. 145–62; Claire E. Gilbert, "The Development of Selected Aspects of Home Demonstration Work in the United States" (Ph.D. diss., Cornell University, 1958), pp. 117, 167–69.

6. *REN,* Jan. 1938, p. 19; and Apr. 1938, pp. 27–28; Robert A. Caro, *The Years of Lyndon Johnson,* vol. 1, *The Path to Power* (New York: Random House, 1982), p. 528; *Report of Rural Electrification Administration, 1938* (Washington, D.C.: GPO, 1939), p. 49.

7. Louisan Mamer, "Electricity Pays Its Way in the Rural Home," conference paper, Mar. 11, 1952, REAA, entry 45, box 3, p. 6.

8. *RL,* Apr. 1959, pp. 7–10, quotations on pp. 7, 8; and Apr. 1960, pp. 10–112.

9. "Strictly Rural and Strictly Cooperative," *RL,* May 1959, pp. 6–8.

10. Thomas R. Ford, "Contemporary Rural America: Persistence and Change," in *Rural U.S.A.: Persistence and Change,* ed. Thomas R. Ford (Ames: Iowa State Univ. Press, 1978), pp. 3–16, on pp. 3–4.

11. Olaf F. Larson, "Values and Beliefs of Rural People," in Ford, *Rural U.S.A.,* pp. 91–112, on p. 110.

12. Ford, "Contemporary Rural America," p. 4.

13. Lizabeth Cohen, *Making a New Deal: Industrial Workers in Chicago* (Cambridge: Cambridge Univ. Press, 1990).

14. Tim Davis, "Rural America Enters the Space Age," *Journal of Popular Culture,* 21 (fall 1987): 117–42; Kenneth T. Jackson, *Crabgrass Frontier: The Suburbanization of the United States* (New York: Oxford Univ. Press, 1985), pp. 261–63; James A. Wren and Genevieve Wren, *Motor Trucks of America* (Ann Arbor: Univ. of Michigan Press, 1979).

15. I witnessed this use of the cellular phone when visiting a high-school friend, Gene Ferguson, at his place near Mound Valley, Kansas, in the summer of 1998.

Bibliographical and Methodological Note

All archives cited in this book appear in the list of abbreviations; all sources are cited fully in the notes. Here, I discuss key secondary sources and the book's methodology. As stated in the introduction, the book bridges two historiographic traditions: a constructivist history of technology focusing on consumers and the social history of rural life.

Technology

I build on recent work in the history of technology that shifts the field's traditional focus from the "producers" of technology (e.g., inventors, engineers, and manufacturers) to the "users" of technology (e.g., laborers, factory owners, home workers, and consumers). Examples of this work that are especially relevant to the present study are Ruth Schwartz Cowan, *More Work for Mother: The Ironies of Household Technology from the Open Hearth to the Microwave* (New York: Basic Books, 1983); Carolyn Marvin, *When Old Technologies Were New: Thinking about Electric Communication in the Late Nineteenth Century* (New York: Oxford Univ. Press, 1988); and Claude S. Fischer, *America Calling: A Social History of the Telephone to 1940* (Berkeley: Univ. of California Press, 1992). My interest is in how users have helped to shape the artifact or system itself. Susan Douglas, *Inventing American Broadcasting, 1899–1922* (Baltimore: Johns Hopkins Univ. Press, 1987), has demonstrated that radio amateurs helped change the dominant interpretation of radio from point-to-point communication to broadcasting. Michele Martin, *'Hello, Central?': Gender, Technology, and Culture in the Formation of Telephone Systems* (Montreal: McGill–Queen's Univ. Press, 1991), and Fischer have argued that the actions of telephone callers eventually convinced the industry to regard the telephone as a social, as well as a business, instrument. And David E. Nye, *Electrifying America: Social Meanings of a New Technology, 1880–1940* (Cambridge: MIT Press, 1990) has described the manifold social meanings given to electric streetlights, trolleys, factories, appliances, and electricity itself by a wide variety of users.

My user-oriented approach extends the social construction of technology method developed in the mid-1980s by Trevor Pinch and Wiebe Bijker to include power relations between social groups and reciprocal relations between social groups and the use of artifacts. See Ronald Kline and Pinch, "Users as Agents of Technological Change: The Social Construction of the Automobile in the Rural United States," *Technology and Culture*, 37 (1996): 763–95, which forms the basis for chapter 2; and Pinch and Bijker, "The Social Construction of Facts and Artifacts: Or How the Sociology of Science and the Sociology of Technology Might Benefit Each Other," in *The Social Construction of Technological Systems: New Directions in the Sociology and History of Technology*, ed. Bijker, Thomas Hughes, and Pinch (Cambridge: MIT Press, 1987), pp. 17–50. My focus on users and mediators is indebted to Fischer, who aptly calls his method a "user heuristic" (*America Calling*, pp. 17–21), and to Ruth Schwartz Cowan, "The Consumption Junction: A Proposal for Research Strategies in the Sociology of Technology," in Bijker, Hughes, and Pinch, *Social Construction of Technological Systems*, pp. 261–80. The reciprocal aspect of my approach is indebted to Peter L. Berger and Thomas Luckmann, *The Social Construction of Reality: A Treatise in the Sociology of Knowledge* (New York: Doubleday, 1966).

Wiebe Bijker has developed a reciprocal model of technology and society, the convergence model, in which technological frames link artifacts and social groups. See Bijker, *Of Bicycles, Bakelite, and Bulbs: Toward a Theory of Sociotechnical Change* (Cambridge: MIT Press, 1995). I also adopt some of the language of "sociotechnical networks," but my usage of the term *network* is closer to the systems theory of Thomas Hughes than to the actant-network approach of Michel Callon and Bruno Latour, in which no analytical distinction is made between human and nonhuman "actants." See Thomas P. Hughes, *Networks of Power: Electrification in Western Society, 1880–1930* (Baltimore: Johns Hopkins Univ. Press, 1983); Hughes, "The Evolution of Large Technological Systems," in Bijker, Hughes, and Pinch, *Social Construction of Technological Systems*, pp. 51–82; Michel Callon, "Society in the Making: The Study of Technology as a Tool for Social Analysis," in Bijker, Hughes, and Pinch, *Social Construction of Technological Systems*, pp. 83–106; and Bruno Latour, *Science in Action: How to Follow Scientists and Engineers through Society* (Cambridge: Harvard Univ. Press, 1987).

For recent work on mediators of technology, see Lizabeth Cohen, *Making a New Deal: Industrial Workers in Chicago* (Cambridge: Cambridge Univ. Press, 1990); Olivier Zunz, *Making America Corporate, 1870–1920* (Chicago: Univ. of Chicago Press, 1990); Mark Rose, *Cities of Light and Heat: Domesticating Gas and Electricity in Urban America* (University Park: Pennsylvania State Univ. Press, 1995); and Carolyn Goldstein, "Part of the Package: Home Economists in the Consumer Products Industry, 1920–1940," in *Rethinking Home Economics: Women and the History of a Profession*, ed. Sarah Stage and Virginia B. Vincenti (Ithaca: Cornell Univ. Press, 1997), pp. 271–96. Thomas J. Misa, *A Nation of Steel: The Making of Modern America, 1865–1925* (Baltimore: Johns Hopkins Univ. Press, 1995), develops a producer-consumer methodology along intra-industry lines. The recognition of the blurred boundaries between producer and consumer dates at least to the nineteenth century. See, for example, Karl Marx, *Grundrisse: Foundations of the Critique of Political Economy (Rough Draft)*, trans. Martin Nicolaus (1939; reprint, New York: Penguin, 1993), pp. 90–94.

For early work on the culture of consumption, see Richard Wightman Fox and T. J. Jackson Lears, eds., *The Culture of Consumption: Critical Essays in American History, 1880–1980* (New York: Pantheon, 1983); and Roland Marchand, *Advertising the American Dream: Making Way for Modernity, 1920–1940* (Berkeley: Univ. of California Press, 1985). Lizabeth Cohen has recently critiqued the corporate-hegemony view of Fox and Lears in her article, "The Class Experience of Mass Consumption: Workers as Consumers in Interwar America," in *The Power of Culture: Critical Essays in American History*, ed. Richard Wightman Fox and T. J. Jackson Lears (Chicago: Univ. of Chicago Press, 1993), pp. 135–60.

In studying interactions among users, mediators, and producers of technology, I have benefited from several works on the related issues of technological progress, technological determinism, gender and technology, resistance to technology, and modernization. For critical histories of the ideology of technological progress and technological determinism, see Thomas J. Misa, "How Machines Make History and How Historians (and Others) Help Them to Do So," *Science, Technology, and Human Values*, 13 (1988): 308–31; Merritt Roe Smith and Leo Marx, eds., *Does Technology Drive History?: The Dilemma of Technological Determinism* (Cambridge: MIT Press, 1994); Howard P. Segal, *Technological Utopianism in American Culture* (Chicago: Univ. of Chicago Press, 1985); and John Jordan, *Machine Age Ideology: Social Engineering and American Liberalism, 1911–1939* (Chapel Hill: Univ. of North Carolina Press, 1994). I use *ideology* in the nonpejorative sense of cultural codes advocated by Clifford Geertz in *The Interpretation of Cultures* (New York: Basic Books, 1973), chap. 8. On the progressive ideology of electricity, see James W. Carey and John J. Quirk, "The Mythos of the Electronic Revolution," *American Scholar*, 39 (1970): 395–424; Thomas P. Hughes, "The Industrial Revolution That Never Came," *American Heritage of Invention and Technology*, 3 (winter 1988): 58–64; and Ronald Kline, *Steinmetz: Engineer and Socialist* (Baltimore: Johns Hopkins Univ. Press, 1992), chap. 9.

Langdon Winner has argued that technology, broadly defined, is more deterministic than these accounts and the present book maintain. See Winner, *Autonomous Technology: Technics-out-of-Control as a Theme in Political Thought* (Cambridge: MIT Press, 1977). In an influential article, "Do Artifacts Have Politics?," Winner argues that some technologies can be flexible enough to be designed to achieve political ends (such as low bridges on the Long Island Expressway preventing inner-city buses from going to Jones Beach), whereas others, such as nuclear power, seem to require a certain type of political system (centralized in this case). See Winner, *The Whale and the Reactor: A Search for Limits in an Age of High Technology* (Chicago: Univ. of Chicago Press, 1986), chap. 2. Like other scholars, I hold that the socioeconomic context of the bridges, not just their height, resulted in their apparent agency—for a time.

My constructivist view of gender and technology is indebted to Judy Wajcman, *Feminism Confronts Technology* (University Park: Penn State Univ. Press, 1991). I use the analytical framework of gender structure, identity, and symbolism in discussing gender relationships. See Joan W. Scott, "Gender: A Useful Category of Historical Analysis," *American Historical Review*, 91 (1986): 1053–75; Sandra Harding, *The Science Question in Feminism* (Ithaca: Cornell Univ. Press, 1986), chap. 2; and Nina E. Lerman, Arwen Palmer Mohun, and Ruth Oldenziel, "The Shoulders We Stand On and the View from Here: Historiography and Directions for Research," in "Gender Analysis and the History of Technology," a special issue of *Technology and Culture*, 38 (1997): 9–30.

On the protean subject of resistance to technology, see David B. Danbom, *The Resisted Revolution: Urban America and the Industrialization of Agriculture, 1900–1930* (Ames: Iowa State Univ. Press, 1979); Adrian J. Randall, "The Philosophy of Luddism: The Case of the West England Woolen Workers, ca. 1790–1809," *Technology and Culture*, 27 (1986): 1–17; James C. Scott, *Weapons of the Weak: Everyday Forms of Peasant Resistance* (New Haven: Yale Univ. Press, 1985); Donald B. Kraybill, *The Riddle of Amish Culture* (Baltimore: Johns Hopkins Univ. Press, 1989); Charles B. Dew, *Bond of Iron: Master and Slave at Buffalo Forge* (New York: Norton, 1994); and Martin Bauer, ed., *Resistance to New Technology: Nuclear Power, Information Technology, and Biotechnology* (Cambridge: Cambridge Univ. Press, 1995).

Resistance is a constitutive element of Michel Foucault's analysis of the exercise of power in disciplinary networks; Foucault, however, does not leave much room for changes effected by resistance. See his *Discipline and Punish: The Birth of the Prison*, trans. Alan Sheridan (New York: Vintage, 1979); *Power/Knowledge: Selected Interviews and Other Writings, 1972–1977*, ed. Colin

Gordon, trans. Gordon, Leo Marshall, John Mepham, and Kate Souper (New York: Pantheon, 1980), especially pp. 162–65; and Hubert Dreyfus and Paul Rabinow, *Michel Foucault: Beyond Structuralism and Hermeneutics, with an Afterword and Interview with Michel Foucault*, 2d ed. (Chicago: Univ. of Chicago Press, 1982), pp. 146–47, 206–7, 211.

In contrast, I view resistance to new technology as a means of social transformation. I believe that this concept applies equally well to groups other than those we call "rural" and that it provides a fruitful way to talk about changes in technology and society without adopting the language of autonomous technology, technological determinism, or corporate hegemony.

Some influential theories of modernity are presented in Richard D. Brown, *Modernization: The Transformation of American Life, 1600–1865* (New York: Hill and Wang, 1976); T. J. Jackson Lears, *No Place of Grace: Antimodernism and the Transformation of American Culture, 1880–1920* (1983; reprint, Chicago: Univ. of Chicago Press, 1993), chap. 1; and Anthony Giddens, *The Consequences of Modernity* (Stanford: Stanford Univ. Press, 1990). For a critique of the interpretation that a wholesale shift from community to society occurred in the modern era, see Thomas Bender, *Community and Social Change in America* (1978; reprint, Baltimore: Johns Hopkins Univ. Press, 1982). On the use of modernization theory in studies of rural America, see Robert P. Swierenga, "Theoretical Perspectives on the New Rural History: From Environmentalism to Modernization," *Agricultural History*, 56 (1982): 495–502. For examples of Brown's influence on rural historians, see Warren J. Gates, "Modernization as a Function of an Agricultural Fair: The Great Grangers' Picnic Exhibition at Williams Grove, Pennsylvania, 1873–1916," *Agricultural History*, 58 (1984): 262–79; and Larry Hasse, "Watermills in the South: Rural Institutions Working Against Modernization," *Agricultural History*, 58 (1984): 280–95. For a less formal usage of the concept, see Richard S. Kirkendall, "The Agricultural Colleges: Between Tradition and Modernization," *Agricultural History*, 60 (1986): 3–21.

Several recent books provide background for the general development of the telephone, automobile, radio, and electric light and power in the United States, mostly from an urban perspective. On the general history of the telephone industry, see Robert W. Garnet, *The Telephone Enterprise: The Evolution of the Bell System's Horizontal Structure, 1876–1909* (Baltimore: Johns Hopkins Univ. Press, 1985); Kenneth Lipartito, *The Bell System and Regional Business: The Telephone in the South, 1877–1920* (Baltimore: Johns Hopkins Univ. Press, 1989); and Charles A. Pleasance, *The Spirit of Independent Telephony* (Johnson City, Tenn.: Independent Telephone Books, 1989). Excellent social studies are Fischer's *America Calling* and Lana F. Rakow, *Gender on the Line: Women, the Telephone, and Community Life* (Urbana: Univ. of Illinois Press, 1992). On rural telephony, see Don F. Hadwiger and Clay Cochran, "Rural Telephones in the United States," *Agricultural History*, 58 (1984): 221–38; Claude Fischer, "The Revolution in Rural Telephony, 1900–1920," *Journal of Social History*, 21 (1987): 5–26; and Fischer, "Technology's Retreat: The Decline of Rural Telephony in the United States," *Social Science History*, 11 (1987): 295–327. On an earlier form of rural communications, see Wayne E. Fuller, *RFD: The Changing Face of Rural America* (Bloomington: Indiana Univ. Press, 1964).

The literature on the automobile is vast. The following have been particularly helpful for understanding the general history of the car in American life: Warren J. Belasco, *Americans on the Road: From Auto Camp to Motel, 1910–1945* (Cambridge: MIT Press, 1979); David L. Lewis and Laurence Goldstein, eds., *The Automobile and American Culture* (Ann Arbor: Univ. of Michigan Press, 1983); Bruce E. Seely, *Building the American Highway System: Engineers as Policy Makers* (Philadelphia: Temple Univ. Press, 1987); James J. Flink, *The Automobile Age* (Cambridge: MIT Press, 1988); Virginia Scharff, *Taking the Wheel: Women and the Coming of the Mo-*

tor Age (New York: Free Press, 1991); and Clay McShane, *Down the Asphalt Path: The Automobile and the American City* (New York: Columbia Univ. Press, 1994).

On the auto in rural life, see Reynold M. Wik, *Henry Ford and Grass-Roots America.* (Ann Arbor: Univ. of Michigan Press, 1972); Michael Berger, *The Devil Wagon in God's Country: The Automobile and Social Change in Rural America, 1893–1929* (Hamden, Conn.: Archon, 1979); Joseph Interrante, "You Can't Go to Town in a Bathtub: Automobile Movement and the Reorganization of Rural American Space, 1900–1930," *Radical History Review,* 21 (1979): 151–68; and Hal S. Barron, "And the Crooked Shall Be Made Straight: Public Road Administration and the Decline of Localism in the Rural North, 1870–1930," *Journal of Social History,* 26 (1992): 81–103. An earlier form of rural transportation is described in George W. Hilton and John F. Due, *The Electric Interurban in America* (Stanford: Stanford Univ. Press, 1960).

The history of household technology has received much attention. See Joann Vanek, "Time Spent in Housework," *Scientific American,* 231 (Nov. 1974): 116–20; Susan Strasser, *Never Done: A History of American Housework* (New York: Pantheon, 1982); Cowan, *More Work for Mother;* Suellen Hoy, *Chasing Dirt: The American Pursuit of Cleanliness* (Oxford: Oxford Univ. Press, 1995); and Maureen Ogle, *All the Modern Conveniences: American Household Plumbing, 1840–1890* (Baltimore: Johns Hopkins Univ. Press, 1996). For divergent views on the history of home economics, see Barbara Ehrenreich and Deirdre English, *For Her Own Good: 150 Years of the Experts' Advice to Women* (New York: Doubleday, 1978); and Stage and Vincenti, *Rethinking Home Economics.* On the farm home, see Katherine Jellison, *Entitled to Power: Farm Women and Technology, 1913–1963* (Chapel Hill: Univ. of North Carolina Press, 1993).

On the development of the early radio industry, see Hugh Aitken, *Continuous Wave: Technology and American Radio, 1900–1932* (Princeton: Princeton Univ. Press, 1985); Susan J. Douglas, *Inventing American Broadcasting, 1899–1922* (Baltimore: Johns Hopkins Univ. Press, 1987); and Susan Smulyan, *Selling Radio: The Commercialization of American Broadcasting, 1920–1934* (Washington: Smithsonian Institution Press, 1994). On the material culture of radios, see Michael B. Schiffer, *The Portable Radio in American Life* (Tucson: Univ. of Arizona Press, 1991). The influence of the radio in rural life is described by John C. Baker, *Farm Broadcasting: The First Sixty Years* (Ames: Iowa State Univ. Press, 1981); Reynold M. Wik, "The Radio in Rural America during the 1920s," *Agricultural History,* 55 (1981): 339–50; and Wik, "The USDA and the Development of Radio in America," *Agricultural History,* 62 (1988): 177–88.

For recent histories of the electrical industry, see Hughes's magisterial *Networks of Power;* Nye, *Electrifying America;* and Ronald C. Tobey, *Technology as Freedom: The New Deal and the Electrical Modernization of the American Home* (Berkeley: Univ. of California Press, 1996). Tobey makes the cogent argument that U.S. census figures mask steps in the usage of electricity that mark purchases of major appliances. On selling electricity in the city, see Harold L. Platt, *The Electric City: Energy and the Growth of the Chicago Area, 1880–1930* (Chicago: Univ. of Chicago Press, 1991), chap. 5; and Rose, *Cities of Light and Heat.*

Most accounts of rural electrification focus on the Rural Electrification Administration. See Mark C. Stauter, "The Rural Electrification Administration: A New Deal Case Study" (Ph.D. diss., Duke University, 1973); Philip J. Funigiello, *Toward a National Power Policy: The New Deal and the Electric Utility Industry, 1933–1941* (Pittsburgh: Univ. of Pittsburgh Press, 1973), chaps. 5–6; D. Clayton Brown, *Electricity for Rural America: The Fight for the REA* (Westport, Conn.: Greenwood, 1980); Jean Christie, *Morris Llewellyn Cooke: Progressive Engineer* (New York: Garland, 1983), chap. 9; R. Douglas Hurt, "REA: A New Deal for Farmers," *Timeline,* 2 (Dec. 1985–Jan. 1986): 32–47; and Keith R. Fleming, *Power at Cost: Ontario Hydro and Rural Elec-*

trification, 1911–1958 (Montreal: McGill–Queen's Univ. Press, 1992). Several state histories are cited in the notes. On agricultural uses of electricity, see Nye, *Electrifying America*, chap. 7.

AGRICULTURE AND RURAL LIFE

Because the literature on the history of American agriculture and rural life is as vast as that on technology, I give only selected references. For recent historiographic issues, see R. Douglas Hurt, "American Agricultural and Rural History," *American Studies International*, 35 (1997): 50–71; the introductions to Steven Hahn and Jonathan Prude, eds., *The Countryside in the Age of Capitalist Transformation: Essays in the Social History of Rural America* (Chapel Hill: Univ. of North Carolina Press, 1985); and Frederick V. Carstensen, Morton Rothstein, and Joseph A. Swanson, eds., *Outstanding in His Field: Perspectives on American Agriculture in Honor of Wayne D. Rasmussen* (Ames: Iowa State Univ. Press, 1993). For a social science perspective, see Frederick H. Buttel and Philip McMichael, "Sociology and Rural History: Summary and Critique," *Social Science History*, 12 (1988): 93–120.

For recent overviews of the agricultural history of the United States, see Willard W. Cochrane, *The Development of American Agriculture: A Historical Perspective* (Minneapolis: Univ. of Minnesota Press, 1979); R. Douglas Hurt, *American Agriculture: A Brief History* (Ames: Iowa State Univ. Press, 1994); and David B. Danbom, *Born in the Country: A History of Rural America* (Baltimore: Johns Hopkins Univ. Press, 1995). On the 1920s, see James Shideler, *Farm Crisis, 1919–1923* (Berkeley: Univ. of California Press, 1957). The agrarian myth is analyzed by Richard Hofstadter, *The Age of Reform: From Bryan to FDR* (New York: Vintage, 1955), chaps. 1–3; and Danbom, "Romantic Agrarianism in Twentieth-Century America," *Agricultural History*, 65 (1991): 1–12.

On the industrialization and urbanization of the American farm in the twentieth century, see John L. Shover, *First Majority, Last Minority: The Transforming of Rural Life in America* (De Kalb: Northern Illinois Univ. Press, 1976); Gilbert C. Fite, *American Farmers: The New Minority* (Bloomington, Indiana Univ. Press, 1981); Pete Daniel, *Breaking the Land: The Transformation of Cotton, Tobacco, and Rice Cultures since 1880* (Urbana: Univ. of Illinois Press, 1985); Jack Temple Kirby, *Rural Worlds Lost: The American South, 1920–1960* (Baton Rouge: Louisiana State Univ. Press, 1987); Robert C. Williams, *Fordson, Farmall, and Poppin' Johnny: A History of the Farm Tractor and Its Impact on America* (Urbana: Univ. of Illinois Press, 1987); Deborah Fitzgerald, *The Business of Breeding: Hybrid Corn in Illinois, 1890–1940* (Ithaca: Cornell Univ. Press, 1990); Jane Adams, *The Transformation of Rural Life: Southern Illinois, 1890–1990* (Chapel Hill: Univ. of North Carolina Press, 1994); Hal S. Barron, *Mixed Harvest: The Second Great Transformation in the Rural North, 1870–1930* (Chapel Hill: Univ. of North Carolina Press, 1997); and R. Douglas Hurt, ed., *The Rural West since World War II* (Lawrence: Univ. Press of Kansas, 1998).

The development of the agricultural complex has been of great concern to historians and policy makers. See Gould P. Colman, *Education and Agriculture: A History of the New York State College of Agriculture at Cornell University* (Ithaca: Cornell Univ. Press, 1963); Charles E. Rosenberg, *No Other Gods: On Science and American Social Thought* (Baltimore: Johns Hopkins Univ. Press, 1976), chaps. 9–10; Jim Hightower, *Hard Tomatoes, Hard Times: The Original Hightower Report, Unexpurgated, of the Agribusiness Accountability Project on the Failure of America's Land Grant College Complex . . .* (Rochester, Vt.: Schenkman, 1978); Theodore Saloutous and John D. Hicks, *The American Farmer and the New Deal* (Ames: Iowa State Univ. Press, 1982); Alan I. Marcus, *Agricultural Science and the Quest for Legitimacy: Farmers, Agricultural Colleges, and Ex-*

periment Stations, 1870–1890 (Ames: Iowa State Univ. Press, 1985); and Jeffrey W. Moss and Cynthia B. Lass, "A History of Farmers' Institutes," *Agricultural History*, 62 (1988): 150–63. For the private side, see Earl W. Hayter, *The Troubled Farmer, 1850–1900: Rural Adjustment to Industrialism* (DeKalb: Northern Illinois Univ. Press, 1968); William L. Bowers, *The Country Life Movement in America, 1900–1920* (Port Washington, N.Y.: Kennikat, 1974); and Zunz, *Making America Corporate*, chap. 6.

The Cooperative Extension Service has been treated by Alfred C. True, *A History of Agricultural Extension Work in the United States, 1785–1923* (Washington, D.C.: GPO, 1928); Gladys Baker, *The County Agent* (Chicago: Univ. of Chicago Press, 1939); Roy V. Scott, *The Reluctant Farmer: The Rise of Agricultural Extension to 1914* (Urbana: Univ. of Illinois Press, 1970); Danbom, *The Resisted Revolution;* and Wayne D. Rasmussen, *Taking the University to the People: Seventy-Five Years of Cooperative Extension* (Ames: Iowa State University Press, 1989). On the agricultural social sciences, see Richard S. Kirkendall, *Social Scientists and Farm Politics in the Age of Roosevelt* (Columbia: Univ. of Missouri Press, 1966); Lowry Nelson, *Rural Sociology: Its Origin and Growth in the United States* (Minneapolis: Univ. of Minnesota Press, 1969); and Harry C. McDean, "Professionalism in the Rural Social Sciences, 1896–1919," *Agricultural History*, 58 (1984): 373–92.

The complexity of the history of farm organizations is indicated by Theodore Saloutous and John D. Hicks, *Twentieth-Century Populism: Agricultural Discontent in the Middle West, 1900–1939* (Lincoln: Univ. of Nebraska Press, 1951); Grant McConnell, *The Decline of Agrarian Democracy* (New York: Athenaeum, 1969); Christiana McFadyen Campbell, *The Farm Bureau: A Study of the Making of National Farm Policy, 1933–1940* (Urbana: Univ. of Illinois Press, 1962); Joseph G. Knapp, *The Rise of American Cooperative Enterprise: 1620–1920* (Danville, Ill.: Interstate, 1969); Thomas A. Woods, *Knights of the Plow: Oliver H. Kelly and the Origins of the Grange in Republican Ideology* (Ames: Iowa State Univ. Press, 1991); Donald B. Marti, *Women of the Grange: Mutuality and Sisterhood in Rural America, 1866–1920* (Westport, Conn.: Greenwood, 1991); and Robert C. McMath Jr., *American Populism: A Social History, 1877–1898* (New York: Hill and Wang, 1993).

On urban-rural tensions in the twentieth century, see Don S. Kirschner, *City and Country: Rural Responses to Urbanization in the 1920s* (Westport, Conn.: Greenwood, 1970); James H. Shideler, "Flappers and Philosophers and Farmers: Rural-Urban Tensions of the Twenties," *Agricultural History*, 47 (1973): 283–99; Robert S. Bader, *Hayseeds, Moralizers, and Methodists: The Twentieth-Century Image of Kansas* (Lawrence: Univ. Press of Kansas, 1988); and Charles W. Eagles, *Democracy Delayed: Congressional Reapportionment and Urban-Rural Conflict in the 1920s* (Ames: Iowa State Univ. Press, 1990).

Several studies of urban history have been particularly useful in comparing technology in urban and rural life. William Cronon, *Nature's Metropolis: Chicago and the Great West* (New York: Norton, 1991), analyses the importance of hinterlands to the development of a major city. The growth of suburbs comes to life in Kenneth T. Jackson, *Crabgrass Frontier: The Suburbanization of the United States* (New York: Oxford Univ. Press, 1985). On the development of urban infrastructures, see Joel A. Tarr and Gabriel DuPuy, eds., *Technology and the Rise of the Networked City in Europe and America* (Philadelphia: Temple Univ. Press, 1988).

There is now a large and sophisticated literature about American farm women. For the late nineteenth and the twentieth centuries, see Deborah Fink, *Open Country Iowa: Rural Women, Tradition, and Change* (Albany: State Univ. of New York Press, 1986); Corlann Gee Bush, "'He Isn't Half So Cranky As He Used to Be': Agricultural Mechanization, Comparable Worth, and the Changing Farm Family," in *"To Toil the Livelong Day": American Women at Work, 1780–1980*, ed. Carol Groneman and Mary Beth Norton (Ithaca: Cornell Univ. Press, 1987), pp. 213–29; Sara Elbert, "Amber Waves of Grain: Women's Work in New York Farm Families," in Groneman and Norton, *"To Toil the Livelong Day,"* pp. 250–68; Deborah Fink and Dorothy Schwieder, "Iowa Farm Women in the 1930s," *Annals of Iowa*, 49 (1989): 570–90; Nancy G. Osterud, *Bonds of Community: The Lives of Farm Women in Nineteenth-Century New York* (Ithaca: Cornell Univ. Press, 1991); Marti, *Women of the Grange;* Fink, *Agrarian Women: Wives and Mothers in Rural Nebraska, 1880–1940* (Chapel Hill: Univ. of North Carolina Press, 1992); Jellison, *Entitled to Power;* Kathleen Babbitt, "The Productive Farm Woman and the Extension Home Economist in New York State, 1920–1940," *Agricultural History*, 67 (1993): 83–101; Adams, *The Transformation of Rural Life;* Sally McMurry, *Transforming Rural Life: Dairying Families and Agricultural Change, 1820–1885* (Baltimore: Johns Hopkins Univ. Press, 1995); Mary Neth, *Preserving the Family Farm: Women, Community, and the Foundations of Agribusiness in the Midwest, 1900–1940* (Baltimore: Johns Hopkins Univ. Press, 1995); and Sandra Schackel, "Ranch and Farm Women in the Contemporary West," in Hurt, *The Rural West since World War II*, pp. 99–118. Fink views rural gender relations as an oppressive rural patriarchy, whereas Osterud, Marti, Jellison, and Neth emphasize mutuality between farm men and women within a flexible male-dominated gender system. For a review of how technology is treated in this literature, see Pamela Riney-Kehrberg, "Women, Technology, and Rural Life: Some Recent Literature," *Technology and Culture*, 38 (1997): 942–53.

Eleanor Arnold has edited several useful oral histories of farm women. See *Voices of American Homemakers* (Hollis, N.H.: National Extension Homemakers Council, 1985); *Party Lines, Pumps, and Privies: Memories of Hoosier Homemakers* ([Indianapolis]: Indiana Extension Homemakers Association, [1984]); and *Buggies and Bad Times: Memories of Hoosier Homemakers* ([Indianapolis]: Indiana Extension Homemakers Association, [1984]).

Index